# Probability, Combinatorics and Control

*Edited by Andrey Kostogryzov and Victor Korolev*

Published in London, United Kingdom

IntechOpen

*Supporting open minds since 2005*

Probability, Combinatorics and Control
http://dx.doi.org/10.5772/intechopen.79802
Edited by Andrey Kostogryzov and Victor Korolev

Contributors
Alberto Besana, Cristina Martinez Ramirez, Igor Goncharov, Alexander Dushkin, Nikita Goncharov, Pavel Parinov, Sergey Kochedykov, Igor Sinitsyn, Shalamov Anatoliy, Nikolay Andreevich Makhutov, Anatoly Tchernyaev, Yury Petrenia, Vladimir Moskvichev, Mikhail Gadenin, Anatoly Lepikhin, Andrey Kostogryzov, Victor Korolev, Dmitry Reznikov, Sergey Maslov, Igor Razumovskiy, Andrey Gorshenin, Alexander Zeifman, Franck Adekambi, Irene Lena Hudson, Susan Kim, Marie Keatley, Carlos Rodríguez Lucatero, Davood Shiri, F. Sibel Salman, Vladimir Nadein, Dmitriy A. Neganov, Martin Tamm



First published in London, United Kingdom, 2020 by IntechOpen
IntechOpen is the global imprint of INTECHOPEN LIMITED, registered in England and Wales, registration number: 11086078, 7th floor, 10 Lower Thames Street, London, EC3R 6AF, United Kingdom
Printed in Croatia

British Library Cataloguing-in-Publication Data
A catalogue record for this book is available from the British Library

Additional hard and PDF copies can be obtained from orders@intechopen.com

Probability, Combinatorics and Control
Edited by Andrey Kostogryzov and Victor Korolev
p. cm.
Print ISBN 978-1-83880-103-8
Online ISBN 978-1-83880-104-5
eBook (PDF) ISBN 978-1-78985-949-2

For EU product safety concerns:
IN TECH d.o.o., Prolaz Marije Krucifikse Kozulić 3, 51000 Rijeka, Croatia, info@intechopen.com or visit our website at intechopen.com.

# Meet the editors

Andrey Kostogryzov is Chief Researcher of the Federal Research Center "Computer Science and Control" of the Russian Academy of Sciences, Moscow, and a professor at Gubkin Russian State University of Oil and Gas. He is an Honored Science Worker of the Russian Federation (RF), winner of the Award of the Government of the RF in the Field of Science and Engineering (2015), a Doctor of Engineering (1994), and a professor (1999). Dr. Kostogryzov is Chairman of Subcommittees "Information Security and Industrial Safety" and "Information Technologies" of the Chamber of Commerce and Industry of the RF. He is a Certified Expert of the Russian Academy of Sciences, the Ministry of Education, Rostechnadzor, "Gazprom" Company. He is also author of more than 100 mathematical models (for analyzing and optimizing quality and risks) and more than 190 scientific works, including 19 books.

Victor Korolev is a professor and head of the Department of Mathematical Statistics, Faculty of Computational Mathematics and Cybernetics, Lomonosov Moscow State University, Russia. He is a leading researcher of the Federal Research Center "Computer Science and Control" of the Russian Academy of Sciences, Moscow. He is also a professor at Hangzhou Dianzi University, Hangzhou, China. Dr. Korlev obtained a PhD in Physics and Mathematics in 1981 and a Doctor of Science in Physics and Mathematics from Lomonosov Moscow State University in 1994. He is a Lomonosov Prize winner (2005), Honorary Professor of Lomonosov Moscow State University (2006), and Leonard Euler medal (2014, Perm State University). Dr. Korolev's main fields of interest include limit theorems of probability theory and mathematical statistics and their applications to risk theory, reliability theory, financial mathematics and modeling real phenomena. He is the author of more than 300 scientific papers and more than 20 books.

# Contents

# Preface

Since ancient times combinatorics and probability theory have been closely interrelated. Combinatorial techniques were used for the calculation of probabilities even when the latter were not directly assumed and were hidden behind the concept of "chances." After the notion of probability appeared, was acknowledged and comprehended, this interrelation became even stronger. Now probabilistic and combinatorial techniques are often used for solving advanced problems. The union of probability and combinatorics becomes more and more actual within the problem of Big Data, which assumes the possibility of detection of latent regularities or relations in new non-structured data types and construction of predictive models. One of the most urgent directions of the development of methods for Big Data analysis is their application in artificial intelligence systems. The problems of adequate modeling involve the development of the mathematical apparatus for the construction of reasonable models of statistical regularities and the study of their analytical and asymptotic properties. Whereas for a layman probability is still associated with divination on daisies, for specialists these methods long ago became powerful tools in predicting successes or failures, preventive management and achieving the desired successes. Some of these applicable techniques are demonstrated in this book. It is worth noting that to a great extent the calculus of probabilities became a mathematical theory due to the findings and works of the representatives of the Russian mathematical school: P.L. Chebyshev (1821–1894), A. A. Markov (1856–1922), A.M. Lyapunov (1857–1918), S.N. Bernstein (1880–1968), A.Ya. Khinchin (1894–1959), A.N. Kolmogorov (1903–1987), B.V. Gnedenko (1912–1995), Yu.V. Prokhorov (1929–2013), who by all means deserve to be remembered along with the famous creators of probability theory J. Bernoulli (1655–1705), P.-S. Laplace (1749–1827), S.D. Poisson (1781–1840), C.F. Gauss (1777–1855) and others.

This book is very unusual. It is not at all a random collection on a topic devoted to a formalization of uncertainties. Considering that because of high complexity and uncertainties the existing probabilistic models can't be used sometimes directly to predict and estimate desired results, the initial concept of the book is as follows:

- to draw the reader's attention to the fact that the same mathematical models for cognitive solving of different problems are useful in a variety of applications (e.g., artificial intelligence systems, high-power hydro turbine, system of flowering synchronization with respect to climate dynamics, organizational-technical-economical systems, offshore platforms, asymptotic approximations, graphs, uncertain costs, equipment of nuclear power plants, trunk oil pipelines, risk processes, social networks, combinatorial cosmology);

- to educate how modern probabilistic and combinatorial models may be created to formalize uncertainties;

- to train how new probabilistic models can be generated for the systems of complex structures;

- to present the correct use of the presented models for rational control in systems creation and operation; and

- to demonstrate analytical possibilities and practical effects for solving different system problems on each life cycle stage.

For example, the described approaches are applicable:

- on concept stage – for a system analysis of expected challenges, risks, usefulness and profits, studying of system creation, rationale of system requirements;
- on development, production and utilization stages – for a system analysis of feasibility, estimation of technical solutions and risks, prediction (on possible data) of reliability, safety, quality, effectiveness and efficiency, for optimization;
- on support stage – for an estimation of challenges and risks, prediction (on real data) of reliability, safety, quality, effectiveness and efficiency, for optimization, planning and rationale of improvements.

Of course, the described approaches do not exhaust all existing views on the problems of "probability, combinatorics and control." Nevertheless, after competent application of the proposed models, we are sure the reader will be able to trace distinctly the following chain of purposeful actions of the authors of this book:

1. From a formalization of uncertainties - to probabilistic modeling;
2. From probabilistic modeling - to reasonable control;
3. From reasonable control - to artificial intelligence;
4. From balanced human and artificial intelligence - to achievable effects;
5. From achievable effects - to sustainable harmony.

We wish you, dear readers, the patience in understanding the book's ideas and their successful implementations in different theoretical and applications areas. It will allow to control effects in time and to achieve sustainable harmony in system engineering, your creative life and activity.

**Andrey Kostogryzov**
Federal Research Center "Computer Science and Control"
of the Russian Academyof Sciences,
Gubkin Russian State University of Oil and Gas,
Russia

**Victor Korolev**
Lomonosov Moscow State University
(Faculty of Computational Mathematics and Cybernetics,
Department of Mathematical Statistics),
Federal Research Center "Computer Science and Control"
of the Russian Academy of Sciences,
Russia

Section 1

# From Reasonable Control to Balanced Human and Artificial Intelligence, Achievable Effects and Sustainable Harmony

Chapter 1

# Probabilistic Methods for Cognitive Solving of Some Problems in Artificial Intelligence Systems

*Andrey Kostogryzov and Victor Korolev*

## Abstract

As a result of the analysis of dispatcher intelligence centers and aerial, land, underground, underwater, universal, and functionally focused artificial intelligence robotics systems, the problems of rational control, due to be performed under specific conditions of uncertainties, are chosen for probabilistic study. The choice covers the problems of planning the possibilities of functions performance on the base of monitored information about events and conditions and the problem of robot route optimization under limitations on risk of "failure" in conditions of uncertainties. These problems are resolved with a use of the proposed probabilistic approach. The proposed methods are based on selected probabilistic models (for "black box" and complex systems), which are implemented effectively in wide application areas. The cognitive solving of problems consists in improvements, accumulation, analysis, and use of appearing knowledge. The described analytical solutions are demonstrated by practical examples.

**Keywords:** artificial intelligence system, method, probability, risk, uncertainty

## 1. Introduction

Today, artificial intelligence (AI) has confidently entered our lives. The first mention of it belongs to the mid-50s of the last century. Under AI, we usually understand it as the branch of computer science devoted to develop data processing systems that perform functions normally associated with human intelligence, such as reasoning, learning, and self-improvement (ISO/IEC 2382-1:1993 Information technology–Vocabulary–Part 1). According to this, over the decades, AI has found its application in expert systems supporting decision-making, in heuristic classification, computer vision, pattern recognition, understanding natural language, etc. [1–14]. Here, under AI systems (AIS), we understand systems that include data processing systems that perform functions by AI, in particular by modeling and logic reasoning.

Note. System is a combination of interacting elements organized to achieve one or more stated purposes (according to ISO/IEC/IEEE 15288 "Systems and software engineering–System life cycle processes").

If the modern human brain already possesses skills of adaptation to conditions of various uncertainties in the world around, artificial intelligence systems require creation of effective methods for cognitive solving actual practical problems. "Cognitive solving" means relating to or involving the processes of thinking and reasoning (Cambridge English Dictionary). The applicable mathematical methods are focused mainly on conditions of actions in the logician "if ..., that ..." according to the gathered information, and on an estimation of traced situations by a man-operator. At increase and expansion of uncertainty conditions, quite often, there are failures and errors because of complexity. It means that search of new methods for advanced solving of AIS practical problems today is very important.

In the present chapter, various AIS for supporting decision-making in intellectual manufacture and robotics systems are analyzed. According to robotics, it is supposed that AIS may be used for solving multiple aerial, land, underground, underwater, universal, and special problems of creation and operation. At the same time, we would like to emphasize that the main efforts of this chapter are not focused on illustrating the capabilities of AIS, but on demonstrating the applicability of author's probabilistic models and methods to improve some of the existing capabilities of AIS.

For this goal, the problem of planning the possibilities of functions performance on the base of monitored information and the problem of robot route optimization under uncertainties limitations are chosen. The choice of these problems in AIS applications is caused on the one hand by increase of quantity and a variety of specific uncertainties conditions, and on the other hand by an urgency and width of areas for their practical use. However, some relevant problems (such as the problems of robotics orientation, localization and mapping, information gathering, the perception and analysis of commands, movement and tactile, realizations of manipulations, and also rational control) for which different probabilistic methods are also applicable have been left out of the scope of work.

For cognitive solving and improvements by the use of probabilistic methods, the chosen problems are transformed more specifically to:

- problem 1 of planning the possibilities of functions performance on the base of monitored information about events and conditions, and

- problem 2 of robot route optimization under limitations on risk of "failure" in conditions of uncertainties.

The proposed methods for cognitive solving AIS problems are based on theoretical and practical author's researches [15–37] and need to be used either in combination or in addition to existing methods. There, where often it is required prognostic analysis or where the used approaches are not effective, the proposed methods can be used as rational basis or alternative.

The proposed and referred author's methods and models can be used in AIS life cycle to form system requirements, compare different processes, rationale technical decisions, and estimate reliability, quality, and risks. The decisions, scientifically proved by the offered models and software tools, can provide purposeful essential improvement of quality and mitigation of risks and decrease expenses for created and operating systems. The spectrum of the explored systems by these methods includes systems (not only AIS) operated by government agencies, manufacturing structures (including power generation, coal enterprises, oil and gas systems), food storage, space industry, emergency services, municipal economy, etc. [15–19, 22–37]. The supporting software tools are original Russian creations registered by Rospatent [38–44]. They have been presented at seminars, symposiums, conferences, ISO/IEC working groups, and other forums since 2000 in Russia, Australia, Canada, China, Finland, France, Germany, Italy, Kuwait, Luxembourg, Poland, Serbia, the USA, etc. The software tools were awarded by the Golden Medal of the International Innovation and Investment Salon and the International Exhibition "Intellectual Robots," acknowledged on the World's fair of information technologies CeBIT in Germany, noted by diplomas of the Hanover Industrial Exhibition and the Russian exhibitions of software.

Note. The proposed methods below do not replace existing methods for robots actions (for example, the methods of solving the systems of differential equations, the methods of refreshed linear and geometric algebra, geometry, Lie groups, linearization, solving Jacobians and Hessians, Kalman filters, Lyapunov analysis, the methods of biomechanics, graph theory, Laplas transforming for large-scale dynamic systems, etc.) [1–14].

The structure of the chapter research is shown in **Figure 1**. It provides an explanation of the essence of cognitive solving of problems on the base of probabilistic modeling, selection of some author's probabilistic models applicable for cognitive solving problems 1 and 2, the practical steps to solve these problems, and five practical examples demonstrating system planning the possibilities of functions performance by using robot-manipulators (in space), by AIS for a coal company and by AIS used for a security service of floating oil and gas platform, example of forming input for probabilistic modeling from monitored data and example of robot route optimization under limitations on risk of "failure" in conditions of uncertainties. Various areas of the examples' applications have been chosen purposely to demonstrate universality and analytical usefulness of the proposed methods and models. Appendices includes the proof for the proposed model of a quite general technology of periodical diagnostics of system integrity and some short models results to estimate quality of used information.

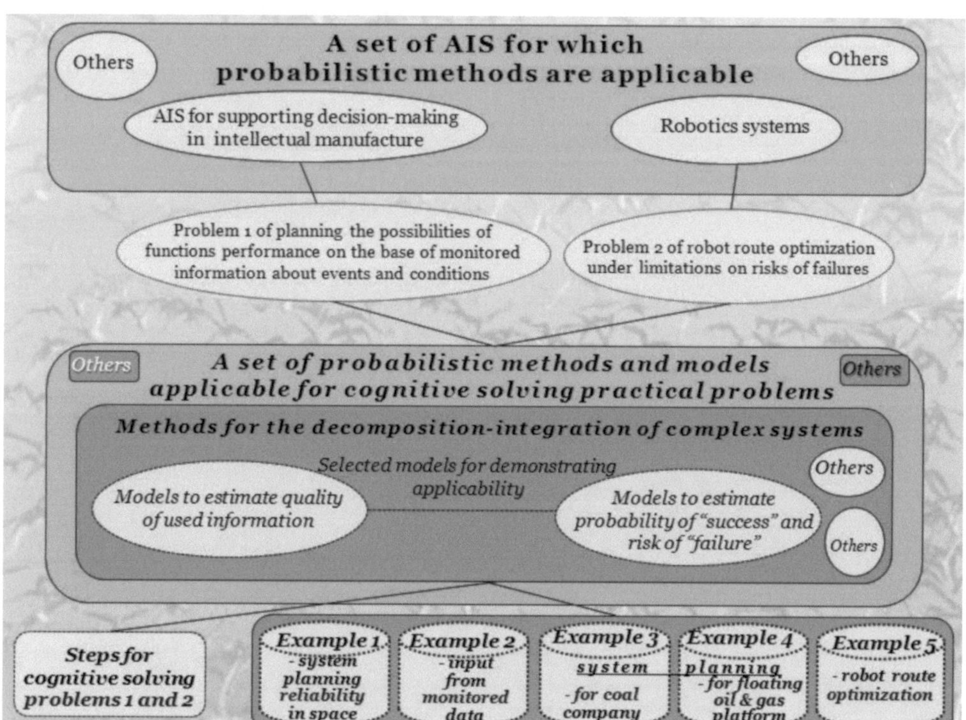


**Figure 1.**
*The structure of the research.*

## 2. The essence of cognitive solving of problems on the base of probabilistic modeling

This section explains the definitions and interpretations which can help to understand the proposed models and results of modeling complex systems in different application areas.

AIS itself can be considered as an interested system (for example, dispatching intellectual center) or as a part of other, more comprehensive interested system (for example, functionally focused robots in safety systems). The current information is processed in real time for performing the set or expected functions of interested system. To meet system requirements, the solutions of considered problems 1 (of planning the possibilities of functions performance) and 2 (of robot route optimization) are initiated along with the solutions of other problems.

The cognitive solving of problems include improvements, accumulation, analysis, and a use of appearing knowledge, see **Figure 2**. Possible uncertainties for the given period (from initial time point $t_1$ to future moment $t_x$) may be considered by using proposed probabilistic modeling, prediction, and optimization.

The solutions for problems 1 and 2 are estimated by probability of "success" and/or "failure" (risk of "failure") during given prognostic time period. Thus, prognostic period should be defined so to be in time to recover capabilities (which can be lost), or to carry out preventive action (with which the initiation of solving the problem is connected). Such behavior means operation in real time.

In each real case of modeling the term "success" should be defined in terms of admissible condition of interested system to operate for the purpose. The term "failure" means "unsuccess." Generally, a "success" of interested system operation during the given time period means an admissible degree of integrity. Accordingly, "failure" for interested system during given time period means inadmissible degree of integrity at least once within this period. System (or system element) integrity is defined as such system (system element) state when system (system element) purposes are achieved with the required quality and/or safety. The risk of "failure"

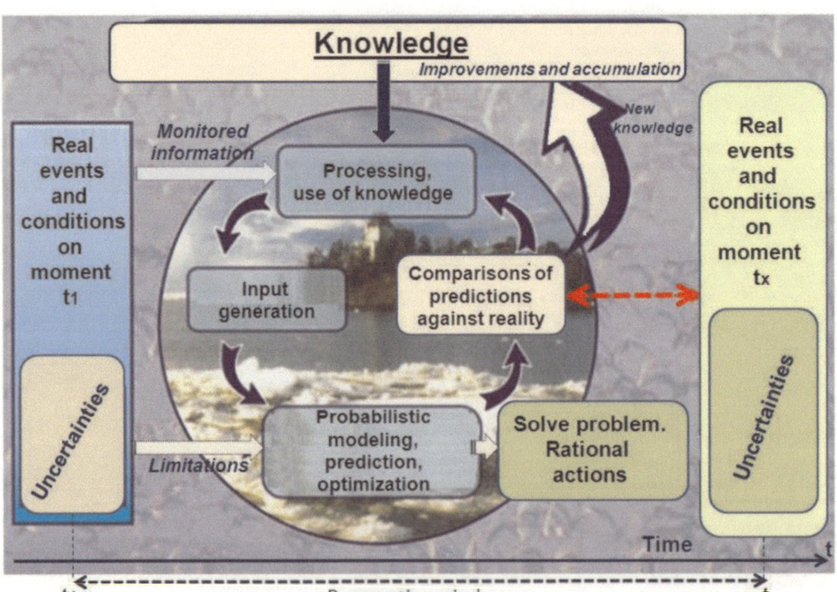


**Figure 2.**
*The essence of cognitive of solving of problems.*

is understood as a probabilistic measure of "failure" considering consequences (according to ISO Guide 73).

Note. For example, an interested system is a dangerous manufacturing object. The object structure includes an AIS, which monitors events and conditions in and/or around its manufacture. Equipment parameters (temperature, pressure, and so forth) which should be in norm limits are traced. The "failure" of interested system operation may mean an incident or accident on object.

Generally, from the point of view of formalization for each estimated variants (for problem 1 or 2), the interested system is logically decomposed to compound subsystems; see **Figures 3** and **4**. Each subsystem is a set of components (elements and/or other subsystems): for problem 1, this set covers the components participating in functions performance; and for problem 2, the set covers compound parts of a possible route of the robot in space. Complete set of these components formally characterizes a variant of decomposed system for solving problem 1 or 2. The analysis and optimization are carried out on complete set of all compared possible variants.

Interpretation of such decomposition is the following:

The subsystem from serial connected elements provides functions performance with admissible level of integrity (quality and/or safety) at given time, if:

- "AND" 1st component, ..., "AND" last element provide admissible level of integrity (quality and/or safety) at given time (for problem 1);
- "AND" 1st compound part of the route, ..., "AND" last compound part of the route are overcame successfully by the robot at given time (for problem 2).

The subsystem from parallel connected elements provides functions performance with admissible level of integrity (quality and/or safety) at given time, if:

- "OR" 1st component, ..., "OR" last component in the subsystem provide admissible level of integrity (quality and/or safety) at given time (for problem 1);

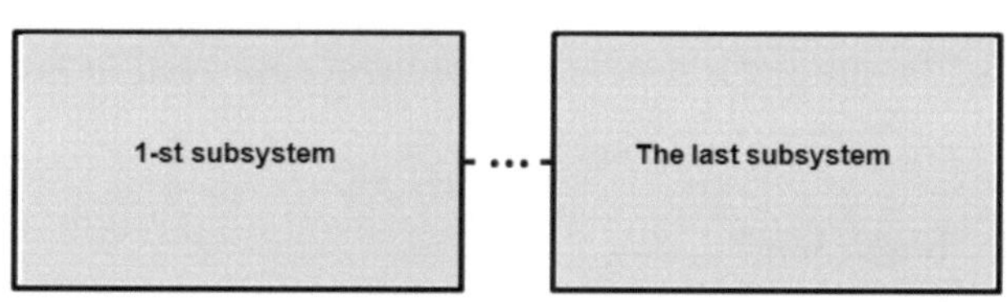


**Figure 3.**
*Variant of system decomposition.*

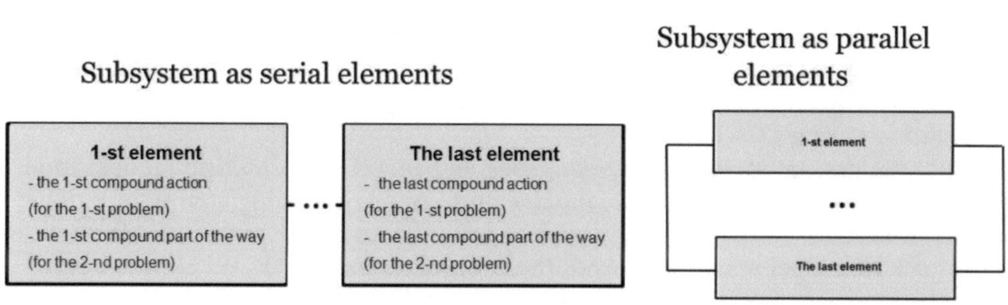


**Figure 4.**
*Variant of subsystem decomposition.*

- "OR" 1st compound part of the route, ..., "OR" last compound part of the route are overcame successfully by the robot at given time (for problem 2).

Each component after system decomposition is presented as a "black box." For each "black box," various probabilistic models can be applied for calculations and for building required probabilistic distribution function (PDF) of time between the next deviations from an established norm. A norm is connected with definitions of "success" and "failure," it may be connected with the precondition to "failure" (to prevent "failure"—see Example 2). Focus on processes' description allows to use only time characteristics (mean time or frequency of events), the dimensionless or cost characteristics peculiar for various applications.

Appropriate calculated probabilities of "success" and/or "failure" (risk of "failure") in comparisons to real events during the prediction periods represent the knowledge of admissibility borders for probabilities of "success" and acceptability borders for risks of "failure." The process of cognitive solving of problems 1 and 2 means not only the formation and use of this knowledge for interested system, but also the estimated quality of monitored and used information (including definition of input for continuous modeling).

## 3. Selection of the models

The proposed probabilistic methods for cognitive solving of problems 1 and 2 are based on selected probabilistic models which are implemented effectively in wide application areas. The main principle at a selection of models consists that useful knowledge should be result of their application in conditions of various uncertainties. Knowledge is understood as the form of existence and ordering of results of cognitive activity of human. In the applications to solve problems 1 and 2, useful knowledge (received as a result of probabilistic modeling in time) is an output information of admissible quality or cognitive conclusion that allows to solve a specific applied problem. As the results of selection, the author's models to estimate the probabilistic measures of a quality of used information and the probabilities of "success" and risks of "failure" for "black box" and for complex structures are proposed for AIS. The models are widely tested and approved in practice [15–19, 22–37].

### 3.1 Selection for "black box"

Selected models for every system element, presented as "black box," allow to estimate probabilities of "success" and/or "failure" during given prognostic period. A probabilistic space ($\Omega$, $B$, $P$) for estimation of system operation processes is traditional [15–21], where $\Omega$ is a limited space of elementary events; $B$ is a class of all subspace of $\Omega$-space, satisfied to the properties of $\sigma$-algebra; and $P$ is a probability measure on a space of elementary events $\Omega$. Such space ($\Omega$, $B$, $P$) is built and proposed for using [22–37].

Not considering uncertainty specificities, in general case, intellectual operation of AIS component aims to provide reliable and timely producing complete, valid and/or, if needed, confidential information; see **Figure** 5. The gathered information is used for its proper specificity. And, the proposed models [18–19] allow to estimate the intellectual operation processes on a level of used information quality, which is important for every AIS (information may be used by technical devices, "smart" elements, robotics, users, etc.).

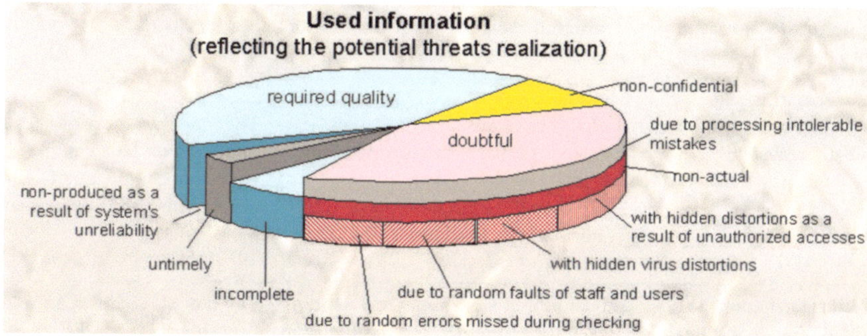


**Figure 5.**
*Quality of used information (abstraction).*

The proposed analytical models ("The model of functions performance by a complex system in conditions of unreliability of its components," "The models complex of calls processing for the different dispatcher technologies," "The model of entering into system current data concerning new objects of application domain," "The model of information gathering," "The model of information analysis," "The models complex of dangerous influences on a protected system," and "The models complex of an authorized access to system resources") allow to estimate the probability of "success" and risks to lose quality of intellectual operation during given prognostic period considering consequences; see **Table 1**. Required limits on probability measures are recommended as produced knowledge for the best AIS practice (estimated on dozens practical estimations for various application areas).

The next probabilistic model is devoted to estimate a probability of "success" and risk of "failure" on high meta level. This is based on studying the general AIS technology of periodical diagnostics of system integrity. Some general technologies were researched for "The models complex of dangerous influences on a protected system," see **Table 1**. Here, the general case for AIS is presented.

For system element allowing prediction of risks to lose its integrity during given prognostic period, there is studied the next general AIS technology of providing system integrity.

Technology is based on the periodical diagnostics of system integrity (without the continuous monitoring between diagnostics). Diagnostics are carried out to detect danger sources occurrence from threats into a system or consequences of negative influences (for example, these may be destabilizing factors on dangerous enterprise). The lost system integrity can be detected only as a result of diagnostics, after which system recovery is started. Dangerous influence on system is acted step-by step: at first, a danger source occurs into a system, and then after its activation may be a loss of integrity; see **Figure 6**. Occurrence time is a random value that can be distributed by PDF of time between neighboring occurrences of danger $\Omega_{occur}(t) = P(\tau_{occurrence} \leq t) = 1 - \exp(t/T_{occur})$, $T_{occur}$ is mean time, frequency $\sigma = 1/T_{occur}$. Activation time is also random value which can be distributed by PDF of activation time of occurred danger $\Omega_{activ}(t) = P(\tau_{activation} \leq t) = 1 - \exp(t/T_{activ})$, $T_{activ}$ is mean time. System integrity cannot be lost before an occurred danger source is activated. A threat is considered to be realized only after a danger source has activated and influenced on system.

It is supposed that used diagnostics tools allow to provide system integrity recovery after revealing danger sources occurrence or the consequences of influences. Thus, the probability ($P$) of providing system integrity within the given

| Threats to AIS operation quality | Evaluated measure (required limits as produced knowledge for the best practice) | Model tittle |
|---|---|---|
| Information is not produced as a result of system unreliability | Probability of providing reliable functions performance during given time (no less than 0.99).<br>Mean time between failures.<br>System availability (no less than 0.9995) | The model of functions performance by a complex system in conditions of unreliability of its components |
| Delayed information producing (i.e., not in real time) | Probability of well-timed processing during the required term (no less than 0.95).<br>Mean response time.<br>Relative portion of all well-timed processed calls.<br>Relative portion of well-timed processed calls of those types for which the customer requirements are met (no less than 95%) | The models complex of calls processing for the different dispatcher technologies |
| Producing of incomplete information | Probability that system contains information about states of all real object and coincides (no less than 0.9) | The model of entering into system current data concerning new objects of application domain |
| Information validity deterioration caused by:<br>• non-actual input information;<br>• errors missed or made during information verification;<br>• incorrectness of processing | Probability of information actuality on the moment of its use (no less than 0.9).<br>Probability of errors absence after checking (no less than 0.97).<br>Fraction of errors in information after checking.<br>Probability of correct analysis results obtaining (no less than 0.95) | The model of information gathering.<br>The model of information analysis |
| Violation of information confidentiality | Probability of system protection against unauthorized access during objective period (no less than 0.999) | The models complex of an authorized access to system resources |
| Violation of secure system operation including<br>• random faults of staff and users;<br>• dangerous influences (revealing of software and technical defects, virus influences, violators' influences, terrorist attacks in information environment, psychological influence etc.);<br>• unauthorized access | Probability of faultless (correct) operation during given time (no less than 0.95).<br>Mean time between errors.<br>Probability of system protection against unauthorized access (no less than 0.99) | The models complex of dangerous influences on a protected system.<br>The models complex of an authorized access to system resources |

**Table 1.**
*The proposed analytical models to estimate AIS operation quality [18, 19].*

prognostic period $T_{given}$ (i.e., probability of "success") may be estimated as a result of the use of the next probabilistic model. Risk to lose integrity (R) addition to 1 for probability of providing system integrity R = 1 − P.

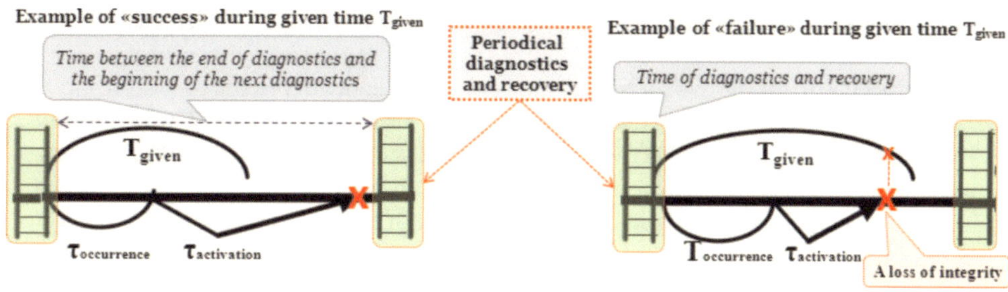


**Figure 6.**
*Some random events for technology: left—correct operation to provide system integrity; right—a loss of integrity during prognostic period $T_{given}$.*

There are possible the next variants:

variant 1—given prognostic period $T_{given}$ is less than the established period between neighboring diagnostics ($T_{given} < T_{betw.} + T_{diag}$);
variant 2—given prognostic period $T_{given}$ is more than or equal to the established period between neighboring diagnostics ($T_{given} \geq T_{betw.} + T_{diag}$).

Here, $T_{betw.}$ is the time between the end of diagnostics and the beginning of the next diagnostics and $T_{diag}$ is the diagnostics time.

For the given period $T_{given}$, the next statements are proposed for use, see in detail [18, 19, 35–37].

Under the condition of independence of considered characteristics, the probability of providing system integrity (probability of "success") is equal to

1. for variant 1

$$P_{(1)}\left(T_{given}\right) = 1 - \Omega_{occur} * \Omega_{activ}\left(T_{given}\right), \tag{1}$$

2. for variant 2

measure (a)

$$\begin{aligned} P_{(2)}\left(T_{given}\right) = {} & N\left(\left(T_{betw} + T_{diag}\right)/T_{given}\right)P_{(1)}{}^{N}\left(T_{betw} + T_{diag}\right) \\ & + \left(T_{rmn}/T_{given}\right) P_{(1)}\left(T_{rmn}\right), \end{aligned} \tag{2}$$

where $N = [T_{given}/(T_{betw.} + T_{diag.})]$ is the integer part, $T_{rmn} = T_{given} - N(T_{betw} + T_{diag})$;

measure (b)

$$P_{(2)}\left(T_{given}\right) = P_{(1)}{}^{N}\left(T_{betw} + T_{diag}\right)P_{(1)}\left(T_{rmn}\right) \tag{3}$$

The probability of success within given prognostic period $P_{(1)}(T_{given})$ is defined by (1).

The modification of this model allows to use different values of diagnostics and recovery time [35–37]; for formulas (1)–(3), recovery time is equal to diagnostics time.

All these models, supported by various versions of software tools, registered by Rospatent, may be applied and improved for solving quality and safety problems, connected with intellectual system presented as "black box" [18, 19, 38–44].

Summaries for the last model are as follows:

- The input for modeling include: frequency of the occurrences of potential threats (or mean time between the moments of the occurrences of potential threats which equals to 1/frequency); mean activation time of threats; mean recovery time; time between the end of diagnostics and the beginning of the next diagnostics; diagnostics time; and given prognostic period.

- The calculated results of modeling include: the probability of providing system integrity within given prognostic period (i.e., probability of "success"); and risk to lose integrity (i.e., probability of "failure") as addition to 1 for probability of "success."

### 3.2 Integration for complex structures on the level of probability distribution functions

If probability of providing system integrity within given prognostic period for all points $T_{given}$ from 0 to ∞ are computed, it means a trajectory of the PDF depending on characteristics of threats, periodic diagnostics, and recovery. And, the building of PDF is the real base to predict probabilistic metrics for given time $T_{given}$. In analogy with reliability, it is important to know a mean time between neighboring losses of integrity (MTBLI) like mean time between failures in reliability (MTBF), but in application to concepts of quality, safety, etc.

For complex systems with serial or parallel structure, new models with known PDF can be developed by the next method [17–21]. Let us consider the elementary structure from two independent parallel or serial elements (**Figures 3** and **4**). Let the PDF of time between losses of i-th element integrity be $B_i(t)$, i.e., $B_i(t) = P(\tau_i \le t)$, then:

1. time between losses of integrity for system combined from serial connected-independent elements is equal to minimum from two times $\tau_i$: failure of first or second elements (it means the system goes into a state of lost integrity when either first, or second element integrity is lost). For this case, the PDF of time between losses of system integrity is defined by the expression

$$B(t) = P(\min(\tau_1, \tau_2) \le t) = 1 - P(\min(\tau_1, \tau_2) > t) = 1 - P(\tau_1 > t)P(\tau_2 > t) \\ = 1 - [1 - B_1(t)][1 - B_2(t)] \quad (4)$$

2. time between losses of integrity for system combined from parallel connected independent elements (hot reservation) is equal to a maximum from two times $\tau_i$: failure of first and second elements (it means the system goes into a state of lost integrity when both first and second elements have lost integrity). For this case, the PDF of time between losses of system integrity is defined by the expression

$$B(t) = P(\max(\tau_1, \tau_2) \le t) = P(\tau_1 \le t)P(\tau_2 \le t) = B_1(t)B_2(t) \quad (5)$$

By applying recurrently expressions (4) and (5), it is possible to build PDF of time between losses of integrity for any complex system with parallel and/or serial structure.

As summary, the calculated results of modeling are: PDF of time between losses of integrity for system and each compound subsystems and elements; mean time between losses of integrity for system and each compound subsystems and elements (MTBLI as analog of MTBF).

For example, integrated complex system, combined from intellectual structures for modeling interested system including AIS (**Figure** 7), can be analyzed by formulas (1)–(5) and probabilistic models described above and allowing to form PDF by (4) and (5). The correct operation of this complex system during the given period means: during this period both first and second subsystems (left and right) should operate correctly according their destinations, i.e., integrity of complex system is provided if "AND" integrity of first system left "AND" integrity of second system right are provided.

All these ideas of analytical modeling operation processes are supported by the software tools [18, 19, 21, 23, 38–44].

What about new knowledge by using the proposed methods and models for cognitive solving of problems 1 and 2 of the chapter? A use of these methods and models on different stages of AIS life cycle (concept, development, utilization, support stages) allows to produce cognitive answers for the following questions:

- What about different risks to lose integrity in operation?
- What about the justified norms for values of monitored parameters?
- What requirements should be specified to MTBLI and to repair time for different possible scenarios of operation?
- Which information operation processes should be duplicated and how?

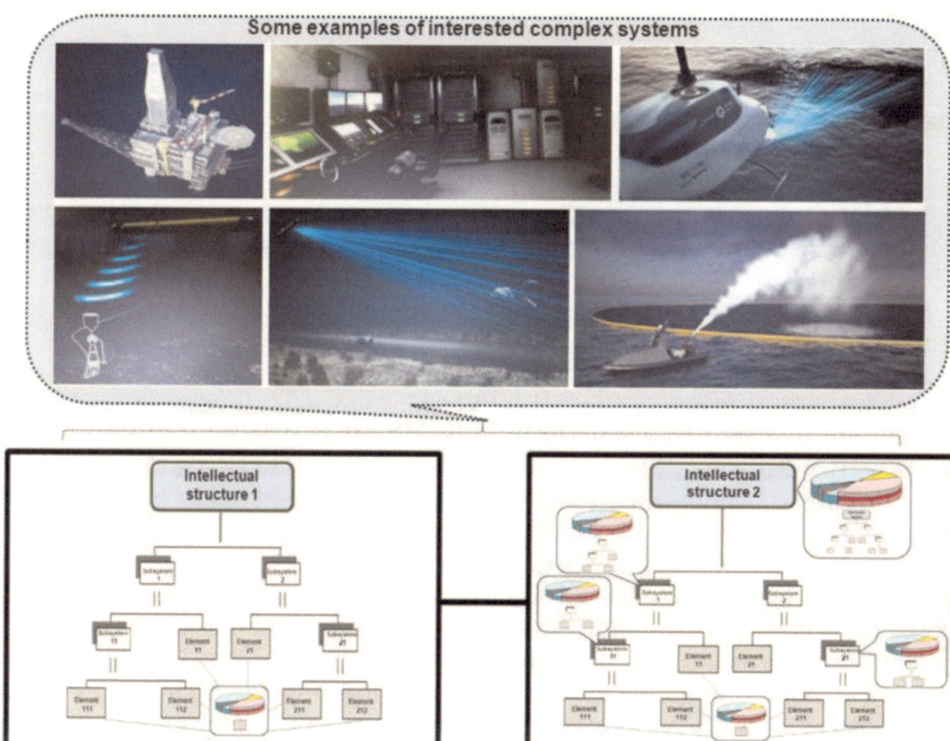


**Figure 7.**
*An integrated complex system of two serial subsystems (abstraction).*

- What processing devices and technologies should be used to achieve the necessary level of system integrity (quality, safety, etc.)?
- What is the system tolerance to data flows changing?
- What data flows and functional tasks may be the main causes of "bottlenecks"?
- What data gathering technologies and engineering solutions can guarantee the completeness and actuality of used information?
- What information verification and validation control should be used?
- What qualification requirements should be for the users of AIS (from the AIS effectiveness and efficiency points of view)?
- How dangerous are scenarios of environment influences and what protective technologies will provide the required security?
- How the use of integrity diagnostics and security monitoring will worsen time-probabilistic characteristics of system?
- What protection system effectiveness should be to prevent an unauthorized access?
- What are the information security risks? etc.

The rationale answers allow to improve and accumulate knowledge concerning AIS.

The proposed methods and models provide the next approach for cognitive solving problems 1 and 2.

## 4. Problem 1 of planning the possibilities of functions performance on the base of monitored information about events and conditions

It is supposed that the terms "success" and accordingly "failure" are defined in terms of admissible condition of interested system to operate for the purpose.

Note. For example, for each parameter of equipment, the ranges of possible values of conditions may be estimated as "Working range inside of norm" and "Out of working range, but inside of norm" ("success") or "Abnormality" ("failure"), interpreted similarly light signals—"green," "yellow," and "red." For this definition, a "failure" of equipment operation characterizes a threat to lose system norm integrity after danger influence (on the logic level this range "Abnormality" may be interpreted analytically as failure, fault, losses of quality, or safety etc.). But the definition may be another: for example, a "failure" may be defined as incident or accident. For this definition, short-time being in the range "Abnormality" is not "failure," because the incident or accident may not happen.

There are four steps proposed for cognitive solving of problem 1 of planning the possibilities of functions performance on the base of monitored information about events and conditions; see **Figure 8**.

**Step 1**. The complete set of variants for actions, and for each variant—a set of components is defined. Each use case may be characterized by an expected gain in comparable conventional units. If the objective value of gain cannot be defined,

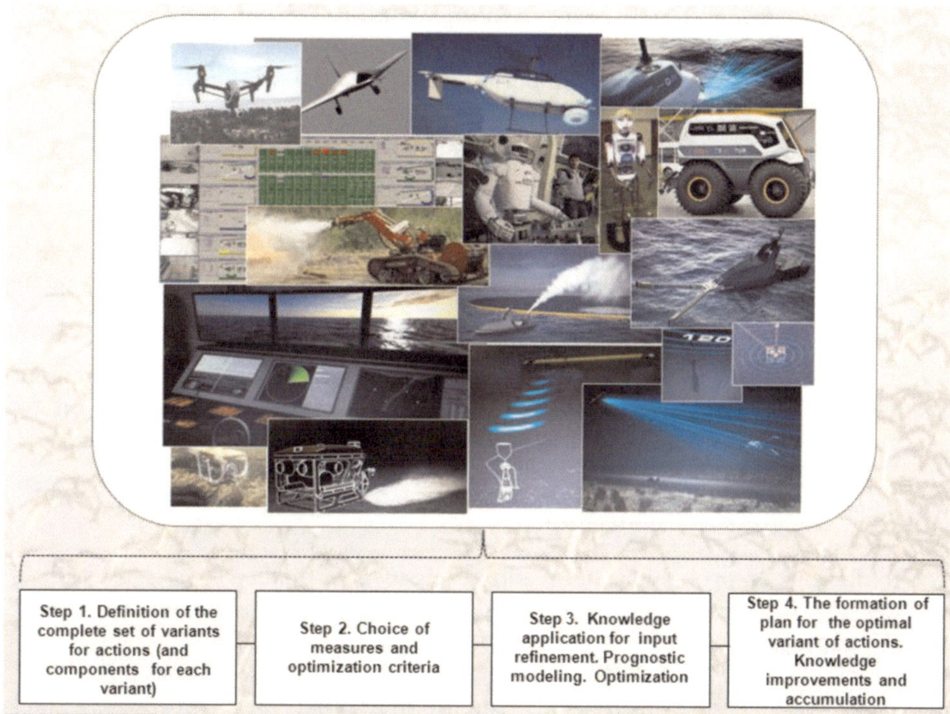


**Figure 8.**
*Steps for cognitive solving of problem 1.*

expert value of expected level of "success" for each variant may be established, for example, on a dimensionless scale from 0 to 100 (0—"no gain", i.e., "failure"; 100—"the maximal gain," i.e., complete "success"). After learning by knowledge base, self-improving AIS uses input and the corresponding results of probabilistic modeling in a form of the solution of previously specific encountered problem 1.

Knowledge base (K-base) is defined as a database that contains inference rules and information about human experience and expertise in a domain (ISO/IEC 2382-1:1993).

**Step 2**. The measures and optimization criteria are chosen. As criteria can be accepted:

- Maximum of gain as a result of the functions performance under the given conditions and limitations on the acceptable risk of failure and/or other limitations

- Maximum probability of "success" or minimum risk of "failure" under limitations

**Step 3**. The accumulated knowledge is used to refine the input for modeling. A quality of used information is estimated by models above considering limitations from **Table 1**. Using the model for each variant, the probabilistic measures are calculated for given prognostic period (see proposed models above and Step 1). From a set of possible variants, the optimal one is chosen, according to Step 2 criterion.

Note. For example, there are proposed the next general formal statements of problems for system optimization:

1. on the stages of system concept, development, production, and support: system parameters, software, technical, and management measures (Q) are the most rationale for the given period if on them the minimum of expenses ($Z_{dev.}$) for creation of system is reached

$$Z_{dev.}\left(Q_{rational}\right) = \min_{Q} Z_{dev.}\left(Q\right), \tag{6}$$

at limitations on probability of an admissible level of quality $P_{quality}(Q) \geq P_{adm.}$ and expenses for operation $C_{oper.}(Q) \leq C_{adm.}$ and under other development, operation, or maintenance conditions;

2. on utilization stage: system parameters, software, technical, and management measures (Q) are the most rational for the given period of operation if on them the maximum of probability of correct system operation is reached

$$P_{quality}\left(Q_{rational}\right) = \max_{Q} P_{quality}\left(Q\right), \tag{7}$$

at limitations on probability of an admissible level of quality $P_{quality}(Q) \geq P_{adm.}$ and expenses for operation $C_{oper.}(Q) \leq C_{adm.}$ and under other operation or maintenance conditions.

For limitation on $P_{quality}(Q)$ K-base is used; for example, see **Table 1**. For calculation probabilistic measures for given prognostic period, the proposed models are used.

These statements (6), (7) may be transformed into the problems of expenses or risk minimization in different limitations. There may be a combination of these formal statements in system's life cycle.

**Step 4**. A plan for the optimal variant of actions (defined in Step 3) is formed. To support the efficiency and/or effectiveness of the functions, the achievable gain calculated at Step 3 is recorded. New knowledge is improved, accumulated, and systematized in K-base by comparing it with reality (for example, by a specific method considering AIS capabilities for self-improving).

Note. A solution that meets all conditions may not exist. In this case, there is no optimal variant of planning the possibilities of functions performance on the base of monitored information. Additional systems analysis, adjustment of the criteria, or limitations is required (see, for example, ISO/IEC/IEEE 15288).

## 5. Problem 2 of robot route optimization under limitations on risk of "failure" in conditions of uncertainties

For a robot, the concept of "failure" under uncertainty is defined as the "unsuccess" to achieve the goal within a given time. It is assumed that there are several possible routes to achieve the goal, and uncertainties may include both the conditions for robot operation (including random events in orientation, localization, and mapping in cooperation with drone for gathering actual data). The minimum risk of failure under the existing conditions and limitations is used as a criterion of optimization.

The next four steps are proposed for cognitive solving of problem 2 of robot route optimization under limitations on risk of "failure" in conditions of uncertainties, see **Figure 9**.

**Step 1**. The complete set of route variants to achieve the goal within the given time, and for each variant—a set of components is defined (redefined). Data characterizing every part of route for each of the variants are gathered (refined) for

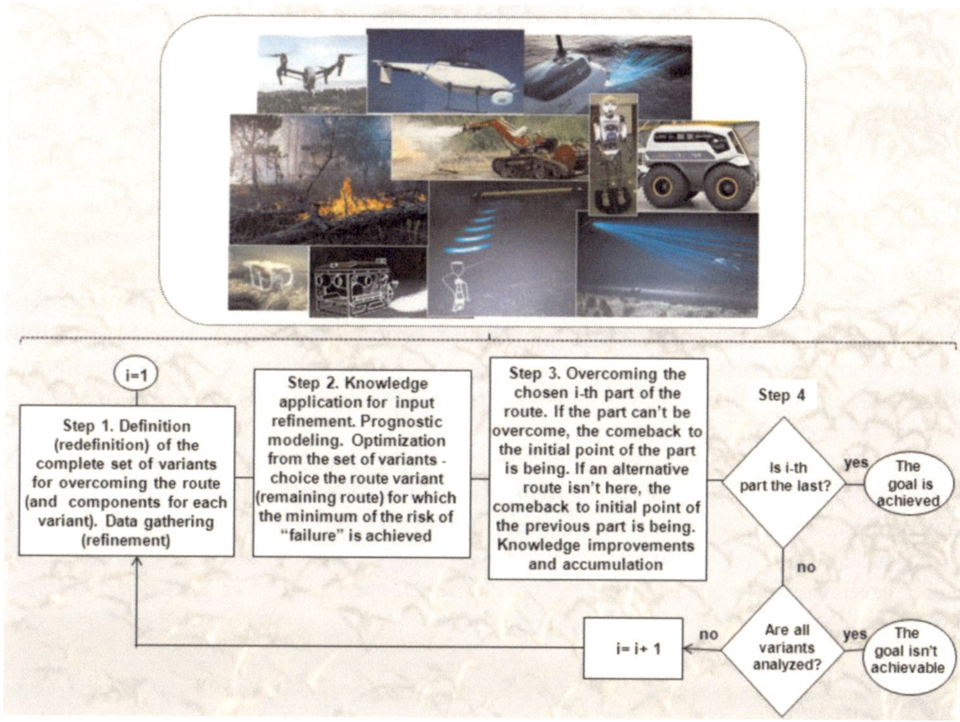


**Figure 9.**
*Steps for cognitive solving of problem 2.*

modeling. To do this, the robot can use data from various sources (for example, from air drones, intelligent buoys on the water or sensors under water, etc.). If necessary, possible damages are taken into account. For example, each use case may be characterized by an expected damage in comparable conventional units. If the objective value of a damage cannot be defined, expert value of expected level of "failure" for each variant may be established, for example, on a dimensionless scale from 0 to 100 (0—"no damages", i.e., "success"; 100—"the maximal damage"). After learning by K-base, self-improving AIS also uses input and the corresponding results of probabilistic modeling in a form of the solution of previously specific encountered problem 2.

The index i of the first part of the selected route is set to the initial value i = 1.

**Step 2**. The accumulated knowledge is used to refine the input for prognostic modeling. A quality of used information is estimated by models above considering limitations from **Table 1**. Using probabilistic model, a calculation of the probability of failure is carried out for each variant. From the set of remaining route variants, the optimal one is chosen (for it is the minimum probability of failure that is achieved).

**Step 3**. The robot overcomes the i-th part of the selected route. If the part cannot be overcome successfully according to probabilistic modeling and/or actual data, the comeback to the initial point of the part is being. If an alternative route is not here, the comeback to initial point of the previous part is being. The input for modeling every part of possible route for each of the variants is updated. New knowledge is improved, accumulated, and systematized in K-base by comparing it with reality (using a specific method considering AIS capabilities for self-improving).

**Step 4**. If, after overcoming the i-th part, the robot arrived at the intended point of route (i.e., the last part of the route is overcome and the goal is achieved), then the solution of task 2 for optimizing the route is complete. If the robot has not yet arrived at the intended point (i.e., the last part of the route is not overcome), then the complete set of different route variants for achieving the goal is redefined (similar to step 1). The input for modeling every part of possible route for each of the variants is updated. i = i + 1. Then, Steps 2–4 are repeated until the last part of the route is overcome on the set of possible variants (i.e., it means the goal is achieved and problem 2 is solved).

If the set of possible options is exhausted and the goal is not achieved, it is concluded that the goal is unattainable with the risk of "failure" less than the acceptable risk (i.e., it means an impossibility of solving problem 2 in the defined conditions).

Thus, for optimizing robot route in space (i.e., for the "successful" solution of problem 2) in real time, information gathering, probabilistic predictions for possible route variants, their comparison, the choice of the best variant, the implementation of further actions, the improvement, accumulation, systematization, and use of knowledge are being, see **Figure 9**.

Note. The proposed methods of solving problems 1 and 2 are essentially identical approaches based on the use of the same probabilistic models (Section 3). The only difference is that for the system planning the possibilities of functions performance (problem 1), the concept of "success" is used; and for the robot route optimization under limitations on risk of "failure" (problem 2), the concept of "failure," which is defined as the lack of "success," is used.

## 6. Examples

### 6.1 Example 1 of system planning the possibilities of functions performance in space by using robot-manipulators

Here, problem 1 (of planning the possibilities of functions performance) is solved by the proposed approach on the base of information gathered from different similar projects, accumulated and systematized in K-base including history. Applicability of the proposed probabilistic methods and models on development stage is demonstrated to improve some of the existing capabilities of robot-manipulator. It is required to predict the possible period of robot-manipulator use in space. When planning the possibilities of performing the functions of the cosmonaut-operator, two variants were compared: first variant–without a use of AIS; second–by using some AIS for supporting decision-making and monitoring the status of the operator's console, power units, central controller, and control handle for manipulator means.

A robot-manipulator as a system is composed on subsystems: an operator's console, a power unit, a central controller with a handle of control and manipulator means. There are supposed that a frequency of anomalies is in average 1 times a year, mean activation time from anomaly occurrence to failure is about 3 days. Time between the end of diagnostics and the beginning of the next diagnostics is about 2 months, and the recovery time is about 2 days.

System decomposition is presented on **Figure 10**. We do STEPS 1–4 (**Figure 8**) and use formulas (1)–(3) for solving the problem for complex structure composed by elementary variants decompositions presented on **Figures 3** and **4**. Here, probability of "success" ($P$) covers the following:

- Probability of reliable operation of robot-manipulator as a system
- Probability of reliable operation of every subsystem

Risks of "failure" (R) means addition to 1 for probability of "success."

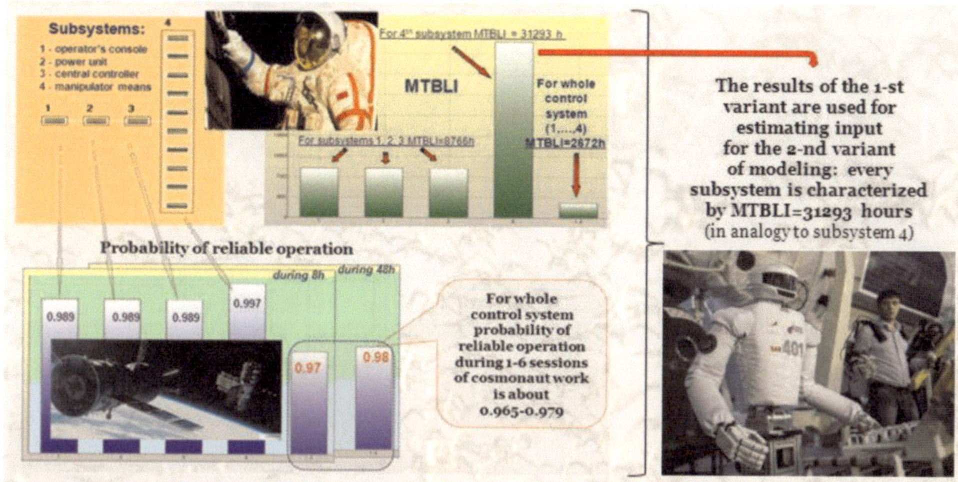


**Figure 10.**
*Results of probabilistic modeling robot-manipulator operation.*

Results of modeling the first variant of project have shown the following (**Figure 10**): for operator's console (first subsystem), power unit (second subsystem) and central controller with a handle of control (third subsystem) MTBLI = 8766 h, for manipulator means (including a hinge of roving of key, a hinge of shoulder, a hinge of roving of elbow, a hinge of elbow, a hinge of roving of brush, a hinge of brushes, a hinge of brush rotation, a device for grasping, videocamera—united as subsystem 4, which can operate if one of these means is available) MTBLI = 31,293 h, for all complex 1,...,4 MTBLI = 2672 h; probability of reliable operation of complex 1,...,4 during 8 h is equal to 0.979; probability of reliable operation of complex 1,...,4 during 48 h is equal to 0.965.

The maximum probability of "success" and minimum risk of "failure" under limitations on the successful functions performance are used as a criterion.

The results of first variant are used for estimating input for the second variant of modeling: every subsystem for second variant (for subsystems equipped by AIS) is characterized by MTBLI = 31,293 h in analogy to the subsystem 4 of first variant. Owing to AIS, the frequency of anomalies is about 0.28 year$^{-1}$ (it is equal to 1/MTBLI), but the conditions of anomalies activation time are more strong: the mean time is 30 min. The time between the end of diagnostics and the beginning of the next diagnostics is 1 month, and the recovery time is about 1 day.

What about the risks of "failure" during period from 0.05 to 2 years?

Analysis of modeling results proves: risks are very high despite the use of AIS with the described characteristics, see **Figure 11**.

For a robot-manipulator used in space, new knowledge for accumulating and improving K-base is as follows:

1. The input (used for modeling) characterizes inadmissible conditions for functions performance by robot-manipulator.

2. The probability of "success" on level 0.98 or risk of "failure" on level 0.02 during six sessions of cosmonaut work is inadmissible for reliable robot-manipulator operation more than 1–2 weeks in space.

3. For a robot-manipulator used in space, the level 31,293 h of MTBLI is inadmissible level for every compound subsystem equipped by considered AIS.

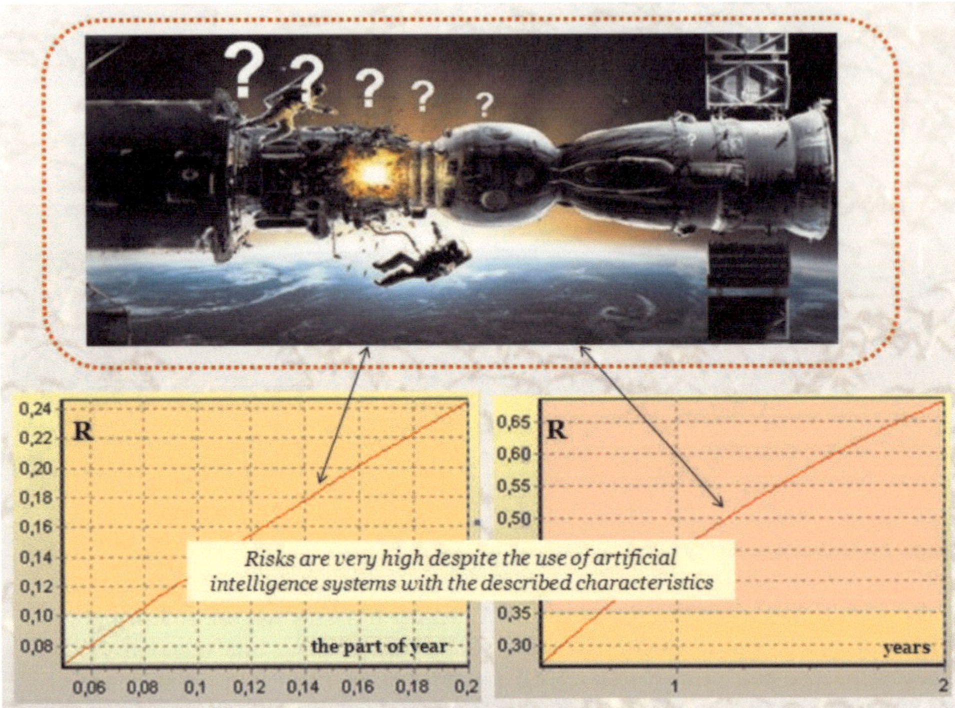


**Figure 11.**
*Risks of "failures" depending on the prognostic period of use (from 0.05 to 2 years).*

4. Analyzed project of robot-manipulator operation effectiveness can be added to K-base history as precedent of "unsuccess."

5. For analyzed project, new research for decreasing risks with the proof of its efficiency on the basis of modeling is strongly required after improving characteristics for every subsystem of robot-manipulator.

### 6.2 Example of forming input for probabilistic modeling from monitored data

In practice, many devices proper to intelligent manufacturing are sources of data monitored. This example explains how monitored data can be tailored in AIS for probabilistic modeling to solve both problems 1 and 2.

The approach to form specific input for modeling is demonstrated on example of mean time $T_{occur}$ for PDF $\Omega_{occur}(t)$ and mean time $T_{activ}$ for PDF $\Omega_{activ}(t)$ from random values $\tau_{occurrence}$ and $\tau_{activation}$ (**Figures 6** and **12**).

The elementary ranges for monitored parameters from quality or safety point of view should be set. For each parameter, the ranges of possible values of conditions are set: "Working range inside of norm," "Out of working range, but inside of norm," and "Abnormality," The condition "Abnormality" characterizes a threat to lose system integrity after danger influence (on the logic level this range "Abnormality" may be interpreted analytically as failure, fault, losses of quality, or safety etc.). The construction on **Figure 12** allows to extract data for probabilistic modeling: time between moments of the occurrences of dangers (potential threats), activation time of occurred dangers, and recovery time.

For example, from **Figure 12**:

Mean time between moments of the occurrences of dangers (potential threats) $T_{occur} = (\tau_{occurrence\ 1} + \tau_{occurrence\ 2})/2$

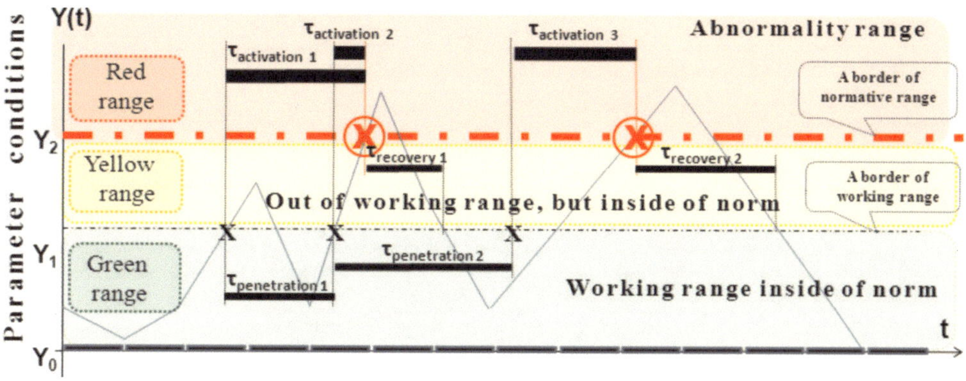


**Figure 12.**
*The universal elementary ranges for monitored parameters.*

Mean activation time $T_{activ} = (\tau_{activation\ 1} + \tau_{activation\ 2} + \tau_{activation\ 3})/3$
Mean recovery time for $T_{recovery} = (\tau_{recovery\ 1} + \tau_{recovery\ 2})/2$

This example is auxiliary to understand some sources of input for the proposed models (Sections 3–5) used for the next examples.

### 6.3 Example of system planning the possibilities of functions performance by AIS for a coal company

Applicability of the proposed probabilistic methods and models is demonstrated to improve some of the existing capabilities of AIS for a coal company. This subsection contains an explanation how problem 1 (of planning the possibilities of functions performance) may be solved for intelligent manufacturing by the proposed approach on the base of data monitored. This demonstrates AIS possibilities for a coal company on its operation stage.

Let a coal company (as system) is decomposed on 9 subsystems for studying efficiency. Of course, every subsystem also may be considered as complex system, for example, see **Figure** 7. Components from 1 to 6 united by multifunctional safety system of the mine, component 7 is associated with the washing factory, component 8 is associated with transport, and component 9 with port, see **Figure 13**: 1—the control system of ventilation and local airing equipment; 2—the system of modular decontamination equipment and compressed air control; 3—the system of air and gas control; 4—the system of air dust content control; 5—the system of dynamic phenomena control and forecasting; 6—the system of fire-prevention protection; 7—the safety system of washing factory; 8—the safety system for transport; and 9—the safety system of port. Information is monitored from different sources, accumulated in a database of dispatcher intelligence center, processed, and systematized (including systematization described in Example 2 to get input for modeling).

For planning possibilities of functions performance by AIS in this example, the probabilistic modeling is being to answer the next two questions:

- How every responsible worker can know a residual time before the next parameters abnormalities?
- What risks to lose system integrity may be for a year, for 10 and 20 years if all subsystems are supported by AISs that transform all system components to the level which is proper to skilled workers (Optimistic view on dangerous coal intelligent manufacturing)?

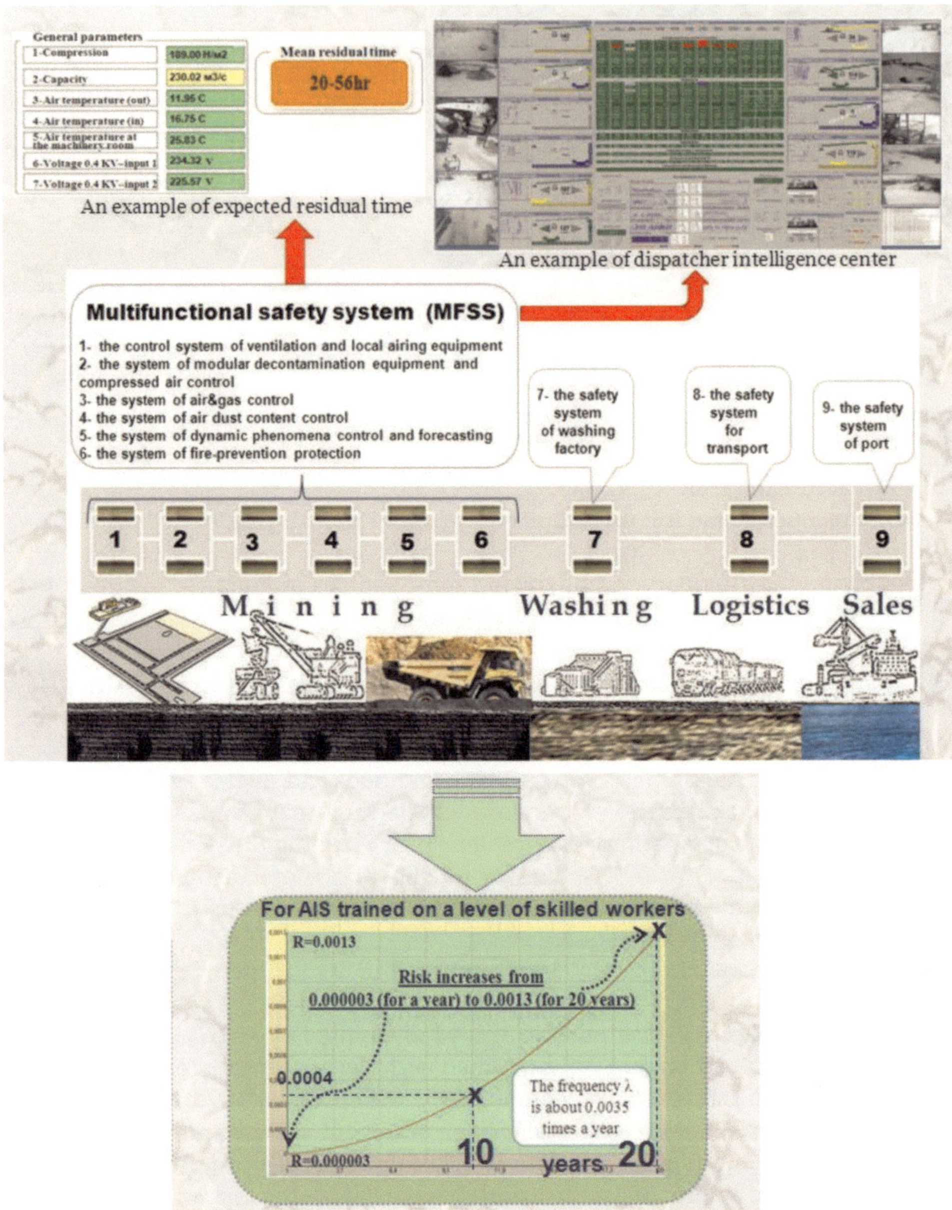


**Figure 13.**
*An example of a coal company with AISs that transformed all system components to the level which is proper to skilled workers.*

To answer the first question, the ranges of possible values of conditions are established: "Working range inside of norm," "Out of working range, but inside of norm," and "Abnormality" for each separate critical parameter of equipment. It is interpreted similarly by light signals—"green," "yellow," and "red," as it is reflected on **Figure 12**. Some examples of parameters may include compression, capacity, air temperature (out, in, at machinery room), voltage, etc. The information from Example 6.2 and additional time data of enterprise procedures are used by AIS as input for using formulas (1) and (3) and Steps 1–4 (from **Figure 8**) in real time of company operation activity. Here, risks to lose the system integrity during the given period $T_{given}$ means risks to be at least once in state "Abnormality" within $T_{given}$. The functions of modeling is performed on special servers (centralized or mapped);

see details in [27, 36]. If virtual risks are computed by formulas (1) and (3) for all points $T_{given}$ from 0 to ∞, the calculated values form a trajectory of the PDF. The mathematical expectation of this PDF means the mean residual time to the next state "Abnormality." It defines MTBLI from this PDF. This output of probabilistic modeling can be transmitted to interested workers. Requirements to AIS operation quality are: quality measures of used information by AIS should meet admissible level recommended in **Table 1**.

Thus, the answer on the first question "How responsible worker can know a residual time before the next parameters abnormalities?" is: the calculated mean residual time to the next state "Abnormality" (MTBLI for "red" range on **Figure 12**) can be transmitted in real time to responsible worker immediately after parameter value cross the border from "Working range inside of norm," "Out of working range, but inside of norm" (from "green" to "yellow" range on **Figures 12** and **13**). It is possible as a result of implementation of the proposed approach—see example of implementation in [27, 36].

To answer the second question, let the next input be formed from data monitored.

Let for every system component, a frequency of occurrence of the latent or obvious threats is equal to once a month and the mean activation time of threats is about 1 day. The system diagnostics are used once for work shift 8 h, a mean duration of the system control is about 10 min, and the mean recovery time of the lost integrity of object equals to 1 day. The workers (they may be robotics, skilled mechanics, technologists, engineers, etc.) are supported by capabilities of an AIS and a remote monitoring systems allowing estimating in real time the mean residual time before the next parameters abnormalities considering the results of probabilistic modeling. Formally they operate as parallel elements with hot reservation (structure on **Figure 4**, right). Owing to AIS support workers are capable to revealing signs of a critical situation after their occurrence. Workers can commit errors on the average not more often once a year (it is proper to skilled workers).

To answer the question we do Steps 1–4 (from **Figure 8**) and use formulas (1)–(3) for solving the problem for complex structure, see **Figure 13**. Here, risks to lose system integrity means risks of "failure" for every subsystem which can be detailed to the level of every separate critical parameter of equipment.

The fragments of built PDF on **Figure 13** show: risk of "failure" increases from 0.000003 for a year to 0.0004 for 10 years and to 0.0013 for 20 years. Thus, the mean time between neighboring losses of integrity (MTBLI) equals to 283 years.

These are some estimations for example assumptions.

Thus, the answer on second question "What risks to lose system integrity may be for a year, for 10 and 20 years if all subsystems are supported by AISs that transform all system components to the level which is proper to skilled workers?" is: risks to lose system integrity may be 0.000003 for a year, 0.0004 for 10 years and 0.0013 for 20 years, herewith (MTBLI) is equal to 283 years. These are the Optimistic estimations for dangerous coal intelligent manufacturing that make sense to take over a desired level of AIS operation effectiveness.

New knowledge for accumulating and improving K-base is as follows:

1. The input (used for modeling) characterizes admissible conditions for functions performance by AIS for a coal company.

2. The probability of "success" on levels 0.99997 for a year, 0.9996 for 10 years and 0.9987 for 20 years or risk of "failure" on levels 0.000003 for a year to 0.0004 for 10 years and 0.0013 for 20 years (with predicted risks levels for discovered "bottlenecks") are admissible.

3. Expected term in average 283 years and more is admissible systemic aim for providing safe company operation.

4. Analyzed project of AISs operation effectiveness (that transform all system components to the level which is proper to skilled workers of coal company) can be added to K-base history as a precedent of "success."

## 6.4 Example of system planning the possibilities of functions performance by AIS used for a security service of floating oil and gas platform

This subsection continues an explanation on how problem 1 (of planning the possibilities of functions performance) may be solved for intelligent manufacturing by the proposed approach on the base of data monitored. This demonstrates the

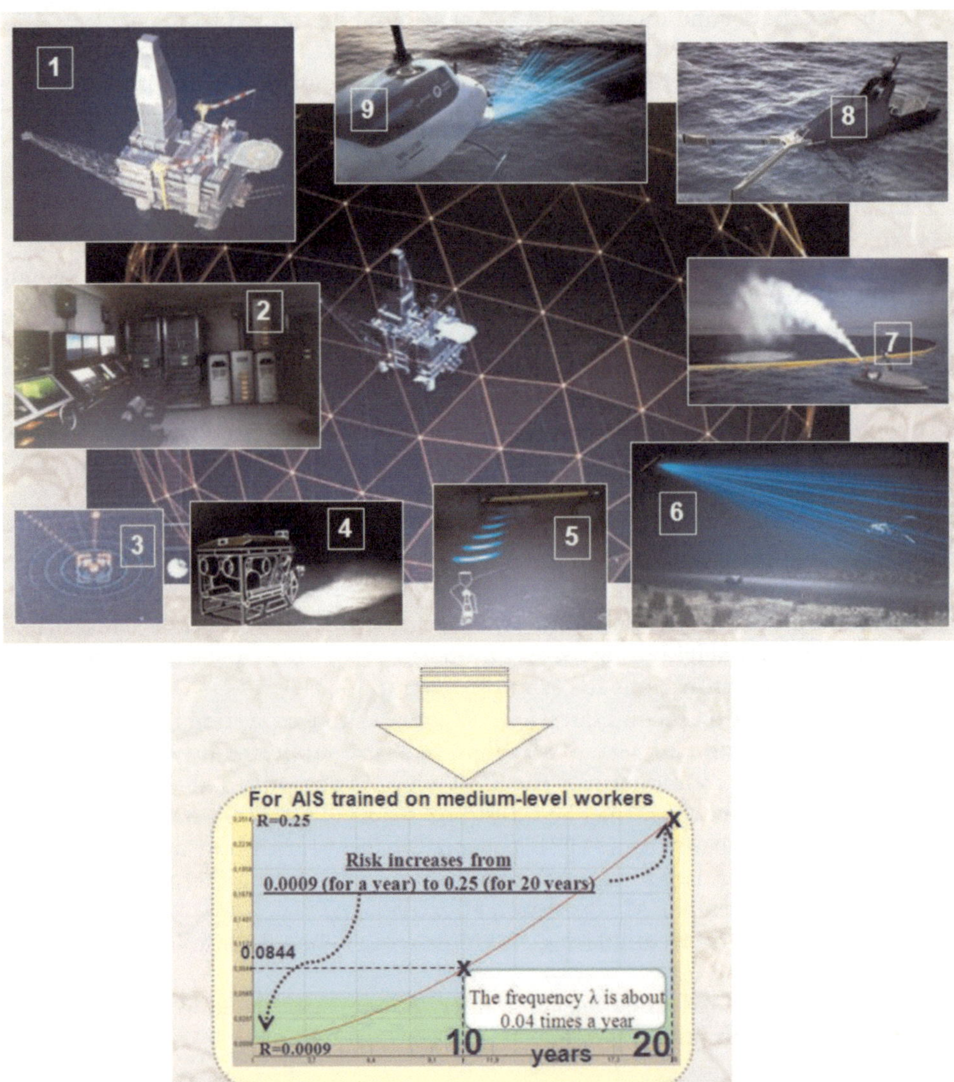


**Figure 14.**
*An example of a floating oil and gas platform with AISs that transform all system components to the level which is proper to medium-level workers.*

capabilities of AIS used for a security service of floating oil and gas platform on its operation stage. The difference from previous example is in more degree of uncertainties (because of high complexity) that allows to transform all system components to the level which is proper to medium-level workers of floating oil and gas platform. The same approach, structure, and formulas for probabilistic modeling are used.

Let a floating oil and gas platform is also decomposed on nine subsystems. Every subsystem is enumerated on **Figure 14**, and operates as parallel elements with hot reservation.

Components are: 1—a construction of platform; 2—an AIS on platform for robotics monitoring and control; 3—an underwater communication modem; 4—a remote controlled unmanned underwater robotic vehicle; 5—a sonar beacon; 6—an autonomous unmanned underwater robotic vehicle; 7—non-boarding robotic boat, a spray of the sorbent; 8—non-boarding robotic boat, a pollution collector; and 9—an unmanned aerial vehicle.

And let input for modeling is the same as in Example 6.3. Only one difference is because of complexity characteristics are proper to medium-level workers of floating oil and gas platform. For this example, it means workers and AIS can commit errors more often in comparison with skilled workers, for one element it is equal to 1 time a month instead of once a year.

For planning possibilities of functions performance by AIS in this example, the probabilistic modeling is being to answer the question:

What risks to lose system integrity may be for a year, for 10 and 20 years if all subsystems are supported by AISs that transform all system components to the level which is proper to medium-level workers (realistic view on dangerous oil and gas intelligent manufacturing)?

To answer the question, we do Steps 1–4 (from **Figure 8**) and use formulas (1)–(3) for solving the problem for complex structure, see structure on **Figure 13**. Here, risks to lose the system integrity mean risks of "failure" for every subsystem. The fragments of built PDF on **Figure 14** show: from 0.0009 for a year to 0.0844 for 10 years and 0.25 for 20 years. Thus, MTBLI equals to 24 years. It is 11.4 times less often against the results of Example 6.3.

These are some estimations for example assumptions.

Thus, the answer on question is: risks to lose system integrity may be 0.0009 for a year, 0.0844 for 10 years and 0.25 for 20 years; herewith, mean time between neighboring losses of integrity is equal to 24 years. These are the realistic estimations for dangerous oil and gas intelligent manufacturing.

New knowledge for accumulating and improving K-base is as follows:

1. The input (used for modeling) characterize possible complex conditions for functions performance by AIS used for a security service of floating oil and gas platform.

2. The probability of "success" on levels 0.9991 for a year, 0.9156 for 10 years and 0.75 for 20 years or risk of "failure" on levels 0.0009 for a year, 0.0844 for 10 years and 0.25 for 20 years (with possible consequences) and expected term in average 24 years as estimation of mean time between neighboring losses of integrity are realistic view on dangerous floating oil and gas platform intelligent manufacturing.

3. For analyzed project new research to improve characteristics for the security service of floating oil and gas platform for decreasing risks with the proof of its efficiency on the basis of modeling is required.

4. Analyzed project of AISs operation effectiveness (that transform all system components to the level which is proper to medium-level workers of floating oil and gas platform) can be added to K-base history as precedent.

## 6.5 Example of robot route optimization under limitations on risk of "failure" in conditions of uncertainties

Applicability of the proposed probabilistic methods and models is demonstrated to improve some of the existing capabilities of rescue robot for route optimization. This subsection contains an explanation on how problem 2 may be cognitively solved. Similar problems of specific robot route optimization from point A (Start) to point F (Finish) can arise on water, under water (**Figure 15**), in burning wood (**Figure 16**), in the conditions of a city or in mountains (**Figure 17**), and in other situations in conditions of uncertainties. Specific cases of uncertainties can be connected additionally with complex conditions of environment and necessity of robotics orientation, localization, and mapping that influences on input for the proposed probabilistic models.

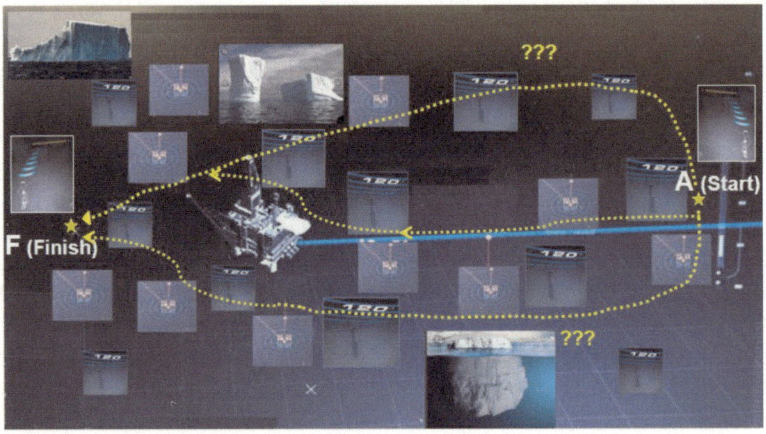


**Figure 15.**
*A system view on situation for robot route from point A (Start) to point F (Finish) on water and under water.*

**Figure 16.**
*A system view on situation for robot route from point A (Start) to point F (Finish) in burning wood.*

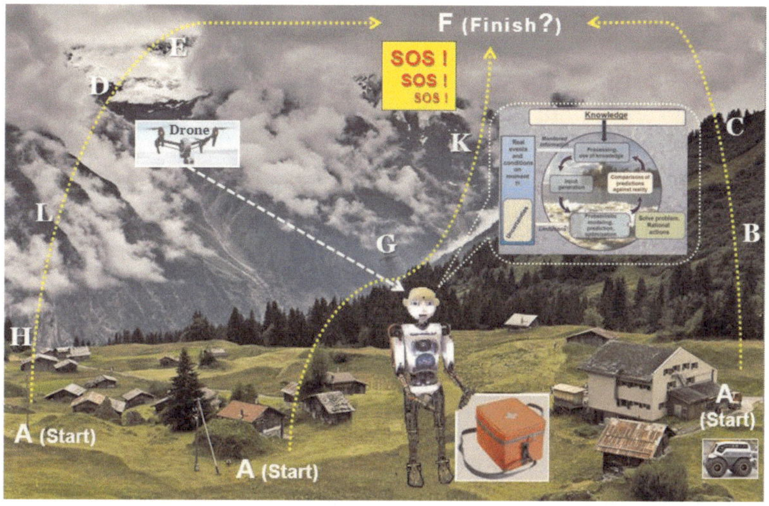


**Figure 17.**
*A system view on situation for robot route from point A (Start) to point F (Finish) in mountains.*

Here, we demonstrate the proposed approach by a simplified example of moving a special rescue robot from point A to the final point F of the route (from where the SOS signals from tourists are following). It is required to optimize the route of the robot in space under uncertainty of weather, complex snow conditions in mountains to achieve the goal in 2 h with an acceptable risk of failure less than 0.1 (i.e., a probability of success should be more than 0.9). Interaction with the drone-informant is supposed, see **Figure 17**.

The applications to cognitive solving the problem of robot route optimization are demonstrated by the next steps.

**Step 1**. The complete set of route variants to achieve the goal within about 2 h: first route is ABCF, second route AGKF, third route if AHLDEF, and fourth variant is a combination of routes 1–3. Points A, B, C, G, K, H, L, D, E, F mean that they may change the route (including return to the previous point). Respectively, it may be a refinement of the further route at these points. Robot speed allows to overcome any route in time.

For each variant, a set of system compared by modeling is defined: there are ABCF, AGKF, AHLDEF, and possible combinations. Inputs characterizing every part of route for each of the variants are formed by K-base and gathered data from drone-informant:

- Frequencies of the occurrences of potential threats are for route ABCF $\sigma$ = 1 time at 10 h, AGKF $\sigma$ = 1.5 times at 10 h, AHLDEF $\sigma$ = 2 times at 10 h (since 8.00 a.m. to 8.00 p.m.)
- Mean activation time of threats $T_{activ}$ = 30 min
- Time between the end of diagnostics and the beginning of the next diagnostics of robot availability $T_{betw.}$ = 2 min
- Diagnostics time of robot availability $T_{diag}$ = 30 s
- Recovery time of robot availability = 10 min (for modified model [42–44])
- Given prognostic period $T_{given}$ = 2 h

i = 1.

**Step 2** (i = 1). Using probabilistic model, a calculation of the probability of failure is carried out for each variant. From the set of variants ABCF, AGKF, and AHLDEF, the shorter variant ABCF for which risk is equal to 0.034 is chosen (for the route AGKF risk = 0.051, for route AHLDEF risk = 0.067), see **Figure 18**. The relevant data from the drone about the forecasted conditions and the weather on the part CF to 8.30 a.m. are taken into account.

**Step 3** (i = 1). The robot overcomes the part AB of route. For the new initial point B, the input for modeling every part of possible route is updated in real time for routes BCF, BGKF, and BGHLDEF.

**Step 4** (i = 1). The robot has not yet arrived at the intended point F (i.e., the last part of the route is not overcome).

i = i + 1 = 2.

**Step 2** (i = 2 for variants BCF, BGKF, and BGHLDEF). Input for modeling is not changed. Risks are the same. From the route variants BCF, BGKF, and BGHLDEF, the shorter one BCF (with minimal risk) is chosen.

**Step 3** (i = 2 for variant BCDEF). The robot overcomes the part BC. For the new initial point C, the input for modeling every part of possible route is updated in real time: bad weather on the CF part does not allow further movement. And weather improvements in the next 2 h are not expected. Part CF is impassable. The come-back to the initial point B of the part is being.

**Step 2** (i = 2 for two remaining variants). From variants BGKF and BGHLDEF, the shorter one BGKF (with minimal risk 0.051) is chosen.

**Step 3** (i = 2 for variant BGKF). The robot overcomes the part BG. For the new initial point G, the input for modeling every part of possible route is updated in real time: according drone from 9.00 a.m. on parts GK and KF the imminent avalanche are detected. The accumulated knowledge is used to clarify the input for modeling, namely: the frequency threats in the part GKF increases from 1.5 to 2.5 times at 10 h. Using a probabilistic model for each variant, a recalculation of the risk of failure is carried out. Of the variants GKF and GHLDEF, the variant GHLDEF is chosen (risk is equal to 0.067, for the route GKF risk equals 0.083).

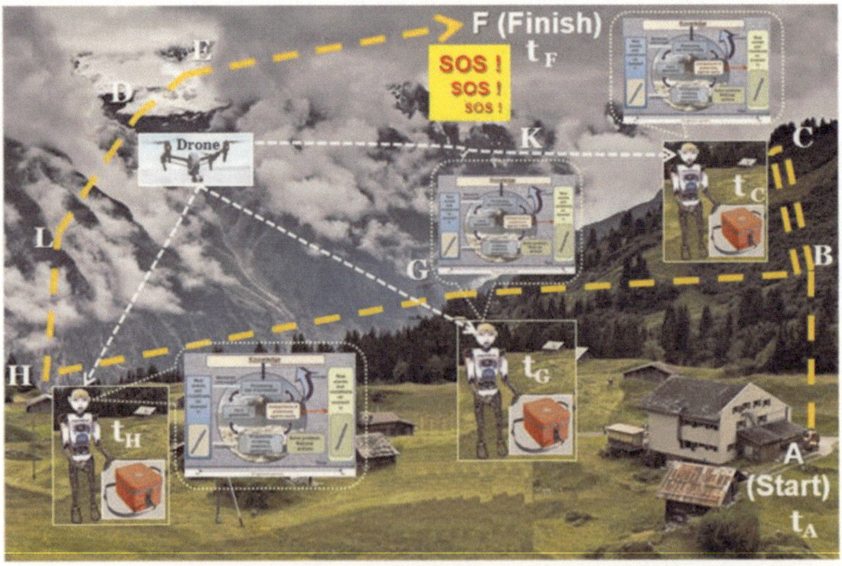


**Figure 18.**
*The risk of "failure" in dependence on prognostic period during the robot route from point A (Start) to point F (Finish).*

**Step 4**. After overcoming the part GHLDEF, the robot arrived at the intended point F of route in time.

Thus, the way ABCBGHLDEF is the result of optimization. The robot purpose was achieved owing to preventive measures which were defined by using risk control on the way (with probability of "success" more than 0.9).

New knowledge for accumulating and improving K-base is as follows:

1. The input (used for modeling) characterizes possible complex conditions for rescue robot route optimization under limitations on risk of "failure" in conditions of uncertainties. In particular, the information updates every 2 min for robot route optimization under limitations on risk of "failure" less than 0.1 is admissible for considered situation.

2. The acceptable risk 0.1 is justified; the predicted risks for all variants of the routes did not exceed 0.1.

3. Analyzed project can be added to K-base history as precedent.

## 7. Conclusion

The proposed approach to build and implement the probabilistic methods and models is demonstrated by application to cognitive solving:

- The problem of planning the possibilities of functions performance on the base of monitored information about events and conditions

- The problem of robot route optimization under limitations on risk of "failure" in conditions of uncertainties

There is proposed to carry out probabilistic prediction of critical processes in time so that not only to act according to the prediction, but also to compare predictions against their coincidence to the subsequent realities.

The described analytical solutions are demonstrated by practical examples such as:

System planning the possibilities of functions performance in space by using robot-manipulators, by AIS for a coal company and for a floating oil and gas platform

Forming input for probabilistic modeling from monitored data

Robot route optimization under limitations on risk of "failure" in conditions of uncertainties

A cognitive solving of the chosen problems consists in improvements, accumulation, analysis, and use of appearing knowledge.

## Appendix

Proofs for formulas (1)–(3)

According to the proof of formula (1): because between diagnostics system is not protected from threats an influence (a loss of integrity) will take place only after danger occurrence and activation during given time before the next diagnostic (**Figure 6**). A risk to lose integrity (i.e., probability of "failure") is equal to

$\Omega_{penetr}{}^{*}\Omega_{activ}(T_{req})$ because these PDF are independent. The found probability of providing system integrity (probability of "success") is equal to addition to 1.

The proof of formula (1) is complete.

For the special case, if $\Omega_{occur}(t) = 1 - \exp(\sigma t)$, $\sigma = 1/T_{occur}$, $\Omega_{activ}(t) = 1 - \exp(t/\beta)$, $\beta = T_{activ}$

$$P_{(1)}\left(T_{given}\right) = \begin{cases} \left(\sigma - \beta^{-1}\right)^{-1}\left\{\sigma e^{-T_{given}/\beta} - \beta^{-1}e^{-\sigma T_{given}}\right\}, if\ \sigma \neq \beta^{-1}, \\ e^{-\sigma T_{given}}\left[1 + \sigma T_{given}\right], if\ \sigma = \beta^{-1}. \end{cases}$$

Note. This formula (1) is used also for the estimation of system operation without diagnostics. There is supposed that before the beginning of period $T_{given}$ system integrity is provided.

According to the proofs of formulas (2) and (3), we consider independence. Then formula (2) means measure $P_{(2)}(T_{given}) = P_{mdl} + P_{end}$, where $P_{mdl}$ is the probability of correct operation ("success") within the period $T_{given}$ since beginning to the last diagnostics, $P_{mdl} = N((T_{betw} + T_{diag})/T_{req})P_{(1)}{}^{N}(T_{betw} + T_{diag})$, here $P_{(1)}(T_{betw} + T_{diag})$ is defined by formula (1), but one is calculated only for time $T_{betw} + T_{diag}$; $P_{end}$ is the probability of correct operation ("success") within the assigned period $T_{given}$ after the last diagnostics, i.e. in the last remainder $T_{rmn} = T_{req} - [N(T_{betw} + T_{diag})]$, $P_{end} = (T_{rmn}/T_{req})\ P_{(1)}(T_{rmn})$. Here, $P_{(1)}(T_{rmn})$ is defined by formula (1), but one is calculated only for the remainder time $T_{rmn}$. Really, for this time $T_{rmn}$, the main condition of the first variant is true: $T_{rmn} < T_{betw} + T_{diag}$.

Formula (3) means measure $P_{(2)}(T_{given}) = P_{(1)}{}^{N}(T_{betw} + T_{diag})P_{(1)}(T_{rmn})$. Interpretation is the next: "success" is on all N periods $(T_{betw} + T_{diag})$ AND on remainder time $T_{rmn}$.

The proofs for formulas (1)–(3) are complete.

## Author details

Andrey Kostogryzov[1]* and Victor Korolev[2]

1 Federal Research Center "Computer Science and Control" of the Russian Academy of Sciences, Gubkin Russian State University of Oil and Gas, Russia

2 Lomonosov Moscow State University (Faculty of Computational Mathematics and Cybernetics, Department of Mathematical Statistics), Federal Research Center "Computer Science and Control" of the Russian Academy of Sciences, Russia

*Address all correspondence to: akostogr@gmail.com

## References

[1] Turing A. Computing machinery and intelligence. Mind. 1950;**59**(236):433-460

[2] Ani Hsieh M, Oussama Khatib, Vijay Kumar (Editors). Experimental Robotics. Berlin: Springer; 2016. 913p

[3] Ajoudani A. Transferring Human Impedance Regulation Skills to Robots. Berlin: Springer; 2016. 180p

[4] Jean-Paul Laumond, Nicolas Mansard, Jean-Bernard Lasserre (Editors). Geometric and Numerical Foundations of Movements. Berlin: Springer; 2017. 417p

[5] Henrik I. Christensen, Oussama Khatib (Editors). Robotics Research. Berlin: Springer; 2017. 646p

[6] Radek Silhavy, Petr Silhavy, Zdenka Prokopova (Editors). Cybernetics Approaches in Intelligent Systems. Computational Methods in Systems and Software. Vol. 1. Berlin: Springer; 2017. 405p

[7] Radek Silhavy, Petr Silhavy, Zdenka Prokopova (Editors). Applied Computational Intelligence and Mathematical Methods. Computational Methods in Systems and Software. Vol. 2. Berlin: Springer; 2017. 393p

[8] Valencia R, Andrade-Cetto J. Mapping, Planning and Exploration with Pose SLAM. Berlin: Springer; 2018. 124p

[9] Matthew Spenko, Stephen Buerger Karl Iagnemma (Editors). The DARPA Robotics Challenge Finals: Humanoid Robots to the Rescue. Berlin: Springer; 2018. 692p

[10] Antonelli G. Underwater Robots. Berlin: Springer; 2018. 374p

[11] Daniel Sebastian Leidner. Cognitive Reasoning for Compliant Robot Manipulation. Berlin: Springer; 2019. 190p

[12] Venture G, Laumond J-P, Watier B. Biomechanics of Anthropomorphic Systems. Berlin: Springer; 2019. 304p

[13] Santamaria-Navarro A, Solà J, Andrade-Cetto J. Visual Guidance of Unmanned Aerial Manipulators. Berlin: Springer; 2019. 144p

[14] Tadokoro S. Disaster Robotics. Berlin: Springer; 2019. 532p

[15] Gnedenko BV, Korolev VY. Random Summation: Limit Theorems and Applications. Boca Raton: CRC Press; 1996

[16] Korolev VY, Sokolov IA. Mathematical Models of Non-Homogeneous Flows of Extremal Events. Moscow: Torus-Press; 2008

[17] Bening V, Korolev V. Generalized Poisson Models and Their Application in Insurance and Finance. Utrecht: VSP; 2002

[18] Kostogryzov A, Nistratov G. Standardization, Mathematical Modelling, Rational Management and Certification in the Field of System and Software Engineering. Moscow: Armament. Policy. Conversion; 2004

[19] Kostogryzov AI, Stepanov PV. Innovative Management of Quality and Risks in Systems Life Cycle (Modern Standards and Ideas of System Engineering, Mathematical Models, Methods, Techniques and Software Tools Complexes for System Analysis, Including Modelling through Internet, 100 Examples with an Explanation of Logic of the Reached Results, Useful Practical Recommendations). Moscow: Armament. Policy. Conversion; 2008

[20] Enrico Z. An Introduction to the Basics of Reliability and Risk Analysis. Singapore: World Scientific Publishing Co. Pte. Ltd.; 2006

[21] Kolowrocki K, Soszynska-Budny J. Reliability and Safety of Complex Technical Systems and Processes. London: Springer-Verlag Ltd.; 2011

[22] Kostogryzov A, Nistratov G, Nistratov A. Some applicable methods to analyze and optimize system processes in quality management. In: Total Quality Management and Six Sigma. London: InTech; 2012. pp. 127-196

[23] Grigoriev L, Guseinov C, Kershenbaum V, Kostogryzov A. The methodological approach, based on the risks analysis and optimization, to research variants for developing hydrocarbon deposits of Arctic regions. Journal of Polish Safety and Reliability Association. Summer Safety and Reliability Seminars. 2014;**5**(1–2): 71-78

[24] Akimov V, Kostogryzov A, Mahutov N, et al. Security of Russia. Legal, Social & Economic and Scientific & Engineering Aspects. The Scientific Foundations of Technogenic Safety. Moscow: Under the Editorship of Mahutov N.A. Znanie; 2015

[25] Kostogryzov A, Nistratov A, Zubarev I, Stepanov P, Grigoriev L. About accuracy of risks prediction and importance of increasing adequacy of used adequacy of used probabilistic models. Journal of Polish Safety and Reliability Association. Summer Safety and Reliability Seminars. 2015;**6**(2): 71-80

[26] Eid M, Rosato V. Critical infrastructure disruption scenarios analyses via simulation. In: Managing the Complexity of Critical Infrastructures. A Modelling and Simulation Approach. Berlin: Springer Open; 2016. pp. 43-62

[27] Artemyev V, Kostogryzov A, Rudenko J, Kurpatov O, Nistratov G, Nistratov A. Probabilistic methods of estimating the mean residual time before the next parameters abnormalities for monitored critical systems. In: Proceedings of the 2nd International Conference on System Reliability and Safety (ICSRS); Milan, Italy; 2017. pp. 368-373

[28] Kostogryzov A, Stepanov P, Nistratov A, Nistratov G, Klimov S, Grigoriev L. The method of rational dispatching a sequence of heterogeneous repair works. Energetica. 2017;**63**(4):154-162

[29] Kostogryzov A, Stepanov P, Nistratov A, Atakishchev O. About probabilistic risks analysis during longtime grain storage. In: Proceedings of the 2nd Internationale Conference on the Social Science and Teaching Research (ACSS-SSTR), Volume 18 of Advances in Social and Behavioral Science. Singapore: Singapore Management and Sports Science Institute Pte. Ltd.; 2017. pp. 3-8

[30] Kostogryzov A, Stepanov P, Grigoriev L, Atakishchev O, Nistratov A, Nistratov G. Improvement of existing risks control concept for complex systems by the automatic combination and generation of probabilistic models and forming the storehouse of risks predictions knowledge. In: Proceedings of the 2nd International Conference on Applied Mathematics, Simulation and Modelling (AMSM); Phuket, Thailand. Lancaster: DEStech Publications Inc.; 2017. pp. 279-283

[31] Kostogryzov A, Atakishchev O, Stepanov P, Nistratov A, Nistratov G, Grigoriev L. Probabilistic modelling processes of mutual monitoring operators actions for transport systems. In: Proceedings of the 4th International Conference on Transportation Information and Safety (ICTIS); Banff, Canada; 2017. pp. 865-871

[32] Kostogryzov A, Panov V, Stepanov P, Grigoriev L, Nistratov A, Nistratov G. Optimization of sequence of performing heterogeneous repair work for transport systems by criteria of timeliness. In: Proceedings of the 4th International Conference on Transportation Information and Safety (ICTIS); Banff, Canada; 2017. pp. 872-876

[33] Kostogryzov A, Nistratov A, Nistratov G, Atakishchev O, Golovin S, Grigoriev L. The probabilistic analysis of the possibilities to keep "organism integrity" by continuous monitoring. In: Proceedings of the International Conference on Mathematics, Modelling, Simulation and Algorithms (MMSA); Advances in Intelligent Systems Research, Vol. 159; Chengdu, China. Paris: Atlantis Press; 2018. pp. 432-435

[34] Kostogryzov A, Grigoriev L, Golovin S, Nistratov A, Nistratov G, Klimov S. Probabilistic modeling of robotic and automated systems operating in cosmic space. In: Proceedings of the International Conference on Communication, Network and Artificial Intelligence (CNAI); Beijing, China. Lancaster: DEStech Publications, Inc.; 2018. pp. 298-303

[35] Kostogryzov A, Grigoriev L, Kanygin P, Golovin S, Nistratov A, Nistratov G. The experience of probabilistic modeling and optimization of a centralized heat supply system which is an object for modernization. In: International Conference on Physics, Computing and Mathematical Modeling (PCMM); Shanghai. Lancaster: DEStech Publications, Inc.; 2018. pp. 93-97

[36] Artemyev V, Rudenko J, Nistratov G. Probabilistic Modeling in System Engineering. Probabilistic Methods and Technologies of Risks Prediction and Rationale of Preventive Measures by Using "Smart Systems". Applications to Coal Branch for Increasing Industrial Safety of Enterprises. London: InTech; 2018. pp. 23-51

[37] Kershenbaum V, Grigoriev L, Kanygin P, Nistratov A. Probabilistic Modeling in System Engineering. Probabilistic Modeling Processes for Oil and Gas Systems. London: InTech; 2018. pp. 55-79

[38] Kostogryzov AI, Bezkorovainy MM, Lvov VM, Nistratova EN, Bezkorovainaya IV. Complex for evaluation of information systems operation quality—"Know-How" (CEISOQ). Registered by Rospatent No. 2000610272 from April 06, 2000

[39] Kostogryzov AI, Nistratov GA, Nistratova EN, Nistratov AA. Mathematical modeling of system life cycle processes (Mathematical modeling of processes)—"Know-How". Registered by Rospatent No. 2004610858 from April 08, 2004

[40] Kostogryzov AI, Nistratov GA, Nistratov AA, Nistratova EN. Software tools complex for evaluation of information systems operation quality by internet—"Know-How" (CEISOQ-Internet). Registered by Rospatent No. 2008612348 from May 15, 2008

[41] Kostogryzov AI, Nistratov GA, Nistratova EN, Nistratov AA. Complex for evaluating quality of production processes. Registered by Rospatent No. 2010614145 from June 25, 2010

[42] Kostogryzov AI, Nistratov GA, Nistratov AA. Remote analytical support of informing about the probabilistic and time measures of operating system and its elements for risk-based approach. Registered by Rospatent No. 2018617949 from July 05, 2018

[43] Kostogryzov AI, Nistratov GA, Nistratov AA. Remote rationale of requirements to means and conditions

for providing "smart" systems operation quality. Registered by Rospatent No. 2018618572 from July 16, 2018

[44] Kostogryzov AI, Nistratov GA, Nistratov AA. Remote probabilistic prediction of informatized systems operation quality. Registered by Rospatent No. 2018618686 from July 17, 2018

Chapter 2

# Laboratory, Bench, and Full-Scale Researches of Strength, Reliability, and Safety of High-Power Hydro Turbines

*Nikolay Makhutov, Yury Petrenia, Anatoly Lepikhin, Vladimir Moskvichev, Mikhail Gadenin and Anatoly Tchernyaev*

## Abstract

Large hydropower plants (HPPs) are categorized as critically and strategically important infrastructure facilities in industrialized countries. Therefore, the issues of ensuring HPPs safety are of paramount importance. In this chapter, the basic aspects of the safety analysis of HPPs, calculation and experimental substantiation of the strength, and resource and reliability of the main equipment are discussed. The scientific and technical measures to ensure safety of HPPs are presented. As a defining measure of safety, it is proposed to ensure the protection of HPPs from severe accidents and disasters according to risk criteria. The main provisions of the risk assessment are presented on the basis of a sequential analysis of loads, features of stress-strain states, characteristics of mechanical properties, and limit states of hydraulic equipment of HPPs. The issues of calculation and experimental evaluation of hydro turbine's resource, which limit the safety of HPPs, are considered. The features of technical diagnosis of hydraulic turbines are considered; characteristic defects and damages are described. The main provisions of the estimated residual life of hydro turbines are presented. The results of the risk estimates of HPPs and hydro turbine resource are given.

**Keywords:** hydropower plants, hydro turbines, safety, risk, protection, strength, resource, experimental studies, technical diagnostics, operational state

## 1. Introduction

Hydropower plants are among the most important elements of the life support infrastructure of many countries. Currently, there are over 100 hydropower plants with a capacity of over 100 MW with a total installed capacity of about 45 GW in Russia. This number includes 10 large hydropower plants with a capacity of more than 1000 MW and 5 largest hydropower plants with a capacity of over 3000 MW. The latter include the Sayano-Shushenskaya HPP with a capacity of 6400 MW, with a dam height of 245 m and a length of 1074 m; Krasnoyarskaya HPP with a capacity of 6000 MW, with a dam height of 128 m and a length of 1072 m; Ust-Ilimskaya

HPP with a capacity of 3840 MW, with a dam height of 105 m and a length of 1475 m; Bratskaya HPP with a capacity of 4500 MW, with a dam height of 124 m and a length of 924 m; and Boguchanskaya HPP with a capacity of 3000 MW, with a dam height of 96 m and a length of 2690 m (**Figure 1**). In the world, there are 7 large hydropower plants with capacity from 5000 to 14,000 MW.

In the presence of extensive national and international experience in the design, construction, and operation of large hydropower plants, accidents of various scales occur on them, including great economic losses and human losses. The largest in the history of hydropower is the Sayano-Shushenskaya HPP disaster, accompanied by the destruction and flooding of the machine room, damage to hydraulic units, and the death of people (**Figure 2**). In this regard, the development of measures and means to ensure the safety of hydropower facilities is of paramount importance.

**Figure 1.**
*Sayano-Shushenskaya (a), Krasnoyarskaya (b), Ust-Ilimkaya (c), and Bratskaya (d) hydropower plants.*

**Figure 2.**
*Disaster of Sayano-Shushenskaya HPP.*

For many years, the safety of technical systems was based on the assumption that a technical object is sufficiently reliable and safe if it meets the requirements of the current regulatory documents. However, the operating experience of such objects showed that compliance with the design, manufacturing, and operating regulations does not exclude the possibility of emergency situations, accidents, and disasters. From this it follows that the security system for hydropower plants should be based not only on traditional approaches but also on new, scientifically based methods of computational-experimental analysis within the concepts of "security—protection—risk—safety—survivability—reliability—resource—strength." Such an analysis requires adjustment of the existing traditional methods of design, construction, and operation of hydropower plants, with the solution of problems of strength and service life of structures and equipment from the standpoint of ensuring the lowest possible risk of accidents. This chapter outlines the main provisions of such an approach for hydro turbines, which are the main equipment of hydropower plants. The scientific and technical tasks, considered here, reflect the experience of research and ensuring the strength, resource, and safety of the technical objects [1–3].

## 2. Concept of computational and experimental substantiation of hydropower plant safety

Taking into account the consequences of accidents and disasters, hydropower plants with a capacity of 1000–5000 MW can be attributed to critical infrastructure objects. Hydropower plants with a capacity of more than 5000 MW can be considered as strategical objects of infrastructure. For such objects, along with the provision of generally accepted standards and requirements of strength, resource, and reliability, the problems of protection from severe accidents and disasters with survivability and risk analysis should be considered [4]. General requirements for ensuring the protection of hydropower plants from severe accidents were formulated in [5]. For water power plants, along with abidance of generally accepted requirements of technical regulations and standards, protection against the most severe catastrophes (design, beyond design and hypothetical) and terrorist impacts must be considered. When solving safety problems, the following should be analyzed:

- Types of disasters
- Scenarios and sources of occurrence
- Critical elements, critical zones, and critical points of the most critical nodes
- Probability characteristics of disasters
- Consequences of the emergence and development of disasters
- Methods and systems of protection against disasters (rigid, functional, natural, combined)
- Measures to counter disasters and analyze and reduce the risks of disasters at water power plants for the region and the country

The results of the investigation into the causes of the Sayano-Shushenskaya HPP catastrophe [2, 4, 5] indicate the need for conducting special studies of the causal complex of accidents at similar facilities to create scientifically based regulations of risk analysis, survivability, and safety criteria. These researches should include next computational and experimental works [5]:

1. Develop a fundamentally new methodology for assessing and improving the protection of hydropower plants, as critical facilities, from severe disasters according to risk criteria.

2. Conduct a computational modeling and experimental analysis of the new parameters of the resource, survivability, safety, and risks in the conditions of a severe disaster of hydroelectric power plants.

3. Develop a methodology for refined estimation of the dynamics, hydrodynamics, and aerodynamics of the occurrence and development of a severe catastrophe on typical hydraulic units.

4. Develop a methodology for constructing a special control system and automated protection of hydro turbines and hydroelectric power plants in the transition from standard to emergency and catastrophic situations.

In order to form a general regulatory framework for the protection of hydropower plants, it is necessary to implement the following measures with the preparation of the relevant regulatory documents:

1. Develop standards and carry out categorization of hydropower plants as critically and strategically important infrastructure facilities according to the risk levels of national, regional, and local disasters.

2. Develop a nomenclature of emergency and catastrophic situations at hydropower plants and levels of protection from them.

3. Build scenarios for the development of severe disasters; identify the damaging factors and the degree of vulnerability of hydropower stations in severe accidents.

4. Develop a methodology for assessing the strategic risks of severe disasters at hydropower plants, taking into account all stages of the life cycle.

5. Develop principles, methods, and systems for protecting hydropower plants from accidents and disasters.

6. Develop the diagnostic methods of hydropower plants and automated protection systems in the event of emergency and catastrophic situations.

7. Determine the role of human factors and responsibility at the stages of decision making, project implementation, and operation of hydropower plants to prevent severe disasters.

8. Perform the complex computational and experimental studies' survivability, safety, and protection of hydropower plants from severe disasters on models and objects.

9. Develop safety criteria for hydraulic engineering dams and methods for assessing actual safety factors.

10. Develop computational models of dams and computational technologies for analyzing the characteristics of their stress-strain state, taking into account the actual changes in the characteristics of concrete and the presence of cracks and damage.

11. Develop criteria for the performance of hydroelectric equipment based on the models of cavitation processes, fatigue, and corrosion damage.

12. Conduct model calculations of emergency situations and scenarios of their development for all existing hydropower plants of Russia.

13. Develop models and methods for assessing the social, environmental, and economic consequences of accidents of hydropower plants.

To ensure the protection of hydropower plants from severe accidents and disasters, their design, construction, and operation should fully address traditional tasks:

- Carrying out normative calculations for static and cyclic strength

- Conducting bench studies of hydrodynamic processes in the flow part of the hydroelectric station

- Control and repair work on the damaged items of equipment

In addition to this, it is necessary to conduct:

- Calculation and experimental analysis of hazardous mechanical and hydraulic processes in power systems of hydropower plants in regular and emergency situations

- Calculation and experimental analysis of the limiting states of critical elements for normal and extreme loads and impacts

To solve these problems, it is necessary to develop fundamental research in the following areas:

- Study of hazardous processes in the environment and technical systems of hydropower plants, taking into account the role of the human factor

- Development of methods and tools for mathematical (computational) modeling of mechanical, hydrodynamic, and electromagnetic processes that affect the conditions for the occurrence and development of severe accidents and disasters

- Development of new methods and means of prompt diagnosis of emergency situations

- Development of the theory and methods to ensure the protection of hydropower stations from severe accidents and disasters

The scientific and methodological basis for ensuring the protection of hydropower plants from severe accidents and disasters is a risk analysis. Risk assessments can be performed in the classical form:

$$R_\Sigma(t) = \sum_i R_i(t), \tag{1}$$

$$R_i(t) = P_i(t) \times U_i(t+\tau) \tag{2}$$

where $P_i(t)$ is the probability of reaching the $i$th limiting state leading to disaster with damage $U_i(t + \tau)$ and $\tau$ is discounting time.

The probability $P_i(t)$ is determined by the given criteria of the limiting state for the most critical zones of highly loaded elements of hydraulic equipment structures:

$$P_i(t) = F_p\{\sigma_c, e_c, d_c, l_c, t\} \tag{3}$$

The values of critical stresses $\sigma_c$, deformations $e_c$, damage $d_c$, and size of crack-like defects $l_c$ depend on the complex of operating technological and operational loads:

$$Q(t) = F_Q\{Q_M(t), Q_H(t), Q_E(t), Q_V(t), Q_S(t)\} \tag{4}$$

where $Q_M$ is mechanical loads from the weight and installation and welding of elements and from the rotation of the hydraulic turbine; $Q_H$ is hydrodynamic loads from pressure and pressure of water, water hammer, and pressure pulsation; $Q_E$ is electromagnetic load from the interaction of the rotor and the stator of the turbine; $Q_V$ is vibration loads; and $Q_S$ is seismic loads.

The components of stresses $\sigma$ and deformations $e$, which characterize the stress-strain state of the structure, are determined by calculation and experimental methods according to the values of the indicated loads:

$$\{\sigma, e\} = F_\sigma\{Q(t), \alpha_\sigma, E, \mu, m, A, W\} \tag{5}$$

where $E$, $\mu$, and $m$ are modulus of elasticity, Poisson's ratio, and strain hardening coefficient, $A$ and $W$ are sectional areas and moments of resistance of the considered elements of hydro turbines, and $\alpha_\sigma$ is stress concentration factors.

Characteristics $E$, $\mu$, and $m$ of the mechanical properties of materials are determined by laboratory testing and full-scale sample testing. The stress concentration coefficient $\alpha_\sigma$ is determined experimentally or by calculation methods.

Based on Eq. (5), the characteristics of cyclic loading of hydro turbine elements are determined: stress amplitudes $\sigma_a$ and deformations $e_a$, mean stresses $\sigma_m$ and deformations $e_m$, and cycle asymmetry coefficients $r = \sigma_{min}/\sigma_{max}$. Next, determine the number of loading cycles to failure:

$$N_c = F_N\{\sigma_a, e_a, \sigma_m, e_m, r, S_c, \psi_c, \sigma_{-1}, m_N, m_\sigma, \omega\} \tag{6}$$

where $S_c$ and $\psi_c$ are tensile strength and ultimate plasticity of the material, $\sigma_{-1}$ is material fatigue limit, $m_N$ and $m_\sigma$ are the characteristics of the sensitivity of materials to cyclic loading, and $\omega$ is the loading frequency.

The ratio of actual $N_e$ to critical $N_c$ loading cycles establishes damage level:

$$\begin{aligned} d(N,t) &= \sum_i d_i(N,t), d_i(N,t) \\ &= N_e^{(i)}/N_c^{(i)} \end{aligned} \tag{7}$$

If there are crack-like defects in the structural elements, the resources $l(t)$ or $N(l)$ are determined at the crack growth stage from initial $l_0$ to critical $l_c$ sizes:

$$l(t) = F_l\{Q(t), l_0, N_e, \Delta K\} \quad (8)$$

$$N(l) = F_{Nl}\{Q(t), N_e, l_0, l_c, \Delta K\} \quad (9)$$

where $\Delta K$ is the magnitude of the stress intensity factor.

The estimation of probabilities $P_i(t)$ is carried out taking into account Eqs. (3)–(9) under the assumption that the form of the probability functions $F_P$, $F_Q$, $F_\sigma$, $F_N$, $F_l$, and $F_{Nl}$ and their parameters is defined. Damage assessment $U_i(t + \tau)$ is performed by actual losses or calculated by economic methods for the considered scenarios of possible accidents of hydro turbines.

The calculated estimates of the probability of damages to equipment and structures of hydropower plants in accordance with the above principles gave the following accident probability values:

- Normal working conditions of hydropower plant (regulatory loads)

$$P_i = 2.2 \times 10^{-4} - 1.5 \times 10^{-3}$$

- Violation of normal operating conditions of hydropower plant (increased loads)

$$P_i = 6.0 \times 10^{-3} - 3.1 \times 10^{-2}$$

- Emergency situations (extreme loads)

$$P_i \geq 0.1$$

Qualitative estimates of potential damage for the enlarged scenarios of accidents of hydropower plants were given the following values (in rubles):

- Overflow over the dam $10^8$–$10^9$
- The destruction of the dam (breakthrough of the pressure front) $10^9$–$10^{11}$
- Destruction (flooding) of hydropower plant $10^9$–$10^{10}$

Taking into account the indicated probabilities and damages, the following generalized risk assessments (in \$) of accidents for hydropower plants of the Angaro-Yenisei cascade of Russia were obtained:

- Risk of breaking the pressure front

$$R_\Sigma = 3.6 \times 10^5 - 2.5 \times 10^6$$

- Risk of destruction of hydropower plant

$$R_\Sigma = 1.0 \times 10^6 - 5.0 \times 10^6$$

- Risk of terrorist threat

$$R_\Sigma \leq 5.1 \times 10^4$$

The aggregated statistical estimates of major accidents of hydropower plants give probabilities $P_i = 3.3 \times 10^{-2}$–$2.3 \times 10^{-3}$. Direct damages from such accidents reach $5 \times 10^9$ \$, and indirect damages are $(1.8–2.5) \times 10^{10}$ \$.

It should be emphasized that the approach outlined requires statistical information on all parameters, which is included in the calculations. Particular attention should be paid to characteristics of mechanical properties, parameters of stress-strain states, and structural damage. Such information can be obtained by conducting large volumes of tests and experimental studies. At the same time, the most preferable are methods and means allowing to evaluate the determining parameters (stresses, deformations, sizes of defects), taking into account the peculiarities of the micro- and macrostructure of structural materials.

## 3. Computational models and experimental evaluation operational state of hydro turbines

The main source of the most severe HPP accidents and disasters are damage and destruction of hydro turbines. Therefore, the problem assessing resource, diagnosing damage, optimizing the operating modes of hydro turbines, and timing of repair works takes a special place in ensuring HPP safety. Until recently, the hydro turbine resource received little attention, since it was assumed that the hydraulic turbines have sufficient strength for long-term safe operation. However, the statistics of failures of hydro turbines shows [6, 7] that large safety margins do not guarantee long-term safe operation of hydro turbines.

The hydro turbine resource must be justified taking into account the peculiarities of the loading modes and damage accumulation processes. With this in mind, the interest in assessment of the resource of hydro turbines is steadily growing. This is facilitated by the following circumstances [8]:

- An increase in the number of powerful hydro turbines that have fulfilled the standard operating time
- Operating modes of hydro turbines with a high level of power variation
- Constantly increasing design requirements for efficiency, maneuverability, and reliability of hydro turbines
- The use of new methods and means of technical diagnostics, indicating the presence of defects and damage not previously detected
- The emergence of new perspectives for studying the behavior and state of hydro turbines based on the achievements of experimental and computational technologies

The main factors that reduce the life of hydro turbines are fatigue, corrosion-fatigue and cavitation damage, degradation mechanical properties of materials, and redistribution of stress and strain fields in the most loaded local zones. Fatigue damages are caused by a complex loading spectrum of hydro turbines, containing components with different frequencies. Low-frequency loads (with a frequency below or equal to rotation frequency) are dangerous the high amplitudes that cause formation and development of cracks in the most loaded zones. High-frequency components have small amplitudes, but the number of cycles can reach $10^9$–$10^{10}$, which ultimately also leads to the formation and development of cracks. A significant danger is represented by "start-stop" cycles, in which parasitic vortex structures, hydraulic shocks, and flow instability zones with nonoptimal flow around the blades arise. The most dangerous are the loads caused by water pressure pulsations due to the interaction between the stator and the rotor at the blade frequency, as well as the loads caused by Karman vortices. Special attention should be paid to resonance phenomena, when the proximity of the natural frequencies of the elements of hydro turbines and the frequencies of external influences occurs.

The level of resource exhaustion is determined by the results of special calculations. These calculations consist in determining the time $t$ or the number of loading cycles $N$ as a function of amplitudes $\sigma_a$ and average values $\sigma_m$ of the loading cycle, defect sizes $l$, characteristics of the mechanical properties of materials (conditional yield strength $\sigma_{0.2}$, temporary fracture resistance $\sigma_b$, fatigue limit $\sigma_{-1}$, destructive deformations $\varepsilon_f$), and safety factors for stresses $n_\sigma$, for a number of cycles $n_N$ and for size of defects $n_l$. The results of the calculations usually defined the fatigue diagrams of the main elements of hydro turbines, the residual resource, and the probability of failure at a given operating time.

The main elements of hydraulic turbines requiring the design justification of the resource are an impeller, a turbine shaft, a turbine cover with fastening elements, a shoulder blade of guide, and other elements. Calculation justification is carried out on the basis of data on operating modes, acting loads, defects, and damages detected during the diagnostics [9].

One of the main stages of resource assessment is the determination of external loads for equipment components and the corresponding internal stresses. Despite the great interest of this topic and the significant achievements of recent years, the problem of correctly describing the dynamic behavior of hydro turbine under partial power conditions and during transients has not been fully resolved. With this in mind, it is becoming a more common method of computational modeling [10]. These methods are based on mathematical models that include three main elements: geometric model, model of external loads, and model of boundary conditions. The accuracy of each model can have a decisive influence on the results of numerical experiments, including the issues of resource estimation [8].

The main problems of estimate resource for hydro turbines today are:

- The complexity of accounting technological and operational defects, stress concentration, residual stresses in welded joints, and heat-affected zones
- The poverty of database on the characteristics of materials for a reliable assessment of the resource
- The complexity of the damage summation mechanism in condition uncertainty of external loads and non-design modes of operation
- The difficulty of predicting crack growth under the conditions of actual spectra and loading conditions

The trends in the development of hydro turbine resource assessment methods at the present stage are characterized by the following circumstances:

1. Increasing interest for the problem resource assessment in connection with emergence of new technical capabilities
2. The desire to increase the reliability and accuracy of solving problems at all stages the assessment of hydro turbine resource
3. The need to take into account the non-project operation condition influence on the resource

4. Predicting the growth of cracks in the process of operating time for the purpose of determining the optimal time between repairs

5. An increase in the share of the numerical experiment due to partial replacement of the model and natural experiments

6. Resource management due to the choice of optimal operational parameters, taking into account the capabilities of the power system

7. Increased interest of resource estimates of hydraulic turbines in the absence of the recommended calculation methods and regulatory requirements for service life and criteria for the admissibility of operation

8. The lack of systematic studies of the residual life of hydro turbines, similar to how it was done for the turbines of thermal and nuclear power plants

Thus, the problem of calculation and experimental evaluation operational state of hydro turbines has a number of unsolved or difficult tasks that require in-depth basic research on the nature of the stress-strain state of hydraulic units, features of damage development mechanisms, and degradation of mechanical properties of materials.

## 4. Technical diagnostics of hydro turbines

The technical condition assessing of hydro turbines is crucial for the estimated assessment of residual life. Such works are performed in accordance with the provisions of the norms and standards [11, 12]. These works include a wide range of studies of actual state of hydraulic turbines by destructive and nondestructive control methods [13]:

- Analysis of design, maintenance, and repair documentation

- Nondestructive testing of structural elements and welded joints

- Experimental studies of metal and welded joints (measurement of hardness, determination of mechanical properties, conducting metallographic studies, and determination of chemical composition)

- Visual measurement control of geometry, surface defects, and shape defects with the determination of their sizes

- Nondestructive penetration control for substance detection of surface defects

- Nondestructive ultrasonic testing of structural elements and welds for detection of internal defects and cracks

- Ultrasonic thickness gauging of elements and determination of the internal stratification of the metal

In addition to the listed methods of nondestructive testing, stress-strain state studies using strain-gauge methods [14] and optical methods of electronic speckle interferometry [15] are performed.

This complex of methods and means is used in diagnosing the technical condition of hydro turbines with over standard operating life. Such work was carried out at the abovementioned hydropower stations in recent years. The main attention was focused on the most loaded structures: the impeller, the turbine shaft, the turbine cover, and the blade of guide.

The systematization and classification results of nondestructive testing showed that the main defects of the impeller blades of hydro turbines are:

- Cavitation erosion
- Corrosion fatigue cracks of the base metal and fatigue cracks of welded joints
- Corrosion damage
- Technological defects of welds

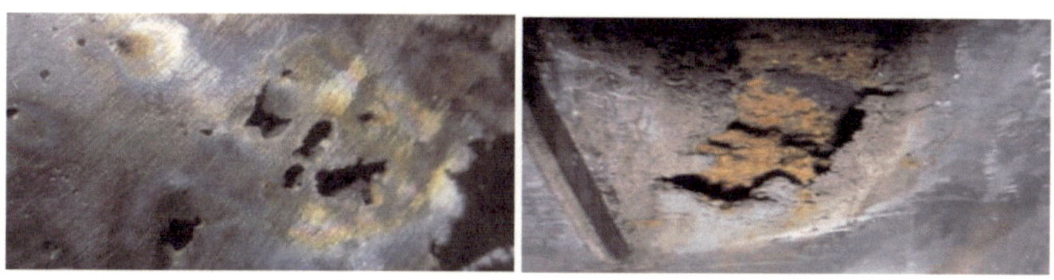

**Figure 3.**
*Corrosion damage and cavitation damage metal of impeller blades.*

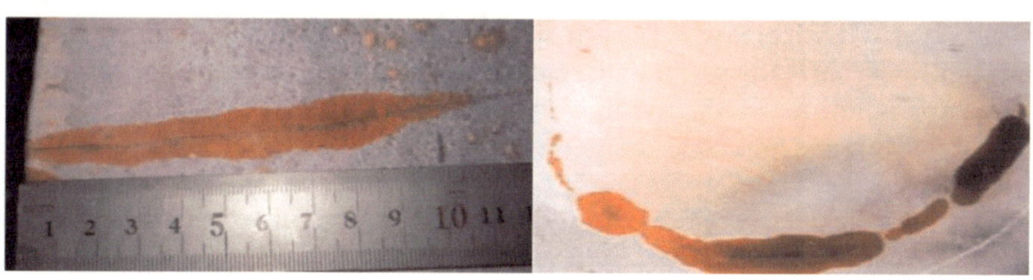

**Figure 4.**
*Cracks in metal of impeller blades.*

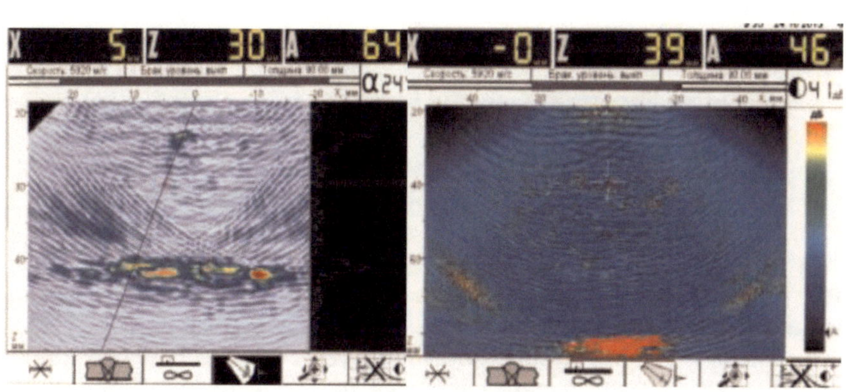


**Figure 5.**
*Internal defects of impeller blades, detected by ultrasound tomography.*

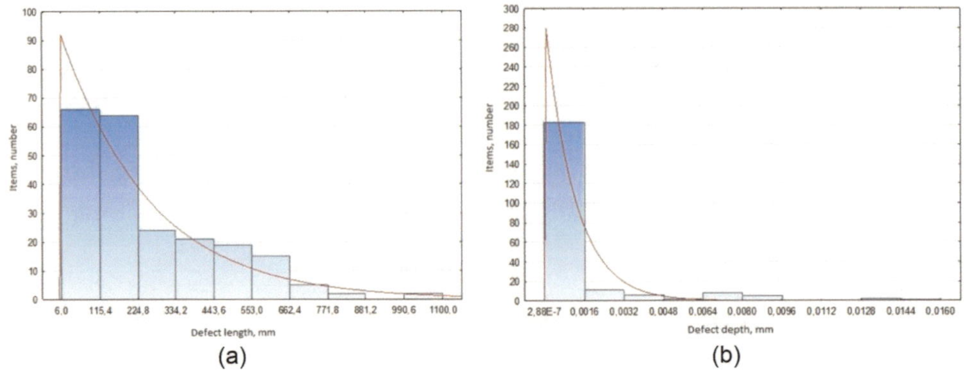


**Figure 6.**
*Distribution corrosion damages of length (a) and depth (b).*

The nature of the defects and damage is presented in **Figures 3–5**. Similar defects were detected and investigated previously in the impellers of the Sayano-Shushenskaya HPP and Krasnoyarskaya HPP [6, 7].

Statistical analysis of the nondestructive testing results for cavitation erosion zones allowed determining the main geometrical parameters for these defects: the length, width, and depth (**Figure 6**).

## 5. Estimated justification resource of hydro turbines

Standard operating life of hydraulic units is established by project documentation and for most units are 30–40 years. Currently, a significant part of hydraulic units of powerful hydroelectric power plants is carried out outside the standard operating time. This leads to a decrease in the overall level of reliability of the structure as a whole and its individual elements and to an increase in the probability of failures and financial costs for technical diagnostics and repairs. The exhaustion of the standard operating life of the elements of hydro turbines raises the question of the assessment of the residual resource as one of the priorities in the field of ensuring the safety of hydraulic structures [7–9]. At the same time, as noted in [8], there are no generally accepted methods for assessing the residual life and the regulatory framework that defines the procedure for extending the service life of the impellers and making decisions about their replacement.

Taking into account the specified circumstances, the authors formulated the basic provisions for the calculation of the residual life of hydraulic turbine elements. The calculation of the resource before the formation of cracks is carried out on the allowable stresses:

$$[\sigma] = \min\left\{\frac{R_{02}}{n_{02}}; \frac{R_{\mathrm{m}}}{n_m}\right\} \tag{10}$$

where $R_{02}$ and $R_m$ are regulatory design resistances of the metal and $n_{02}$ and $n_m$ are dimensionless safety margins.

The estimated allowable number of loading cycles of the elements of the turbine at the stage of the occurrence of cracks, for the "start-stop" mode and mode control with the transition through the "forbidden" zone of operation modes, is determined by the formula:

$$[N] = \frac{1}{4n_N}\left[\frac{Ee_c}{\sigma_a - \sigma_{r0}/(1+\sigma_{r-1}r)}\right]^2, \quad \sigma_{r0} = \frac{\gamma\varepsilon\beta}{K_\sigma}\sigma_{-1}\sqrt{1-\sigma_m/[\sigma]} \tag{11}$$

where $\sigma_{-1}$ is fatigue limit of given loading mode; $\sigma_m$ is stress average of cycle; $\sigma_a$ is stress amplitude of cycle; $r$ is the asymmetry coefficient of the loading cycle; $\gamma$, $\varepsilon$, and $\beta$ are dimensionless coefficients that take into account the influence of the medium, scale factor, and surface quality; and $K_\sigma$ is stress concentration factor.

The estimated allowable number of loading cycles at blade and blade frequencies is determined by the following formula:

$$[N] = \frac{N_0}{n_N}\left(\frac{\sigma_{r0}}{\sigma_a}\right)^m \tag{12}$$

where $N_0$ is the base number of loading cycles and $m$ is dimensionless fatigue pattern indicator.

The influence of the multifrequency component of the loading mode from Karman vortices is taken into account through the reduced stress amplitude:

$$\sigma_a = \sum_{i=1}^{n} \sigma_{ai}(\omega_i/\omega_1)^\alpha \tag{13}$$

where $\sigma_{ai}$ is stress amplitude at the frequency $\omega_i$, $\omega_1$ is the frequency of reduction, $\alpha$ is dimensionless coefficient taking into account the influence of the multifrequency nature of loading, and $n_N$ is the safety factor by the number of loading cycles.

The total accumulated fatigue damage for the considered loading modes is defined as the sum of the ratios of the actual $N_{ei}$ and the calculated loading cycles:

$$d = \sum_i \frac{N_{ei}}{[N]} \leq 1 \tag{14}$$

The number of cycles $N_l$ at the stage of crack growth is determined by the following formulas:

For single-frequency loading mode

$$N_l = \frac{1}{n_l}\frac{2}{(m-2)CY^{\frac{m}{2}}\sigma^m}\left[\frac{1}{l_0^{\frac{m-2}{2}}} - \frac{1}{l^{\frac{(m-2)}{2}}}\right] \tag{15}$$

For multifrequency loading mode

$$\frac{dl}{dN} = q\frac{\pi}{8}\left(\frac{K_{th}}{R_{0.2}}\right)^2\left\{1 + (1-\beta)\frac{\left(1-\widetilde{K}_{th}^2\right)^2 - \left(\widetilde{K}_{max}^2 - \widetilde{K}_{th}^2\right)^2\omega\mu^2}{\left(\widetilde{K}_{max}^2 - \widetilde{K}_{th}^2\right)^2\omega\mu^2}\right\}^{1/(\beta-1)} \tag{16}$$

where $q$, $\beta$, and $\mu$ are parameters of the cyclic crack growth diagram, $\widetilde{K}_{th} = K_{th}/K_C$ and $\widetilde{K}_{max} = K_{max}/K_C$ are relative threshold and maximum stress intensity factors, $K_c$ is the crack resistance characteristic of steel, $\omega$ is relative frequency of loading, and $K_{th}$ is the threshold stress intensity factor.

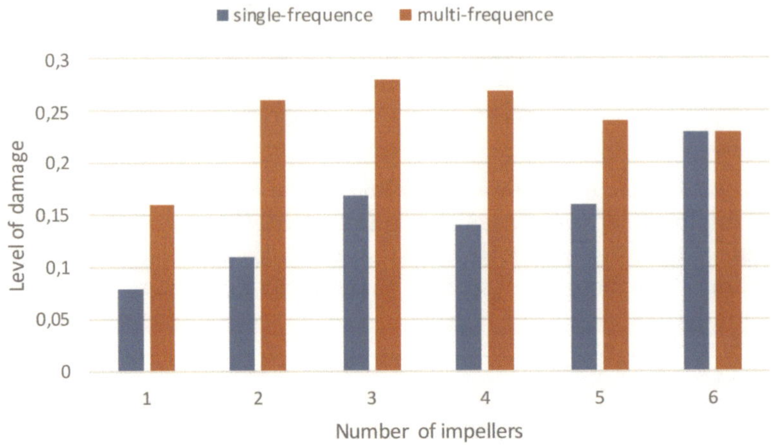


**Figure 7.**
*Estimated cumulative damage for impeller blades.*

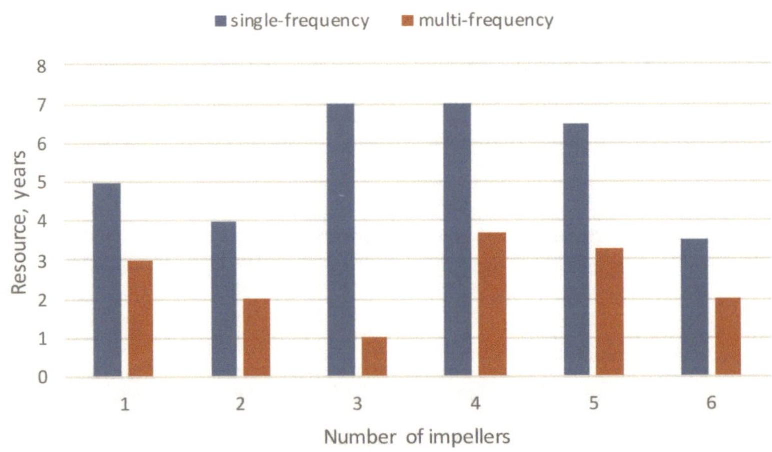


**Figure 8.**
*Estimated residual life of impeller blades by the criterion of cracking.*

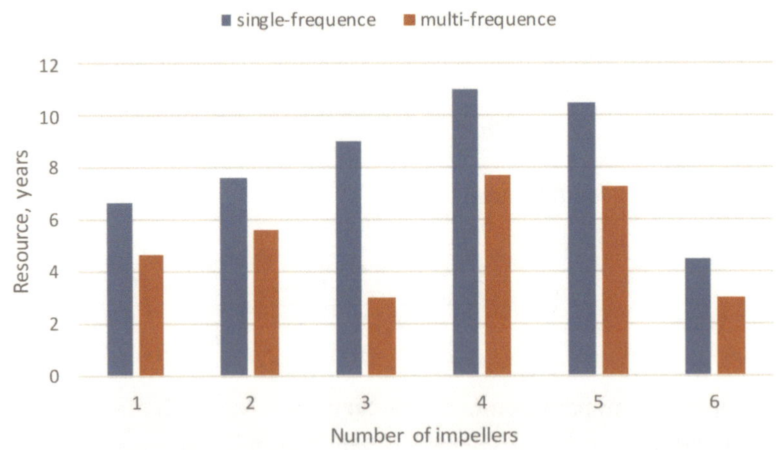


**Figure 9.**
*Estimated residual life of impeller blades at the stage of crack development.*

**Figures 7–9** show the comparison of the results of the resource calculation according to the above procedure for the elements of hydro turbines of the Krasnoyarskaya HPP. The calculations were carried out on the basis of the results of a comprehensive diagnosis of the technical condition, with an assessment of the characteristics of the stress-strain state, the characteristics of the mechanical properties, and the defectiveness of the structural elements. The calculations took into account loading cycles: "start-stop," mode control, on blade frequencies, and at the frequencies of the Karman vortices.

As can be seen from the figures, the resource has a wide range of values. This is due to the different levels of metal damage detected during technical diagnostics and the initial dimensions of crack-like defects in structural elements.

The calculation results show that the hydraulic units surveyed using modern means of technical diagnostics and nondestructive testing have a resource reserve sufficient for planning and carrying out work to replace the impellers with more modern units.

It can also be assumed that an integrated approach to the problem of ensuring the reliability and safety of hydraulic units makes it possible to reliably predict the possibilities, terms, and conditions for their further operation.

## 6. Conclusion

Analysis of domestic and foreign studies and the practice of operating hydraulic equipment of large hydroelectric power plants indicate the need for the development of more advanced computational methods for estimating the life of hydro turbines that have completed their standard (design) service lives. When solving problems of resource assessment, special complex methods of technical diagnostics and modern computational and experimental technologies should be applied. These methods should be based on a combination of engineering design models that take into account the individual characteristics of hydraulic units based on routine monitoring and diagnostics and systems of reasonable safety factors (fatigue, crack length, stress, etc.) reflecting the uncertainty of the task with the required degree of accuracy design loads, material properties, and modes of operation.

It should be emphasized that the purpose, role, and place of technical diagnostics and assessment of the hydraulic equipment resource should be linked to the task of assessing the protection of hydropower stations from severe accidents and disasters according to risk criteria. In technical assignments for the design of hydroelectric power plants, new quantitative safety indicators should be introduced that implement the design-experimental complex "strength—resource—reliability—survivability—safety—risk—security".

## Author details

Nikolay Makhutov[1], Yury Petrenia[2], Anatoly Lepikhin[3*], Vladimir Moskvichev[3], Mikhail Gadenin[1] and Anatoly Tchernyaev[3]

1 Mechanical Engineering Research Institute of RAS, Moscow, Russia

2 Power Machines, Public Stock Company, St-Petersburg, Russia

3 Institute of Computational Technologies SB RAS, Krasnoyarsk, Russia

*Address all correspondence to: aml@ict.nsc.ru

## References

[1] Makhutov NA. Strength and Safety: Fundamental and Applied Research. Novosibirsk: Science; 2008. 528 p

[2] Makhutov NA, Gadenin MM, Moskvichev VV. Formation of the regulatory framework for the safety and security of hydropower plants of Siberia from severe disasters. Security and Emergency Issues. 2011;**4**:28-32

[3] Lepikhin A, Lepikhina I. Probabilistic models for risk assessment of disasters. Safety and Security Engineering IV. 2011:83-90

[4] Makhutov NA, Gadenin MM. Technical diagnostics and risk monitoring for HPS equipment. Destruction of hydro turbine No. 2 of Sayano-Shushenskaya HPP: Reasons and lessons. M: Format-D; 2013. pp. 26-35

[5] Makhutov NA. Proposals for creating a system to prevent and minimize the risk of accidents at all stages of life cycle. Destruction of hydro turbine No.2 of Sayano-Shushenskaya HPP: Reasons and lessons. M: Format-D; 2013. pp. 51-67

[6] Bryzgalov VI, Klyukach AA. The condition of the impellers of hydraulic units of the Sayano-Shushenskaya HPP and the main directions for maintaining their performance. Hydraulic Engineering. 2000;**1**:26-29

[7] Sivkov VG. Problems of cracking in the blades of impellers of hydraulic units of the Sayano-Shushenskaya HPP. Hydraulic Engineering. 2003;**11**:51-52

[8] Vasiliev Yu S, Petrenya Yu K, Georgievskaya EV. On hydro turbine resources: A review of foreign literature. Scientific and Technical Statements SPbPU. Engineering and Natural Sciences. 2017;**2**:184-204

[9] Gavrilov SN, Georgievskaya EV. Ways to improve the reliability of hydro-turbine equipment: An integrated approach to the assessment of residual life. Destruction of hydraulic unit No. 2 of Sayano-Shushenskaya HPP: Reasons and lessons. M: Format-D; 2013. pp. 200-205

[10] Cherny SG, Chirkov DV, Lapin VN, Skorospelov VA, Sharov SV. Numerical Simulation of Currents in Turbomachines. Novosibirsk: Science; 2006. p. 202

[11] STO 17330282.27.140.001-2006. Methods of assessing the technical condition of the main equipment of hydroelectric power plants. UES of Russia; 2006. p. 281

[12] GOST R 55260.3.2-2013. Hydroelectric power. Part 3-2. Water turbines. Techniques for assessing the technical condition. M.: Standartinform; 2015. p. 93

[13] Non-destructive control. Directory. In: 8 volumes. Klyuev VV, editors; M: Mashinostroenie; 2006. p. 840

[14] Aniskovich EV, Moskvichev VV, Makhutov NA, Razumovsky IA, et al. Assessment of residual stresses in the blades of impellers of hydraulic units. Hydraulic Engineering. 2018;**11**:178-184

[15] Makhutov NA, Gadenin MM, Odintsev IN, Razumovsky IA. Development of methods for the calculation and experimental determination of local residual stresses for complex loading spectra. Problems of Engineering and Machine Reliability. 2015;**6**:53-62

Chapter 3

# Mixture Transition Distribution Modelling of Multivariate Time Series of Discrete State Processes: With an Application to Modelling Flowering Synchronisation with Respect to Climate Dynamics

*Irene Hudson, Susan Won Sun Kim and Marie Keatley*

## Abstract

A new approach to assess synchronicity developed in this chapter is a novel bivariate extension of the generalised mixture transition distribution (MTDg) model (we coin this B-MTD). The aim of this chapter is to test MTDg an extended MTD with interactions model and its bivariate extension of MTD (B-MTD) to investigate synchrony of flowering of four Eucalypts species—*E. leucoxylon, E. microcarpa, E. polyanthemos* and *E. tricarpa* over a 31 year period. The mixture transition distribution (MTDg) is a method to estimate transition probabilities of high order Markov chains. Our B-MTD approach allows us the derive rules of thumb for synchrony and asynchrony between pairs of species, e.g. flowering of the four species. The latter B-MTD rules are based on transition probabilities between all possible on and off flowering states from previous to current time. We also apply MTDg modelling using lagged flowering states and climate covariates as predictors to model current flowering status (on/off) to assess synchronisation using residuals from the resultant models via our adaptation of Moran's classic synchrony statistic. We compare these MTDg (with covariates)-based synchrony measures with our B-MTD results in addition to those from extended Kalman filter (EKF)-based residuals.

**Keywords:** multivariate mixed transition distributions, Markov chains, synchrony, climate, eucalypt flowering

## 1. Introduction

Separation or lack of overlap of flowering time in eucalypts has been suggested as a mechanism for maintaining overall 'generic identity' of a plant species. If, for example flowering times and pollinators overlap in sympatric species, hybridization can occur between closely related eucalypts species. Therefore examination of long-term synchrony establishes a baseline of flowering behaviour which may assist in detecting recent or future changes. Although *Eucalyptus* as a genus dominates much of the Australian landscape [1, 2], few studies have quantified eucalypt flowering overlap within or between species, due to the shortage of phenological data in Australia [3, 4, 7]. This chapter examines flowering synchrony over a 31 year period, 1940–1970, at the population level among four eucalypts species—*Eucalyptus leucoxylon, E. microcarpa, E. polyanthemos* and *E. tricarpa* [3–8].

A new approach to assess synchronicity developed in this chapter is a novel bivariate extension of the of the MTDg model [6, 9] (we coin this B-MTD). The aim of this chapter is to test mixture transition distribution (MTD) and an extended MTDg with interactions; and a novel bivariate extension of MTD (B-MTD) to investigate synchrony in phenological data. The MTDg model [6, 9] was the first approach developed to study the multivariate relationship between the probability of flowering with two states of rain and mean temperature via a mixture transition distribution (MTD), assuming, however a different transition matrix from each lag to the present time (our MTDg analysis), thus generalising the MTD approach in [13], (see also [10]) which led to the development of the MARCH software to perform MTD without covariates [11, 12]. The MTDg model is different to MARCH not only in terms of incorporating interactions between the covariates but also in its minimization process, namely using the AD Model BuilderTM [14], which uses auto-differentiation as a minimisation tool. This is shown to be computationally less intensive than MARCH. The assumption Berchtold's MTD model, namely the assumed equality of the transition matrices among different lags, was a strong assumption, so the idea of the mixture transition distribution model was to consider independently the effect of each lag to the present instead of considering the effect of the combination of lags as in pure Markov chain processes. Specifically, an extended model for MTDg analysis which accommodates interactions was developed in [6], and applied to MTDg modelling of the flowering of four eucalyptus species studied in this chapter, as multivariate time series.

This work extends both MARCH and the work in [15, 16] to allow for differing transition matrices among the lags, i.e. our B-MTD method builds on this approach of the MTDg with interaction model [6, 9]. The MTDg model with interactions showed that the flowering of *E. leucoxylon* and *E. tricarpa* behave similarly with temperature (both flower at low temperature) and both have a positive relationship with flowering intensity 11 months ago. The flowering of *E. microcarpa* behaves differently in that *E. microcarpa* flowers at high temperature.

Our B-MTD approach developed in this chapter allows us the derive rules of thumb for synchrony and asynchrony between pairs of species. The latter B-MTD rules are based on transition probabilities between all possible on and off flowering states from previous to current time. Synchronisation is also tested using residuals from the resultant models via an adaptation of Moran's [17, 18] classical synchrony statistic, incorporating MTDg residuals [17–19].

We also apply the earlier MTDg modelling in [6] using climate covariates and lagged flowering states as predictors to model flowering states (on/off) and thus assess synchronisation using an adaptation of the approach of Moran to the resultant MTDg model and fitted residuals. We compare these MTDg (with covariates)-based synchrony measures with our B-MTD results in addition to those using the extended Kalman filter (EKF) [15, 19], based residuals obtained earlier in [21].

## 2. The mixture transition distribution (MTDg) and B-MTD models: mathematical formulations in brief

### 2.1 The MTDg model with interactions between the covariates

The MTD model with covariates was discussed in [6] and developed in [15, 19] to incorporate interactions between the covariates (e.g. rainfall, temperature variants in the case study discussed). The high-order MTD transition probabilities are computed as follows:

$$P(X_t = i_0|X_{t-1} = i_1, ...X_{t-f} = i_f, C_1 = c_1, ...C_e = c_e, M_1 = m_1, ...M_l = m_l) \\ = \sum_{g=1}^{f} \lambda_g q_{i_g i_0} + \sum_{h=1}^{e} \lambda_{f+h} d_{hj_h i_0} + \sum_{u=1}^{l} \lambda_{f+e+u} s_{uv_u i_0} \tag{1}$$

where $\lambda_{f+e+u}$ is the weight for the interaction term, $q_{i_g i_0}$ is the transition probability from modality $i_g$ observed at time $t-g$ and modality $i_0$ observed at time $t$ in the transition matrix $Q$, $s_{uv_u i_0}$ is transition probability between covariate $h_1$ and covariate $h_2$ interaction term ($v_u = d_{h_1 j_{h_1}} \times d_{h_2 j_{h_2}}$) and $X_t$, and where $\sum_{g=1}^{f+e+l} \lambda_g = 1$ and where $\lambda_g \geq 0$.

We refer the reader to [6], and further works in the seminal book by Hudson and Keatley [7] for further mathematical details.

### 2.2 The bivariate mixture transition distribution (B-MTD)

Let $\{X_t\}$ and $\{Y_t\}$ be sequences of random variables (say two (flowering intensity) time series) taking values in the finite set $N = \{1, ..., k\}$. In a $f^{\text{th}}$-order Markov chain, the probability that $\{X_t, Y_t\} = \{i_0, i'_0\}$, ($i_0, i'_0 \in N$) depends on the combination of values taken by $X_{t-f}, ..., X_{t-1}, Y_{t-f}, ..., Y_{t-1}$. In the MTD model, the contributions of the different lags are combined additively. Then a bivariate MTD model, which we denote by B-MTD, has the following formulation:

$$P\left(\{X_t, Y_t\} = \{i_0, i'_0\} | X_{t-1} = i_1, ..., X_{t-f} = i_f, Y_{t-1} = i'_1, ..., Y_{t-f} = i'_f\right) \\ = \sum_{g=1}^{f} \lambda_g q_{i_g, i'_g, i_0, i'_0} \tag{2}$$

where $i_f, ..., i_0, i'_f, ..., i'_0 \in N$, the probabilities $q_{i_g, i'_g, i_0, i'_0}$ are elements of an $m^2 \times m^2$ transition matrix $Q = \left[q_{i_g, i'_g, i_0, i'_0}\right]$, each row of which is a probability distribution (i.e. each row sums to 1 and the elements are nonnegative) and $\lambda = (\lambda_f, ..., \lambda_1)'$ is a vector of lag parameters. To ensure that the results of the model are probabilities, that is, $0 \leq \sum_{g=1}^{f} \lambda_g q_{i_g i'_g i_0 i'_0} \leq 1$ the vector $\lambda$ is subject to the constraints $\sum_{g=1}^{f} \lambda_g = 1$ and $\lambda_g \geq 0$.

Covariates and interaction terms can be added to the bivariate MTD (B-MTD) as follows:

$$P(\{X_{1,t}, ..., X_{n,t}\} = \{i_{1,0}, ..., i_{n,0}\} | X_{t-1} = i_1, ..., X_{t-f} = i_f, Y_{t-1} = i'_1, ..., Y_{t-f} = i'_f, \\ C_1 = c_1, ..., C_e = c_e, M_1 = m_1, ..., M_l = m_l) \\ = \sum_{g=1}^{f} \lambda_g q_{i_{1,g}, ..., i_{n,g}, i_{1,0}, ..., i_{n,0}} + \sum_{h=1}^{e} \lambda_{f+h} d_{hj_h, i_{1,0}, ..., i_{n,0}} + \sum_{u=1}^{l} \lambda_{f+e+u} s_{uv_u i_{1,0}, ..., i_{n,0}} \tag{3}$$

where $\lambda_{f+e+u}$ is the weight for the interaction term, $s_{uv_u i_1, 0, \dots, i_n, 0}$ is the transition probability between covariate $h_1$ and covariate $h_2$ interaction term ($v_u = d_{h_1 j_{h_1}} \times d_{h_2 j_{h_2}}$) and $(X_t, Y_t)$, and where $\sum_{g=1}^{f+e+l} \lambda_g = 1$. For example, if both $X_t$ and $Y_t$ are time series that constitute random realizations of two states {0, 1} and the covariates $C_1, \dots, C_e$ are also defined by bivariate states {0, 1}, then the set of all possible states for $(X_t, Y_t)$ is {(0, 0), (0, 1), (1, 0), (1, 1)}. Hence the transition matrix $Q = \left[q_{i_g i'_g i_0 i'_0}\right]$ is a 4 × 4 matrix as specified below.

| **Previous state $(X_{t-1}, Y_{t-1})$** | | **Current state $(X_t, Y_t)$** | | |
|---|---|---|---|---|
| | **0, 0 (1)** | **0, 1 (2)** | **1, 0 (3)** | **1, 1 (4)** |
| 0, 0 (1) | (1, 1) | (1, 2) | (1, 3) | (1, 4) |
| 0, 1 (2) | (2, 1) | (2, 2) | (2, 3) | (2, 4) |
| 1, 0 (3) | (3, 1) | (3, 2) | (3, 3) | (3, 4) |
| 1, 1 (4) | (4, 1) | (4, 2) | (4, 3) | (4, 4) |

The transition matrices $D_h = \left[d_{hj_h i_0 i'_0}\right]$, $h$ = 1,..., $e$, are 2 × 4 matrices as below.

| | | **Covariate state** |
|---|---|---|
| **$(X_t, Y_t)$** | **0 (1)** | **1 (2)** |
| 0, 0 (1) | (1, 1) | (1, 2) |
| 0, 1 (2) | (2, 1) | (2, 2) |
| 1, 0 (3) | (3, 1) | (3, 2) |
| 1, 1 (4) | (4, 1) | (4, 2) |

## 2.3 Synchrony analysis using Moran's approach

Moran in [17, 18] suggested that if two series $x_t$ and $y_t$ are synchronous, and if $x_t$ can be estimated by a model $f(x)$, the residuals from series $x_t$ fitted to $f(x)$, and the residuals from series $y_t$, fitted with the same model, but with observations, $y_t$, then, $f(y)$ will be positively correlated. The synchrony of two series can then be examined by testing the significance of the correlation of these two series of residuals (using the same model). Moran used an autoregressive integrated moving average (ARIMA) model to test synchrony. Moran's theorem suggests that if two (or more) populations sharing a common linear density-dependence (in a so-called renewal process) are disturbed with correlated noise, they will become synchronised with a correlation matching the noise correlation (see details in [4], and also [6, 15, 21]).

In this chapter we adopt the $k$th order linear stochastic difference to assess synchrony. Goodness of fit of the second order AR ($k$ = 2) model is obtained. The series of residuals can then be found by subtracting the predicted (fitted species) value from the observed series. In summary, synchrony (or otherwise) of two series can be established by performing a test of significance on the correlation coefficient calculated from the two series of residuals as follows:

- Calculate the residuals for say, *E. leucoxylon* (Leu) using its AR (2) model. We denote this residual series by R1.

- Calculate residuals for say, *E. tricarpa* (Tri) using this same model. We denote this residual series by R2.

- Calculate the Pearson correlation coefficient between the residual series R1 and R2 and test for its significance at $p < 0.05$.

Further details on how Moran's method is used and adapted in the case of the MTDg-based models are given in [15] (see also Section 4.8). We use the functionals and parameterisations from the mixture transition distribution (MTD) analysis as the basis of our EKF modelling approach. EKF is likewise a method to estimate the past, present and future status of non-linear time series data by minimising the mean square error. We will also test whether EKF better detects asynchronous species pairs, given EKF estimates the Kalman gain and covariance matrix at each time point [15, 19].

## 3. Data

Flowering data were sourced from the Box-Ironbark Forest near Maryborough, Victoria, in particular the flowering records of *E. leucoxylon*, *E. microcarpa*, *E. polyanthemos* and *E. tricarpa* (1940 and 1971). Flowering intensity was calculated by using a rank score (from 0 to 5) based on the quantity and distribution of flowering [4, 20, 23].

Flowering intensity scores were dichotomised into two discrete states, namely on and off (1/0) flowering (**Figure 1**) as in [6]. One temperature variant, mean monthly diurnal temperature (MeanT), in addition to the monthly rainfall (Rain) were included as climate covariates in the MTDg models; along with the temperature by rain interaction effect. We used discrete state low/high (lower than median temperature *vs* higher than median temperature) for the temperature variable dichotomies and less/more (less than the median rainfall *vs* more than the median

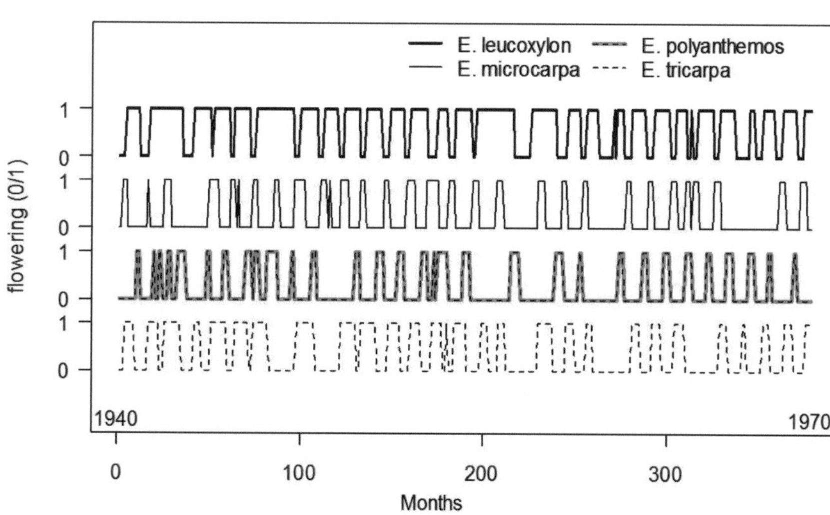


**Figure 1.**
*Flowering of the four eucalypts species.*

| Climate variables | Low (less) | High (more) |
|---|---|---|
| Mean diurnal temp (°C) | ≤13.84 | >13.84 |
| Rain (mm) | ≤40.45 | >40.45 |

**Table 1.**
*Cut-points for climate variables based on medians.*

rainfall) for the rainfall variable. The cut-points for the states or low/high categories of each climate covariate are shown in **Table 1**.

## 4. Results

### 4.1 Bivariate MTD (B-MTD) discrete states results

Four eucalypts species, *E. leucoxylon*, *E. microcarpa*, *E. polyanthemos* and *E. tricarpa* were modelled using the order 1 B-MTD model discussed in Section 2.2—without the inclusion of covariates (such as temperature (variants) and rainfall). These species were paired as follows: *E. leucoxylon* and *E. microcarpa* (LeuMic); *E. leucoxylon* and *E. polyanthemos* (LeuPol); *E. leucoxylon* and *E. tricarpa* (LeuTri) and so on; hence 6 pairs were modelled via B-MTD (see **Table 2**) for the corresponding bivariate transition probabilities (see also **Figure 2**).

The possible states for any pair of species is the set {(0, 0), (0, 1), (1, 0), (1, 1)}, where no flowering is represented as 0 (state = 0 = no flowering) and flowering is represented by a 1 (state = 1 = flowering). Since lag order 1 B-MTD models were used, the mixing probability $\lambda$ is equal to 1.0.

The corresponding transition matrices for the 6 B-MTD models are given in **Table 2**. These transition profiles are also shown schematically as flow diagrams in **Figures 3–4**, and also as transition signatures in **Figures 5–6**. These shall be discussed in more detail later. The transitions to differing states (from **Table 2**) are shown as arrows (transitions A to F) in the schematic diagram of **Figure 2**. The exact probabilities of such transitions are given by the off diagonal elements of **Table 2** and also shown above or below the arrows in **Figures 3** and **4**.

The transitions have the following intuitive interpretation and associated probability (sum), which are derived from the subcomponents of the transition matrices $Q$ (see **Table 2**).

- A: transition of both species off to one species on: q(0, 0;0, 1) + q(0, 0, 1, 0)
- B: transition of both species on to one species off: q(1, 1;0, 1) + q(1, 1, 1, 0)
- C: species switching states: q(0, 1;1, 0) + q(1, 0; 0, 1)
- D: transition of one species off to both species off: q(0, 1;0, 0) + q(1, 0;0, 0)
- E: transition of one species on to both species on: q(0, 1;1, 1) + q(1, 0;1, 1)
- F: transition of one species on/off to both species off/on: q(0, 0;1, 1) + q(1, 1;0, 0)

In this chapter we shall demonstrate that transitions that lead towards both species being off or both species being on (states D, E or F), are considered to be

| Species | Previous state | Current state | | | |
|---|---|---|---|---|---|
| | | (0, 0) | (0, 1) | (1, 0) | (1, 1) |
| LeuMic | (0, 0) | **0.6667** | 0.2280 | 0.1053 | 0.0000 |
| | (0, 1) | 0.0000 | **0.6000** | 0.1333 | 0.2667 |
| | (1, 0) | 0.0845 | 0.0376 | **0.8357** | 0.0423 |
| | (1, 1) | 0.0000 | 0.0612 | 0.4490 | **0.4898** |
| LeuPol | (0, 0) | **0.6970** | 0.0303 | 0.2626 | 0.0101 |
| | (0, 1) | 0.4444 | **0.3889** | 0.0000 | 0.1667 |
| | (1, 0) | 0.0562 | 0.0000 | **0.7921** | 0.1517 |
| | (1, 1) | 0.1309 | 0.0952 | 0.1429 | **0.6310** |
| LeuTri | (0, 0) | **0.6947** | 0.1263 | 0.1053 | 0.0737 |
| | (0, 1) | 0.0455 | **0.3636** | 0.0000 | 0.5909 |
| | (1, 0) | 0.2203 | 0.0085 | **0.7034** | 0.0678 |
| | (1, 1) | 0.0069 | 0.0069 | 0.1736 | **0.8125** |
| MicPol | (0, 0) | **0.7637** | 0.1429 | 0.0879 | 0.0055 |
| | (0, 1) | 0.1818 | **0.6705** | 0.1023 | 0.0455 |
| | (1, 0) | 0.2737 | 0.0000 | **0.6842** | 0.0421 |
| | (1, 1) | 0.0714 | 0.2141 | 0.3572 | **0.3573** |
| MicTri | (0, 0) | **0.7975** | 0.0316 | 0.1329 | 0.0380 |
| | (0, 1) | 0.2232 | **0.7500** | 0.0179 | 0.0089 |
| | (1, 0) | 0.1090 | 0.0182 | **0.5819** | 0.2909 |
| | (1, 1) | 0.0000 | 0.4259 | 0.0000 | **0.5741** |
| PolTri | (0, 0) | **0.7464** | 0.1739 | 0.0797 | 0.0000 |
| | (0, 1) | 0.0719 | **0.7842** | 0.0360 | 0.1079 |
| | (1, 0) | 0.3067 | 0.0400 | **0.6400** | 0.0133 |
| | (1, 1) | 0.0370 | 0.1482 | 0.4074 | **0.4074** |

**Table 2.**
*Transition matrices for the 6 B-MTD models.*

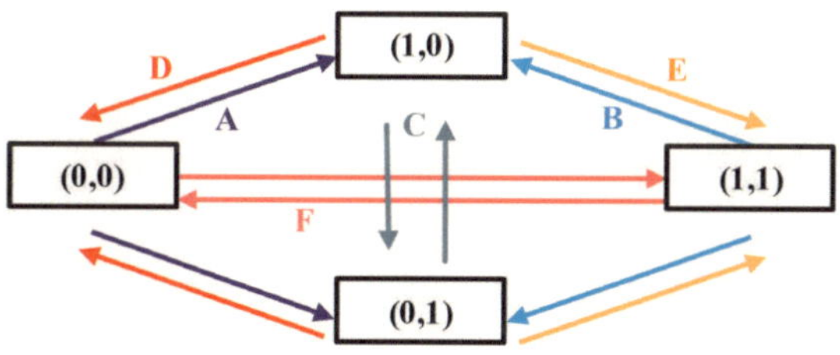


**Figure 2.**
*Subcomponents of possible transitions.*

synchronising. However, transitions that lead towards only one species being on or off (flowering) (A and B) and where within a species pair flowering switches (transitions C) are considered to be asynchronous.

Note that the probabilities of staying in the same state; e.g. both species continuing to be in a non-flowering state (a (0, 0) to (0, 0) transition); one species flowering off and the other species in the pair with flowering on, (a (0, 1) to (0, 1)

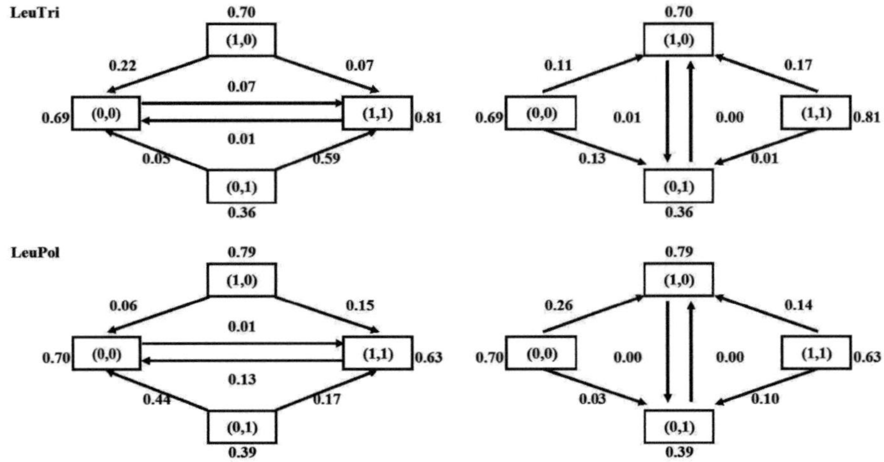


**Figure 3.**
*Diagram of transition probabilities for synchronous pairs: LeuTri and LeuPol.*

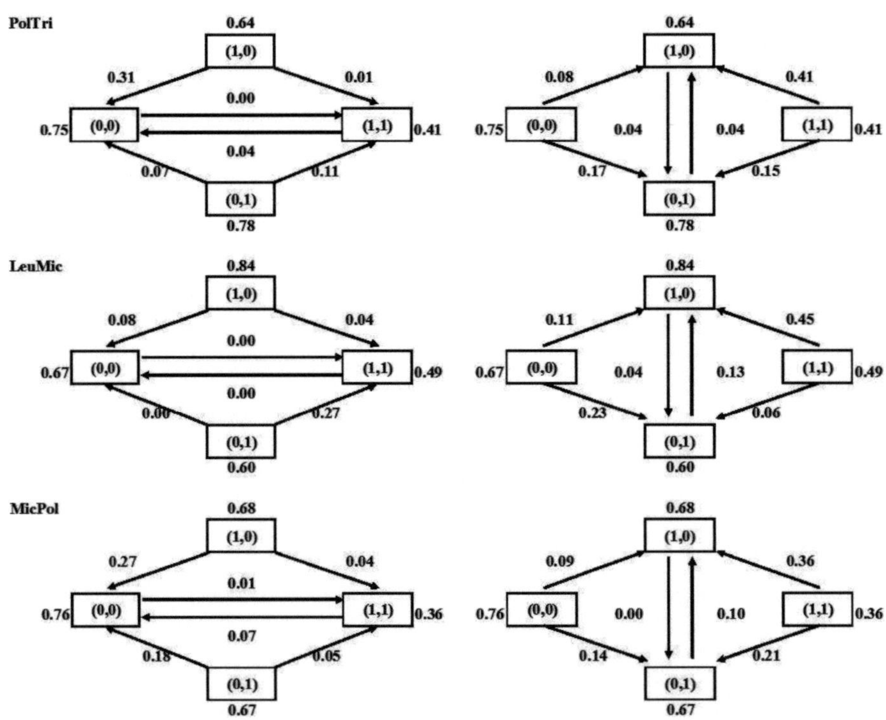


**Figure 4.**
*Diagram of transition probabilities for asynchronous pairs: PolTri, LeuMic and MicPol.*

transition); one species on the other in the pair off (a (1, 0) to (1, 0) transition); and both species continuing to flower (a (1, 1) to (1, 1) transition) are not shown on **Figure 2**. These to same states transitions, are given for each species, by the diagonal elements in the transition matrices (from previous to current states) in **Table 2**; and are also shown in **Figures 3** and **4** as numbers (positioned next to the 4 states as boxes).

An examination of the transition probabilities for the species pairs in **Table 2** shows that there is a significantly high propensity (probability) to remain in the

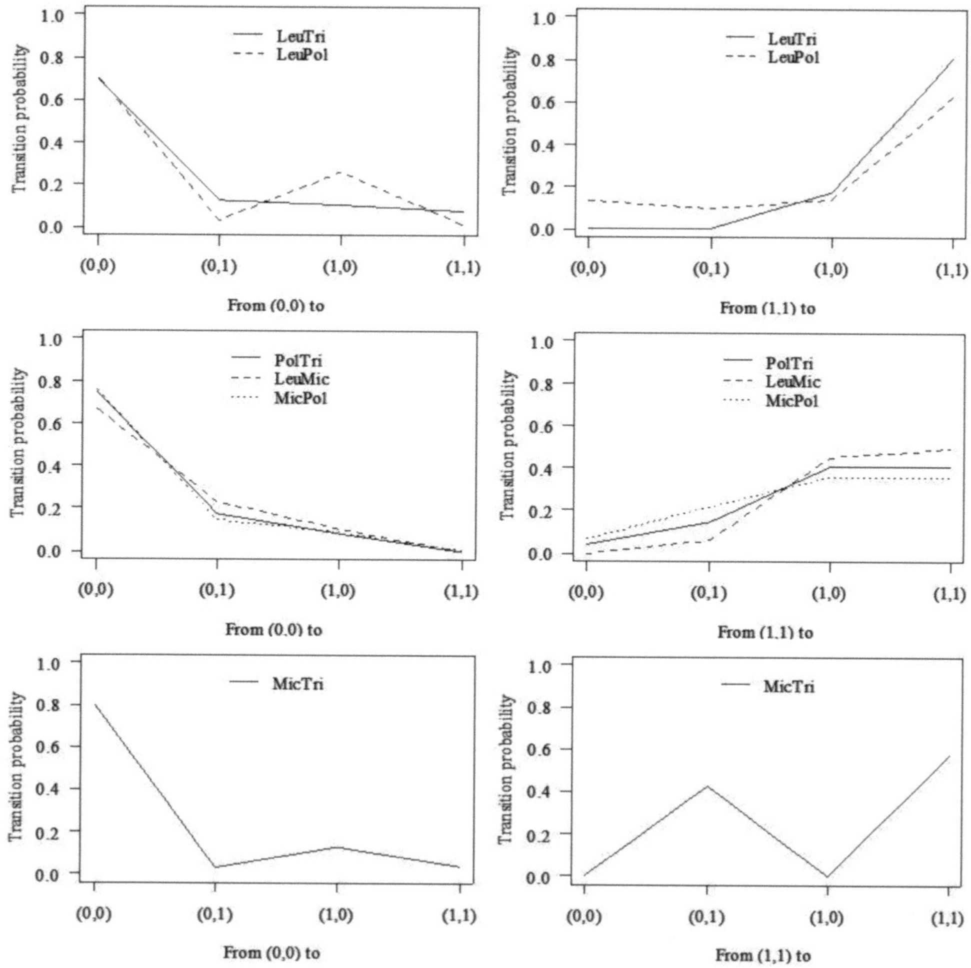


**Figure 5.**
*Transition probabilities from (0, 0) and (1, 1) states for 6 species pairs.*

same (bivariate) state as the previous state (see highlighted transition probabilities on the diagonals). For synchronous species pairs, such as LeuPol, and LeuTri the likelihood of species switching flowering state (states C), i.e. transition from one species flowering in a pair previous state = (0, 1) to the other species flowering, current state = (1, 0) never occurs (transition probability = 0.0000); or the likelihood of the transition from one species flowering to the other species flowering (i.e. a (1, 0) to (0, 1) transition) is rare ($0.0000 \leq$ transition probability $\leq 0.0085$). For asynchronous species pairs such as LeuMic, MicPol, and PolTri, their switching probabilities are significantly higher in that at least one of the transition probabilities from (0, 1) to (1, 0); or from (1, 0) to (0, 1) is greater than 0.036, with associated probability $\geq 0.076$.

Overall for synchronous pairs the probabilities of one species flowering to both or no species flowering, i.e. one off to both off, or one on to both on are high (>0.30). The latter are delineated by D and E transitions in **Figure 2** and **Table 4**. Overall for asynchronous pairs there are high probabilities of both off (or on) to one off (or on). The latter transitions are delineated by A and B in **Figure 2**, with probabilities given in **Tables 3** and **4**.

In summary the transitions that lead to both species being off (no flowering) or both species being on (flowering) (transitions D, E or F), are considered to be

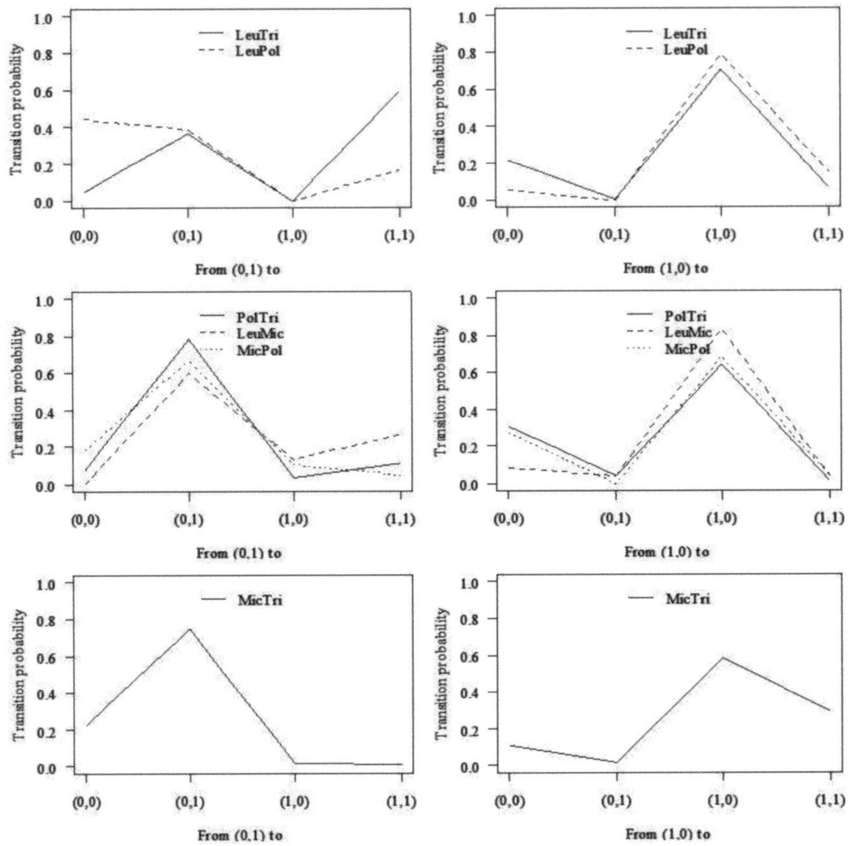


**Figure 6.**
*Transition probabilities from (0, 1) and (1, 0) states for 6 species pairs.*

| Transition names | Description | Probability: sum of subcomponents | Threshold for synchrony | Threshold for asynchrony | Rules |
|---|---|---|---|---|---|
| A | Both off to one off | q(0, 0;0, 1) + q(0, 0;1, 0) | <0.30▼ | ≥0.30▲ | P(A or B) > 0.8 for asynchrony |
| B | Both on to one on | q(1, 1;0, 1) + q(1, 1;1, 0) | <0.50▼ | ≥0.50▲ | |
| C$^{\phi}$ | Switching | q(0, 1;1, 0) + q(1,0;0, 1) | <0.05▼ | ≥0.05▲ | |
| D$^{\phi}$ | One off to both off | q(0, 1;0, 0) + q(1, 0;0, 0) | ≥0.40▲ | <0.40▼ | P(D or E) > 0.65 for synchrony |
| E | One on to both on | q(0, 1;1, 1) + q(1, 0;1, 1) | ≥0.40▲ | <0.40▼ | |
| F | Both on (off) to both off (on) | q(0, 0;1, 1) + q(1, 1;0, 0) | ≥0.08▲ | <0.08▼ | |

$^{\phi}$*Events or transitions C and F do not occur often.*

**Table 3.**
*Descriptions and rules of (a) synchrony based on the transitions A-F.*

synchronizing. However, transitions that lead to only one species being on or off (no flowering) (transitions A and B) and where a species pairs' flowering status switches (transitions C) are considered to be asynchronous.

We now provide a rule for synchrony (or asynchrony) based on subcomponent (sums) of the transition probabilities derived from the B-MTD model:

- Two species are synchronous if P(D or E) > 0.65, i.e. P(one on to both on) + P(one off to both off) > 0.65,

- Two species are asynchronous if P(A or B) > 0.80, i.e. P(both off to one off) + P(both on to one on) > 0.8.

The transitions have the following interpretation and probabilities (**Tables 3** and **4**):

- A: transition of both species off (in the past state) to one species flowering (on) in the current state;

- B: transition of both species on to one species off;

- C: species switching states;

- D: transition of one species off to both species off;

- E: transition of one species on to both species on;

- F: transition of one species on/off to both species off/on.

According to the rules given in **Table 3**, the synchronous pairs are LeuTri and LeuPol (with P(D or E) > 0.65); asynchronous pairs are: PolTri, LeuMic and MicPol (with P(A or B) > 0.80) and a species pair that is neither synchronous nor asynchronous is MicTri.

In summary we have a simple rule for (a) synchrony, which in agreement with the work of [6] (see also [25]), using the synchronisation theory of Moran that:

- *E. leucoxylon* flowering is synchronous with both *E. polyanthemos* and *E. tricarpa*, but asynchronous with *E. microcarpa*.

- *E. microcarpa* is synchronous with none of three species; specifically it is asynchronous with both *E. leucoxylon* and *E. polyanthemos* (and has no relationship with *E. tricarpa*).

| Transition probability sums | P(A) | P(B) | P(A or B) | P(C) | P(D) | P(E) | P(D or E) | P(F) |
|---|---|---|---|---|---|---|---|---|
| **Synchronous (S) pairs** | | | | | | | | |
| LeuTri | 0.232 | 0.181 | | 0.008 | 0.266 | 0.659 | | 0.081 |
| LeuPol | 0.293 | 0.238 | | 0.000 | 0.501 | 0.318 | | 0.141 |
| **Asynchronous (A) pairs** | | | | | | | | |
| PolTri | 0.254 | 0.556 | | 0.076 | 0.379 | 0.121 | | 0.037 |
| LeuMic | 0.333 | 0.510 | | 0.171 | 0.085 | 0.309 | | 0.000 |
| MicPol | 0.231 | 0.571 | | 0.102 | 0.455 | 0.088 | | 0.077 |
| **Neither S nor A** | | | | | | | | |
| MicTri | 0.165 | 0.426 | | 0.036 | 0.332 | 0.300 | | 0.038 |

**Table 4.**
*Transition probabilities of events A to F for each species pair categorised into synchronous and asynchronous (or neither) species pairs.*

- *E. polyanthemos* flowering is synchronous only with *E. leucoxylon*; and asynchronous with both *E. microcarpa* and *E. tricarpa*.

- *E. tricarpa* flowering is synchronous only with that of *E. leucoxylon*; and is asynchronous with *E. polyanthemos* (and has no relationship with *E. microcarpa*).

We can view **Figure 5** as the transition signatures from past states, where both species flowering is off or both species flowering is on, for synchronous pairings (LeuTri or LeuPol) and the asynchronous species pairs (PolTri, LeuMic and MicPol). **Figure 6** likewise delineates transition signatures from past states, where only one species of the pair is flowering. These signatures (**Figures 5** and **6**) distinctly differ according to whether a species pair is synchronous or asynchronous.

For MicTri the associated sum of the probabilities for transitions A and B (both off/on to one off/on) is 0.591 (see **Table 4**), which is close to the threshold for synchrony of 0.65. Note that the more sophisticated MTDg modelling approach in Section 4.2 which incorporates covariates (mean temperature and rainfall) with interactions, shows that indeed *E. microcarpa* and *E. tricarpa* are synchronous (**Tables 6** and 7), wherein the MTDg model allows for prior lag 1 to lag 12 month flowering effects and climate covariates (see also **Table 7** and **Figure 7**).

### 4.2 Moran tests on residuals of the MTDg models incorporating climatic covariates

In this section synchronisation among species pairs is tested using Moran's correlation method on the cross-residuals, based on MTDg models which incorporate both climate covariates and lagged effects of previous flowering. This work is based on [16], where MTDg models allowing interactions were fitted to the same four species. We present here only MTDg models with two covariates, namely, mean temperature and rainfall.

Parameters of the MTDg models are shown in **Table 5**. Significant lag effects of previous flowering states (*lag j*, where $j = 1, \ldots, 12$ months), and of the climatic covariates (*meanT* and *rain*) and their interaction (*meanT*rain*) are also given in **Table 5**. The estimated parameters for the MTDg models generally show a (positive) 1 month lag effect and 9, 11 and 12 months lag effects of previous flowering status (**Table 5**).

From **Tables 5** and **6** we observe that mean diurnal temperature (*meanT*) has a significant effect on flowering for all species; *rain* impacts significantly only on *E. tricarpa* (Tri) and an interaction effect between *rain* and *meanT* exists for *E. polyanthemos* (Pol). Overall, flowering increases as temperature (*MeanT*)

| Species | lag 1 | lag 9 | lag 10 | lag 11 | lag 12 | Temp variable | Rain | Temp × rain |
|---|---|---|---|---|---|---|---|---|
| *E. mic* | 0.534 | - | - | $0.032^{\Phi}$ | 0.275 | 0.136 | - | - |
| *E. poly* | 0.530 | 0.060 | - | 0.160 | 0.105 | 0.091 | 0.009 | 0.045 |
| *E. leu* | 0.611 | - | - | 0.124 | 0.042 | 0.202 | - | - |
| *E. tri* | 0.617 | 0.059 | 0.009 | 0.096 | - | 0.157 | 0.062 | - |

*$^{\Phi}$Covariate effects above 0.03 are considered significant.*
*- indicates cells with zero probabilities.*

**Table 5.**
*MTDg mixing probabilities of MeanT and rain models.*

| Species | Climate effects | Previous flowering | | Temperature | | Rain | | Temperature by rain interaction | | | |
|---|---|---|---|---|---|---|---|---|---|---|---|
| | (temp/ rain) | Off | On | Low[1] | High[2] | Less[3] | More[4] | Low/ less | Low/ more | High/ less | High/ more |
| *E. mic* | (+/−) | 0.00 | 1.00 | 0.00 | 1.00 | 0.39 | 0.28 | - | - | - | - |
| *E. poly* | Inter-action | 0.01 | 1.00 | 0.00 | 0.34 | 0.94 | 0.03 | 0.88 | 0.12 | 0.20 | 0.96 |
| *E. leu* | (−/+) | 0.05 | 1.00 | 1.00 | 0.00 | 0.88 | 0.94 | - | - | - | - |
| *E. tri* | (−/+) | 0.00 | 1.00 | 1.00 | 0.00 | 0.00 | 1.00 | - | - | - | - |

[1]*Cut point for low temperature states: MeanT 13.83°C.*
[2]*Cut point for high temperature states: MeanT >13.84°C.*
[3]*Cut point for less rain: rain 40.44 mm.*
[4]*Cut point for more rain: rain >40.45 mm.*
*Note that '-' indicates cells with zero probabilities.*

**Table 6.**
*Transition probabilities of flowering for the meanT and rain MTDg models.*

| Model species | mic | pol | leu | tri |
|---|---|---|---|---|
| Synchronous fitted species | tri (0.14) | leu (0.14) | pol (0.16) | mic (0.15) |
| | | | tri (0.11) | |
| Asynchronous fitted species | | | mic (−0.14[Φ]) | |

[Φ]*A negative and significant correlation indicates an asynchronous species pair.*

**Table 7.**
*Significant Moran correlations (in brackets) from the MTDg models.*

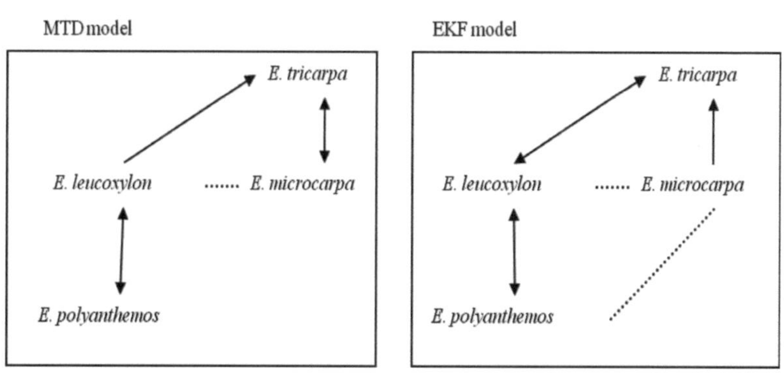


**Figure 7.**
*Synchrony relationships among the four eucalypts species.*

increases for *E. microcarpa*; and flowering decreases as temperature increases for both *E. leucoxylon* and *E. tricarpa*. Rainfall positively impacts the flowering of *E. tricarpa* (i.e. flowering increases with more rainfall). Interestingly *E. polyanthemos* exhibits increased flowering at low *meanT* when there is contemporaneous below average rainfall and at high *meanT* with above average rainfall (see the transition probabilities to flowering for the interaction effect of *E. polyanthemos* (i.e. (0.88, 0.12, 0.20, 0.96)) in **Table 6**.

In what follows we denote the species used to estimate the parameters for the MTD-based equation as the 'Model species' and the species fitted with these estimated parameters as the 'Fitted species'. **Table 7** gives the resultant significant Moran correlations based on the residual series from the MTDg-based model and fitted species equations. Significant Moran correlations from both the MTDg (and the EKF models show that (a)synchronous pairings found via the MTD and EKF models in [15–19] generally agree (**Tables 7** and **8**); refer also to **Figure 7**, where a solid line indicates synchronous pairs and a dashed line indicates asynchronous pairs of species.

**Table 7** shows significant positive MTDg-based correlations ($P < 0.006$) for the following (model species: fitted species) pairs—(LeuPol), (PolLeu), (LeuTri), (MicTri) and (TriMic), indicating that *E. leucoxylon* is synchronous with *E. polyanthemos*, in agreement with the rules of synchrony described earlier (**Tables 3** and **4**). *E. leucoxylon* is synchronous with *E. tricarpa*; and that *E. microcarpa* and *E. tricarpa* are synchronous. The synchrony of the latter species pair (MicTri) however, contrasts the results of Moran-based results on raw intensity profiles which indicate that *E. microcarpa* and *E. tricarpa* were neither synchronous or asynchronous (**Table 4**). It is noteworthy however, that for this species pairing, *E. microcarpa and E. tricarpa* (i.e. MicTri or TriMic), the associated sum of the probabilities for transitions D and E (one species off/on to both species off/on) is 0.591 (**Table 4**), which is close to the threshold for synchrony of 0.65 (**Tables 3** and **4**).

**Tables 7** and **8** shows significant negative-based correlations ($P < 0.001$) for the following (model species: fitted species) pairs; (LeuMic), (PolMic) and (MicLeu) indicating that that *E. leucoxylon* is asynchronous with *E. microcarpa* and *E. microcarpa* is asynchronous with *E. polyanthemos* (only via the EKF-based residuals) (**Figure 7** RHS); in agreement with the rule for asynchrony (**Table 4**) and Moran-based AR analysis of the flowering intensities.

Both the MTDg- and EKF-based models show that *E. tricarpa* is not asynchronous with *E. polyanthemos* (**Tables 7** and **8**). Note that for this species pairing *E. tricarpa* and *E. polyanthemos* (i.e. TriPol and PolTri) the associated sum of the probabilities for transitions P(A) and P(B) (both species off/on to one species off/on) is equal to 0.802 (**Table 4**), which is just above the to the threshold for asynchrony of 0.80.

### 4.3 Principal component analysis on the (λ, Q, d, s) parameters

In this section a novel approach which invokes a principal component analysis (PCA) of the resultant $(\lambda, Q, d, s)$ parameters (Section 2.2) which details the weight $\lambda, q, d$ and $s$ parameters from the MTD ($n = 4$) models) is performed. The resultant two dimensional PCA axis plots (**Figure 8**) of the rotated $(\lambda, Q, d, s)$-based PCs provides an informative visualisation of the synchronous and asynchronous species groupings (of $n > 2$ species) allowing for interpretation of the main climate drivers and climatic profiles (e.g.+/− or ( −/+)) detailed in **Table 6**.

| Model species | mic | pol | leu | tri |
|---|---|---|---|---|
| Synchronous fitted species | tri (0.12) | leu (0.19) | pol (0.18) | leu (0.26) |
| | | | tri (0.33) | |
| Asynchronous fitted species | leu (−0.17$^{\Phi}$) | mic (−0.10$^{\Phi}$) | | |

$^{\Phi}$*A negative and significant correlation indicates an asynchronous species pair.*

**Table 8.**
*Significant Moran correlations (in brackets) from the EKF models.*

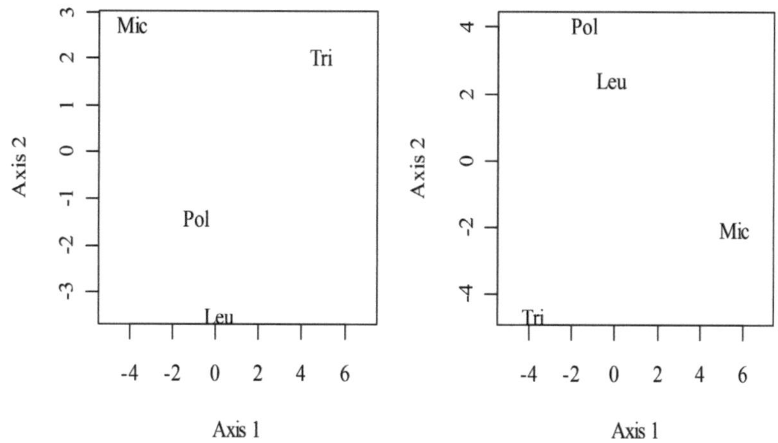


**Figure 8.**
*Distances in the (λ, Q, d, s) parameters among the 4 species—without interaction terms (left) or with interaction terms (right).*

The resulting parameters estimated from the MTDg models with and without interaction terms can be compared among all four species using **Figure 8**, which shows that the separation of *E. tricarpa* (−/+) and *E. microcarpa* (+/−) from other species along the horizontal axis 1, is due to the effect of mean temperature. Although *E. leucoxylon* is affected by the similar lag 1 and 11 month flowering terms as *E. polyanthemos*, *E. leucoxylon* (−/+) commences flowering at low temperature and shuts down at high temperatures. *E. microcarpa* begins flowering at high temperature (+/−). **Figure 8** also displays the similarity (synchronicity) of *E. leucoxylon* and *E. polyanthemos*.

## 5. Discussion and conclusion

The highest degree of synchrony (via the B-MTD rules of synchrony, the MTD models and Moran AR method) occurs between *E. leucoxylon* and *E. tricarpa*; then followed by *E. polyanthemos* and *E. leucoxylon* which indicates the potential for intense competition for potential pollinators, and therefore the prospect for a high level of hybridization. Both these species pairs were shown to be synchronous by Keatley et al., [20]; with *E. leucoxylon* and *E. tricarpa* having 6 years of no overlap (and a long term mean synchrony value of 0.62); and *E. polyanthemos* and *E. leucoxylon* having 5 years of the 31 years (between 1940 and 1970) with no overlap (long term mean synchrony value of 0.51); as quantified in [20]. The degree of synchrony or overlap of flowering was however determined using the method outlined in [22] which measures the extent of overlapping in the flowering periods among pairs of individuals in a population.

*E. leucoxylon* is the only species to synchronise flowering with *E. tricarpa*, as shown by all three methods, namely the B-MTD rules of synchrony, MTD models and Moran's AR method. Synchrony between *E. leucoxylon* and *E. tricarpa*, may be explained in terms of niche/competition and also facilitation may be a factor, due to their different modes of flower production. This agrees with the findings of [20]. Interestingly the MTD models discussed here (see also [6, 16, 24]) show that the climatic drivers or signature of *E. leucoxylon* and *E. tricarpa* is similar with respect to temperature, in that both exhibit decreased flowering with increased temperature.

Likewise *E. leucoxylon* is the only species to synchronise flowering with *E. polyanthemos*. *E. leucoxylon* and *E. polyanthemos* sometimes occur in the same geographical area; and earlier studies have shown they overlap significantly [20]. From the flowering behaviour indices of Keatley and Hudson in [23], *E. leucoxylon* and *E. polyanthemos* were shown to have temporally separated months of peak flowering, September and November, respectively; likewise their flowering commencement months May and October, respectively. These two species can occur in the same geographical area and their flowering period. Differentiation of these two species is based on their differing months of peak flowering as well as their separated months of most probable flowering; October and November, respectively. Likewise their flowering commencement months differ, May and October, respectively [23].

The least degree of synchrony (via the B-MTD rules of synchrony, the MTD models and Moran method) is shown in this chapter to occur between *E. leucoxylon* and *E. microcarpa*; then followed by *E. polyanthemos* and *E. microcarpa*. Our results agree with the findings in [20], which established that a cross between *E. leucoxylon* and *E. microcarpa* is impossible. In terms of climatic signatures: the flowering of *E. microcarpa* behaves differently from *E. leucoxylon* and *E. tricarpa*. *E. microcarpa* flowers at higher temperature and its flowering has a significant and positive relationship with flowering a year ago, refer also to the results reported in [23].

*Eucalyptus tricarpa* and *E. polyanthemos* were shown in this chapter also to be asynchronous (discordant or out of phase). This is in agreement with conclusions reported in [2]. The MTDg model found a significant interaction between two climate variables, mean temperature and rainfall on the flowering of *E. polyanthemos*. As flowering is viewed as either 'off' or 'on' this interaction appears to be delineating *E. polyanthemos*' flowering period. It usually commences flowering in late spring—as mean temperature is increasing and rainfall is decreasing and ceases in early summer; just prior to the warmest mean temperature and lowest rainfall.

Specific temperature thresholds for commencement and for the cessation of flowering for the four species studied here, have been established, see [5, 7, 8]. For example, *E. microcarpa* was shown to flower at high temperatures, and *E. leucoxylon* and *E. tricarpa* both at lower temperatures. The flowering of *E. polyanthemos* was shown to be impacted by both rainfall and temperature, with increased flowering when conditions were either cool and dry, or hot and wet—indicative of a rainfall by temperature interaction.

Moran residual analysis and the B-MTD analysis described in this chapter showed that *E. tricarpa* and *E. microcarpa* did not exhibit a significant synchronous nor an asynchronous relationship. However, for this species pairing, the associated sum of the probabilities for transitions A and B (both off/on to one off/on) is 0.591, which is close to the threshold for synchrony of 0.65. Indeed the more sophisticated MTDg modelling approach which incorporates covariates (mean temperature and rainfall) with interactions, showed that *E. microcarpa* and *E. tricarpa* are synchronous, wherein the MTDg model allows for prior lag 1 to lag 12 month flowering effects and climate covariates.

SOM-based clustering [4] and Moran AR (2) tests also found that *E. polyanthemos* was asynchronous to *E. microcarpa* and *E. tricarpa*, in agreement with the extended Kalman filter (EKF)-based synchrony measures in [15, 21]. Note also it was demonstrated in [20] that *E. polyanthemos* and *E. microcarpa* have 25 years with no overlap (with a long term mean synchrony value of 0.29). Note that the more sophisticated MTDg modelling approach which incorporates covariates (mean temperature and rainfall) with interactions, showed that indeed *E. microcarpa* and *E. tricarpa* are synchronous, wherein the MTDg model allows for prior lag 1 to lag 12 month flowering effects and climate covariates.

Recently synchronisation of eucalypt flowering is shown to be a complex mechanism that incorporates all the flowering elements—flowering duration, timing of peak flowering, and the timing of start and finishing of flowering, as well as possibly specific climate drivers for flowering [4]. The four species studied were shown to be influenced by temperature and rainfall and as a consequence their flowering phenology will change in response to climate change. This in turn will have an impact on species interactions and community [4].

Extensions of the B-MTD models to allow for climate covariates and for the comparison of more than 2 species at a time (a so-called multivariate M-MTD) is the topic of future work. Other forthcoming research is to examine the timing and a/synchronisation of the within species phenostages of both budding and flowering. Refer to earlier work using wavelets [26] and Generalized Additive Model for Location, Scale and Shape (GAMLSS) [27] to model the relationship between climate (mean monthly minimum, maximum temperatures and rainfall) during bud development and the flowering cycles of *Eucalyptus leucoxylon* and *E. tricarpa* from the Maryborough region of Victoria between 1940 and 1962. Monthly behaviour (start, peak, finish, monthly intensity, duration and success) in budding and flowering was assessed using, as in this current chapter, the indices of Keatley in [23].

## Author details

Irene Hudson[1]*, Susan Won Sun Kim[2] and Marie Keatley[3]

1 Department of Mathematical Sciences, Royal Melbourne Institute of Technology, Melbourne, Australia

2 South Australian Health and Medical Research Institute, Adelaide, South Australia

3 School of Ecosystem and Forest Sciences, The University of Melbourne, Melbourne, Australia

*Address all correspondence to: irene.hudson@rmit.edu.au

## References

[1] Brooker MIH, Kleinig DA. South Eastern Australia. In: Field Guide to Eucalypts: Vol. 1. 3rd ed. Hawthorn: Bloomings Books; 2006. 383 p

[2] MacNally R, Horrocks G. Landscape-scale conservation of an endangered migrant: The swift parrot (*Lathamus discolor*) in its winter range. Biological Conservation. 2000;**92**:335-343

[3] Hudson IL. Interdisciplinary approaches: Towards new statistical methods for phenological studies. Climatic Change. 2010;**100**:143-171. DOI: 10.1007/s10584-010-9859-9

[4] Hudson IL, Keatley MR, Lee L. Using self-organising maps (SOMs) to assess synchronies: An application to historical eucalypt flowering records. International Journal of Biometeorology. 2011;**55**: 879-904

[5] Hudson IL, Keatley MR, Kim SW. Climatic influences on the flowering phenology of four eucalypts: A GAMLSS approach. In: Hudson IL, Keatley MR, editors. Phenological Research: Methods for Environmental and Climate Change Analysis. Dordrecht: Springer; 2010. pp. 213-237. DOI: 10.1007/978-90-481-3335-2_10

[6] Hudson IL, Keatley MR, Kim SW. Modelling the flowering of four eucalypt species using new mixture transition distribution models. In: Hudson IL, Keatley MR, editors. Phenological Research: Methods for Environmental and Climate Change Analysis. Dordrecht: Springer; 2010b. pp. 315-340. DOI: 10.1007/978-90-481-3335-2_14

[7] Hudson IL, Keatley MR. Phenological Research: Methods for Environmental and Climate Change Analysis. Dordrecht: Springer; 2010

[8] Hudson IL, Keatley MR, Kang I. Wavelet characterization of eucalypt flowering and the influence of climate. Environmental and Ecological Statistics. 2011;**18**:513-533. DOI: 10.1007/s10651-010-0149-5

[9] Kim SW, Hudson IL, Keatley MR, Anderssen RS, Braddock RD, Newham LTH. Modelling the flowering of four eucalypts species via MTDg with interactions. In: 18th World IMACS Congress and MODSIM09 International Congress on Modelling and Simulation. 13-17 July 2009; Cairns, Australia: Modelling and Simulation Society of Australia and New Zealand and International Association for Mathematics and Computers in Simulation. 2009. pp. 2625-2631

[10] Raftery AE. A model for high-order Markov chains. Journal of the Royal Statistical Society: Series B: Methodological. 1985;**47**:528-539

[11] Berchtold A. March V.2.01. Markovian Models Computation and Analysis Users Guide. 2004. Available from: http://www.andreberchtold.com/march.html

[12] Berchtold A. March v.3.00 Markovian Models Computation and Analysis Users Guide. 2006. Available from: http://www.andreberchtold.com/march.html

[13] Berchtold A, Raftery AE. The mixture transition distribution model for high-order Markov chains and non-gaussian time series. Statistical Science. 2002;**17**:328-356

[14] Fournier DA. AD Model Builder, Version 5.0.1. Canada: Otter Research Ltd.; 2000

[15] Kim SW, Hudson IL, Keatley MR, Agrawal M, Eilers P. Modelling and synchronization of four eucalypt species via mixed transition distribution (MTD)

and extended Kalman filter (EKF). In: Proceedings of the 23rd International Workshop on Statistical Modelling, 23rd International Workshop on Statistical Modelling; 7-11 July 2008; Utrecht, Netherlands. 2008. pp. 287-292

[16] Kim S. Bayesian and non-Bayesian mixture paradigms for clustering multivariate data: Time series synchrony tests. Chapter 4. University of South Australia; 2011. pp. 60-118. Available from: http://researchoutputs.unisa.edu.au/1959.8/138604

[17] Moran PAP. The statistical analysis of the Canadian lynx cycle. I. Structure and prediction. Australian Journal of Zoology. 1953a;**1**:163-173

[18] Moran PAP. The statistical analysis of the Canadian lynx cycle. II. Synchronization and meteorology. Australian Journal of Zoology. 1953b;**1**:291-298

[19] van der Merwe R. Quick-Start Guide for ReBel Toolkit. Oregon Health and Science University; 2004

[20] Keatley MR, Hudson IL, Fletcher TD. Long-term flowering synchrony of box-ironbark eucalypts. Australian Journal of Botany. 2004;**52**:47-54. DOI: 10.1071/BT03017

[21] Kim S. Bayesian and non-Bayesian mixture paradigms for clustering multivariate data: Time series synchrony tests. Chapter 5. University of South Australia; 2011. pp. 119-123. Available from: http://researchoutputs.unisa.edu.au/1959.8/138604

[22] Augspurger CK. Flowering synchrony of neotropical plants. In: WG D'A, Correa MD, editors. The Botany and Natural History of Panama. Saint Louis: Missouri Botanical Garden; 1985. pp. 235-243

[23] Keatley MR, Hudson IL. A comparison of long-term flowering patterns of box-ironbark species in Havelock and Rushworth forests. Environmental Modeling and Assessment. 2007;**12**:279-292. DOI: 10.1007/s10666-006-9063-5

[24] Hudson IL, Kim SW, Keatley MR, Anderssen RS, Braddock RD, Newham LTH. Climatic influences on the flowering phenology of four eucalypts: A GAMLSS approach. In: 18th World IMACS Congress and MODSIM09 International Congress on Modelling and Simulation; 13-17 July 2009. Cairns, Australia: Modelling and Simulation Society of Australia and New Zealand and International Association for Mathematics and Computers in Simulation. 2009. pp. 2611-2617

[25] Hudson IL, Keatley MR, Kim SW, Kang I. Synchronicity in phenology: From PAP Moran to now. In: Australian Statistical Conference/New Zealand Statistical Association (ASC/NZSA) Conference; 3-6 July 2006, Auckland, New Zealand. 2006

[26] Hudson IL, Kang I, Keatley MR, Weber T, Mcphee MJ, Anderssen RS. Wavelet characterization of eucalypt flowering and the influence of climate and budding. In: MODSIM 2015, 21st International Congress on Modelling and Simulation; Modelling and Simulation Society of Australia and New Zealand. 2015. pp. 1813-1819

[27] Hudson IL, Keatley MR, Piantadosi J, Anderssen RS, Boland J. Scoping the budding and climate impacts on eucalypt flowering: Nonlinear time series decomposition modelling. In: MODSIM2013, 20th International Congress on Modelling and Simulation; 1-6 December; Modelling and Simulation Society of Australia and New Zealand. 2013. pp. 1582-1588

Chapter 4

# Hybrid Modeling of Offshore Platforms' Stress-Deformed and Limit States Taking into Account Probabilistic Parameters

*Gennady Yu. Shmal, Vladimir A. Nadein,*
*Nikolay A. Makhutov, Pavel A. Truskov and Viktor I. Osipov*

## Abstract

Offshore platforms should be referred to critically and strategically important objects of a technosphere due to technological and operational challenges, on the one hand, and the danger potential level, on the other hand. Environmental, social and economic losses occurred over several decades of accidents and disasters in unique Great Britain, Norwegian. The Russian and the USA platforms were evaluated in death of dozens of operators, destruction of platforms, environment contamination and hence in multi-bullion losses. All of these indicate insufficiency of currently taken engineering solutions, providing structure strength, operational life and safety. The scientific, design, expert and supervising organizations in Russia and in the world are developing and improving mathematical and physical methods, implementing the probabilistic formulations for accidents and disasters, risk assessment and risks reduction on offshore platforms. The solutions of the following problems are included: extension of the comprehensive computational and experimental strength, operational life and survivability analysis to the cases of nonroutine events, accidental and catastrophic conditions; numerical justification of modelling of critical elements, zones and points with the maximum tension, deformations and damages occurring under impacts of external extreme seismic, ice, wind, low temperature; implementation of comprehensive diagnostic methods for damage states evaluation within nonlinear and probabilistic fracture mechanics; and use of new structural design and technological systems for reduction of negative extreme impacts as well as emergency protection systems. The solution of the specified problems is illustrated by case studies of the Russian specialists for each life cycle stage of the platforms offshore Caspian and Kara Seas and Sea of Okhotsk.

**Keywords:** offshore platform, offshore technologies, safety of engineering systems, design solutions, analysis of the emergency situations, limit states, crash protection, seismic loads, technical diagnostics

## 1. Articulation of issue

Further development of the modern international community is going hand in hand with the intensive growth of fuel and energy raw materials consumption in all spheres of activity. Meanwhile, in the majority of on-land oil-and-gas regions, resources of oil and gas are exhausted and the possibilities of further increase of the discovered and usable economically recoverable reserves are complicated.

With this knowledge in mind, lately we can see special, increasing interest in a problem of the seas and oceans' oil and gas resources development [1, 2].

The gas and oil fields are discovered in 108 countries of the world. Ultimate reserves of gas reached 172 trillion cubic meters, of oil—172 billion tons; at the beginning of the twenty-first century, the world gas production was equal t0 2.6 trillion cubic meters, while oil production was 3.3 billion tons.

Initial recoverable hydrocarbon resources of the World Ocean continental shelf (up to 500 m isobathic line) and the inner continental shelf are estimated equal approx. to 370 billion tons of fuel oil equivalent (TFOE), including more than 200 trillion cubic meters of free gas and about 155 billion tons of oil and condensate.

The primal gas resources in water areas are concentrated within the shelf of the Northern Asia—44.5 trillion cubic meters. Its bigger part is located offshore in the Kara Sea. Offshore gas resources of Eastern Europe, North and South America and the Middle East are also comparable and considerable relative to ones in Northern Asia (21–24 trillion cubic meters).

In the world, since the 1940s, the multiple sea platforms (SP) are engineered and operated with a wide range of parameters and are used for offshore petroleum and gas production. The largest of them are five platforms of the USA, Norway and Russia. They provide production at sea depths up to 2.5 km and well-drilling up to 10–13 km. About 10 platforms are in operation in Russia: on Caspian, Okhotsk seas and on the seas of the Arctic Ocean. The most significant of them are the platforms "Piltun-Astokhskaya-A (former Molikpak)," "Piltun-Astokhskaya-B," "Lunskaya-A," "Orlan," "Berkut" and "Prirazlomnaya" (**Figure 1**). Length of already constructed offshore pipelines is about 300 km. In long term, the need of Russia in offshore projects includes the necessity to provide functioning of about 50 SPs.

In the world history of development of the continental shelf, a number of disasters and serious accidents with catastrophic consequences occurred due to lack of attention to measures for identification and mitigation of threats for safe operation is wrote. The 15 most dramatic accidents on drilling vessels and platforms of various types (semisubmersible, submersible, mobile, stationary) happened during the last 40 years were followed by:

- great loss of lives (up to 164 people) occurred due to limited space on the platform, evacuation difficulties and vulnerability of personnel to thermal fire impact and toxic effects caused by combustion products;
- infilling and destruction of platforms infrastructure;
- offshore areas and airspace pollution; and
- vegetal and animal life demise.

Most economic direct loss suffered after the disaster on the platform in the Gulf of Mexico (USA) and was more than 20 billion dollars, while indirect losses reached

**Figure 1.**
*Large offshore platforms in Russia.*

**Figure 2.**
*Disasters of large scale platforms in USA (a) and Russia (b).*

60 billion dollars; the direct economic loss suffered from flooding of the "Kol'skaya (Kola)" platform (Russia) is about 200 million dollars.

Review of accidents with catastrophic consequences (death of great number of people, large-scale ecological contamination or material losses) occurred on oil and gas production platforms demonstrate reduction in number during recent years. This can be explained by the platforms' technological and design performance improvements and application of modern safety systems (**Figure 2**).

Evaluation of information about accidents and disasters occurred on offshore drilling rigs of various types makes it possible to combine and classify all accidents in accordance with major, internally connected accidents sources (**Figure 3**):

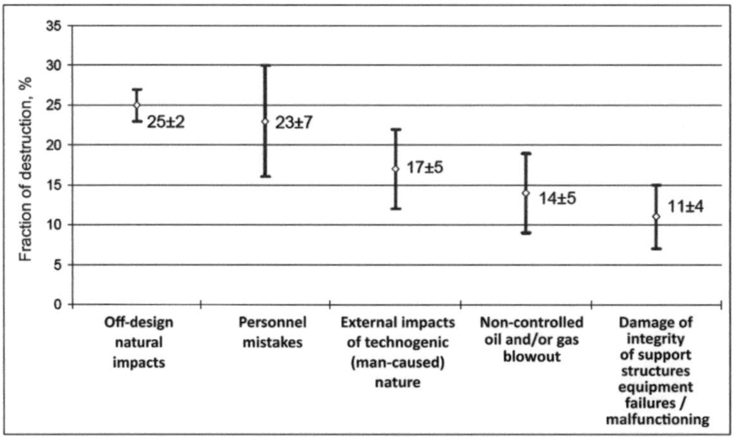


**Figure 3.**
*Characteristic of accidents on oil and gas production platforms.*

uncontrolled release of oil and/or gas from the well; damage of integrity of load bearing (or supporting) structures, as well as equipment failing (or malfunctioning); personnel mistakes; external impacts of technogenic (man-induced) nature (allisions with seagoing vessels, helicopters fall, subversive actions); and off-design impacts of the natural environment.

The probability of accident that may occur during a year on the Unit is in the range from $8 \times 10^{-5}$ up to $1.6 \times 10^{-3}$ per year, and this conclusion is based on the data in the Declaration of Industrial Safety for four Russian production platforms and nine floating drilling rigs.

## 2. Comprehensive issues of industrial safety in the process of the continental shelf development

### 2.1 Risks analysis

One of the first places in the field of strategic planning takes the problem of scientific and methodological frameworks building, while in the field of safe shelf development takes scientifically grounded criterion base. At the same time, it is considered that strategic risks of the Russian continental shelf development can be an essential part of strategic risks of national security.

In view of the foregoing, the main objectives of the Institutes of the Russian Academy of Science (RAS) and the leading security matters sectoral scientific research institutes are as follows [1–3]:

- risks' theorization based on fundamental risk analysis database collected and studied in the process of research works in social, natural and technical science of fundamental base. Risks function R(t) is analyzed in three main spheres of activity—social (N), natural (S) and technogenic (T), forming the uniform complex social-and-natural-and-technogenic system functioning in time $t$

$$R(t) = F_R\{R_N(t), R_S(t), R_T(t)\}; \tag{1}$$

- formulation of the generalized model of the specified complex system with definition of its main components N, S, T role in terms of values of basic risks parameters R(t)—probabilities of occurrence of P(t) negative processes and events (dangers, challenges, threats, crises, disasters and accidents) and consequential losses U(t)

$$R(t) = F_R\{P(t), U(t)\} \quad (2)$$

$$P(t) = F_P\{R_N(t), R_S(t), R_T(t)\} \quad (3)$$

$$U(t) = F_U\{U_N(t), U_S(t), U_T(t)\} \quad (4)$$

- identification of negative events scenarios with regard to a complex system and quantitative risk assessment R(t) through parameters of main triggering and affecting factors—dangerous energies E(t), substances W(t) and information flows I (t)

$$R(t) = F_R\{E(t), W(t), I(t)\}. \quad (5)$$

On the basis of Eqs. (1)–(5), categorization of emergency situations, high-risk objects and dangerous processes in terms of risks R(t) is developed. Objectively, the norm settings, regulation and control in the area of safety provision as per safety and security major components (i.e., social and economic, military, scientific and technical, industrial, environmental and demographic) when using risks nominally comes down to ratio

$$R(t) \leq [R(t)], \quad (6)$$

where $[R(t)]$ is acceptable risks level.

The $[R(t)]$ value is set and defined by bodies of the highest public administration with consideration of abilities and the capacity of the country, level of scientific justifications and domestic and international experience. The realization of the requirement (6) [1–3] will be provided proceeding from the position that the defining risks of R(t) are two groups of risks:

- individual risks (1 per year) of life and health loss caused by abovementioned negative processes and events; and

- economical risks (rubles per year, dollars per year) caused by negative processes and events that are taking into account vulnerability of social *(N)*, natural *(S)* and technogenic *(T)* areas according to Eqs. (1)–(4).

The economic damages due to loss of lives and human health and environmental and technical infrastructure damages are included in the economic risks R(t). Scientific justification of acceptable risks [R(t)] includes development of methodology of definition of critical (limiting, inadmissible) risks Rc (t) and fixing of risks margin $n_R$ in the form of

$$[R_c(t)] = \frac{R_c(t)}{n_R} \quad (7)$$

For quantitative assessment of value of risks Rc (t) relevant to accidents and disasters on SP all basic, Eqs. (1)–(7) can be used while the value of risks margin $n_R$ shall be greater than unity ($n_R \geq 1$). Considering the best domestic and foreign practices, the variation for risks margin can be rather wide ($2 \leq n_R \leq 10$) at the beginning.

Based on (1)–(7), actions to provide enhancement of safety and security with the corresponding economic expenses Z(t) shall be developed. The actions directed to reduction of risks R(t) value to the level [R(t)] have to be effective and correlate with the levels of estimated risks R(t)

$$Z(t) = \frac{R(t)}{m_z}, \tag{8}$$

where $m_z$ is the performance factor of economic costs for reduction of economic risks ($m_z \geq 1$).

The general expression for the analysis and the sea platforms safety provision as per risks criteria based on Eqs. (1)–(8) is the following:

$$R(t) = F_R\{P(t), U(t)\} \leq [R(t)] = \frac{R_c(t)}{n_R} = m_z \cdot Z(t). \tag{9}$$

In the Eq. (9), practically are represented all set above main:

- scientific risks $R(t)$ analysis via its basic components $P(t)$, $U(t)$;
- justification of acceptable risks $[R(t)]$;
- scientific-methodological justification of risks' tolerance $R_c(t)$ and risks' margins $n_R$; and
- development of methodological recommendations on formation and implementation of the actions directed to risks R(t) reduction to the acceptable level [R(t)] providing optimal expenses Z(t) with the set efficiency factor $m_Z$.

### 2.2 Potential hazards characterization in the technical area when developing the sea shelf

With the progress and complication of engineering of technogenic aspects in the field of sea shelf development the analysis of man-caused (technogenic) offshore accidents and disasters becomes one of the most vital tasks of fundamental, interdisciplinary research; applied scientific and technical developments; development of diagnostic and monitoring systems; and designing of barriers and protection means. The ultimate purpose of such research works and development becomes the problem of evidence-based assessment of comprehensive risks and adjusting these risks to acceptable levels by use of expressions (1)–(9).

The analysis and generalization of the numerous data (in the most developed countries, such data bases amount thousands and tens of thousands facts) make it possible to carry out certain classification of technogenic and natural and man-made accidents and disasters [3]. Classification of accidents can be performed on scales of the countries and territories affected by them, on number of the victims and injured persons and on economic and ecological damage; in such classification, seven general groups can be identified: planetary, global, national, regional, local, object-based and local emergency and catastrophic situations (**Figure 4**).

The events resulting in similar serious accidents within technogenic field can also be classified by potential hazard and in this line can be named objects of the nuclear, chemical, metallurgical and mining industry, unique engineer

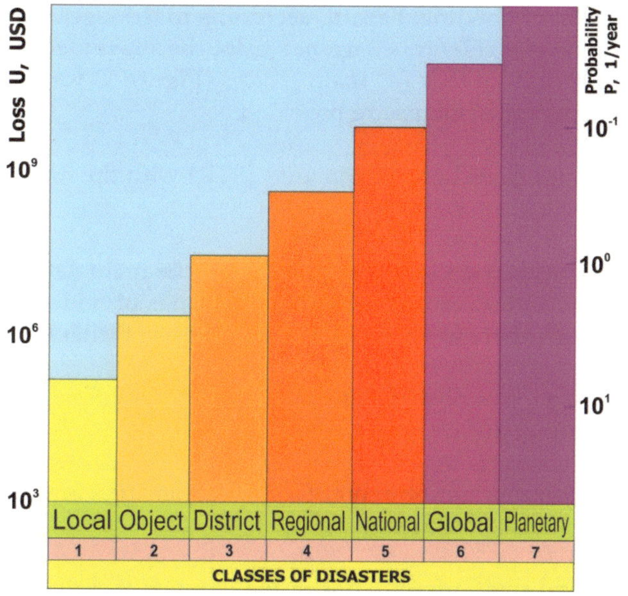


**Figure 4.**
*Losses (damages) and frequency of natural and man-made accidents and disasters.*

constructions (dams, platforms), offshore development objects (sea platforms, hydrocarbons storage tanks, LNG plants), the transport systems (airspace, surface and underwater, on-land) that provide transportation of dangerous cargos, large number of people, main gas-, oil pipelines and product lines. In this line, the hazardous objects of defense industry also shall be mentioned.

At the same time, a majority of accidents and disasters are followed by infringement of stress conditions and depletion of lifetime of the most loaded components in routine situations or in emergencies. The probabilities P(t) characterizing frequency of disaster accidents occurrence in peace time ranges from $(2–3)\cdot10^{-2}$ up to $(0.5–1)\cdot 10^{-1}$ per year, while damages (losses) U(t) ranges from $10^{11}$ to $10^9$ dollars per accident. At the same time, their risks R(t) vary in the limits from $10^4$ dollars per year to $10^{10}$ dollars per year ranging from 104 dollars/year up to 1010 dollars/year.

In view of said above, the new fundamental and applied scientific tasks needed to be set at national and international levels, for instance:

- mathematical theory of disasters and probabilistic theory of risks;
- physics, chemistry and mechanics of emergencies and disasters;
- limit states, strength and lifetime theories taking into account accidental and emergency situations;
- theory of hardware, functional and integral protection in case of emergency of objects, operators and personnel;
- theories of monitoring and forecast of scenarios and technogenic (man-made) disasters consequences (using airspace, airborne and ground-based systems); and
- scientific methods, technologies and hardware for mitigation of consequences of emergency situations of technogenic nature.

Based on the level of potential hazard, according to the legislation requirements and taking into account accidents occurrence risks, the abovementioned objects of a technosphere can be split in four (4) main groups (**Figure** 5) for each of which corresponding safety requirements are provided:

- the objects subject to technical regulation (STR) with the main damages to objects themselves;
- the hazardous production facilities (HPF) with the main damages to production sites and objects which safe operation is provided under the law on industrial safety—there are hundreds of thousands of such facilities;
- the critically important objects (CIO) which damages affect members of the Russian Federation; and
- the strategically important objects (SIO) which damages are followed by losses to the country and the bordering states.

For the continental shelf infrastructures, the number of the objects to be analyzed is reduced by one or two orders.

In the system of initial standards, specifications and guidelines used for design and calculations of SPs were included the following documents:

- Russian regulations database:

GOST 27751-88 "Reliability of structural units and foundations. Basic calculations methodology.", 1988;
SNiP 2.01.07-85 "Loads and impacts", 1996;
SNiP 2.06.04-82*, "Loads and impacts on hydrotechnical structures (waves, ice and sea vessels)", 1995 & 1983;
Marine Registry. FDR/OFR Guidelines, 2001;
VSN 41-88, "Industry Specific Code of Practice for design of offshore ice-resistant fixed platform (OIRFP)", M., 1988;

- Foreign regulations database:

Recommended Practice for Planning, Designing, and Constructing Fixed Offshore Platforms – Load and Resistance Factor Design, API Recommended Practice 2A-LRFD, 1993, Washigton;

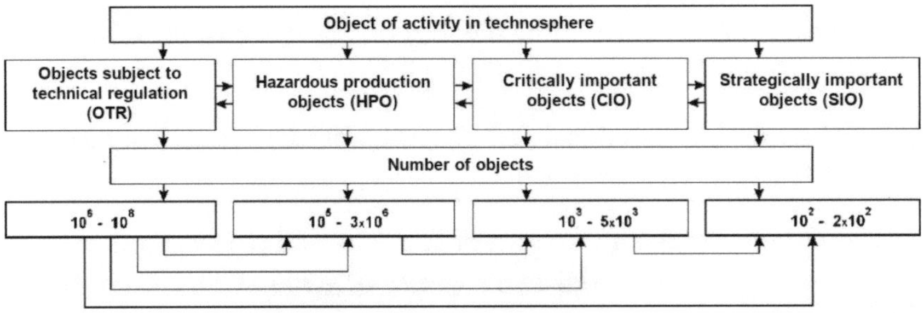


**Figure 5.**
*Diagram of analysis of potentially hazardous objects of the technosphere.*

Recommended Practice for Planning, Designing, and Constructing Fixed Offshore Structures in Ice Environments, API Recommended Practice 2N (RP 2N), 1995, Washington;
CAN/CSA-S471-92, "General Requirements, Design Criteria, the Environment, and Loads", A National Standard of Canada, 1992; Toronto; Commentary to CSA Standard CAN/CSA-S471-92, "General Requirements, Design Criteria, the Environment, and Loads", 1992, Toronto;
CAN CAN/CSA-S473-92, "Offshore Structures", A National Standard of Canada, 1992 CAN CAN/CSA-S16.1-94, "Limit States Design of Steel Structures", A National Standard of Canada, 1992, Toronto;
DnV, "Structural Design, General", Rules for classification of Fixed Offshore Installations, 1993;
DnV, "Structural Reliability Analysis of Marine Structures", 1992. DnV, Offshore Standard OS-C101, Design of Offshore Steel Structures, General, 2001;
ISO 19906, 2010 (ISO/DIS 19906 "Petroleum and natural gas industries - Arctic offshore structures", 2010).
Above documentation was used for definition of the main basic specified characteristic load during design of the sea platforms intended for use at a sea depth from 20 to 70 m to 200–250 m.

### 2.3 Types, design diagrams and cases

Implementation of the proposed recommendations and norms covers the structures with vertical and inclined sides, monopods and multicolumn constructions. In the documents, the rules of definition of the main loads conditioned by action of all potentially dangerous ice features subject to consideration are given. In **Figure 6**, the various structures design versions are presented.

The following loads are subject to analysis:

**Global conventional and extreme loads on conical and vertical constructions:** sheet and rafted ice; ice ridge compression; ultimate moving force (ice field crowding force); global (abnormal loads); ice islands (stopped by a construction).
**Local ice pressure (for vertical and inclined surfaces):** solid ice area; and ice fragments area.
**Ice loads dynamics:** shock actions and interaction "ice—construction" (self-excited); ice load change in time; fatigue ice impact; ice grinding impact; and regelation.

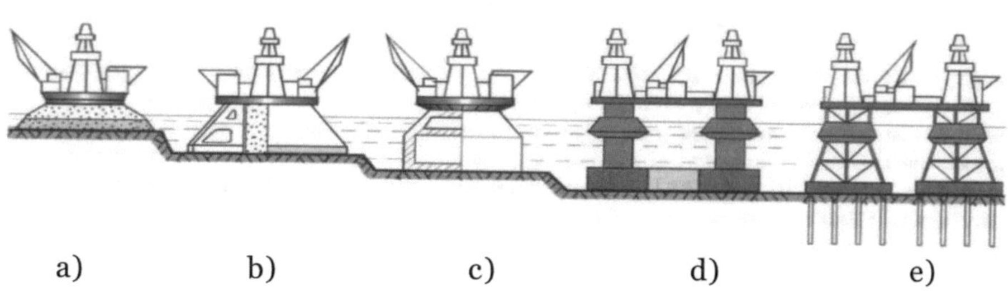


**Figure 6.**
*Types of sea platforms dependent on the sea depth (for standard soil conditions): (a) the artificial pad, depth is up to 5 m; (b) the caisson-island fixed along contour, depth up to 15–20 m; (c) the monopod or monokone, depth is up to 25–30 m; (d) shell support; and (e) the truss-shell type supports, depth 25–30 m and more.*

| Utility | Fixed to the bottom | | | Floating | | Islands | |
|---|---|---|---|---|---|---|---|
| Design | Gravity based | Pile supported | Integrated | With anchor mooring | Dynamic positionable | Outlined | Non-outlined |

**Table 1.**
*Ice resistant oil and gas utilities.*

Broadly speaking, the offshore oil and gas facilities can be classified by the following signs: structural materials; design features; methods of fixing to a bottom; ice resistance indications; and functional features. The design features of offshore oil and gas facilities incorporate the following types: stationary platforms; submersible and semi-submersible platforms; pendulum constructions; tension structures; platforms of SPAR type (with the underwater cylindrical foundation); access bridges and pier sites; and dams and unpaved sites.

Ice-resistant constructions can be grouped as follows (**Table 1**).

## 2.4 Russian shelf specific conditions

The Russian continental shelf area exceeds 6 million sq. km that takes about 25% of a shelf zone of all the World Ocean. The Arctic and Far East shelf areas are the areas of the greatest interest.

With respect to environmental, bathymetric, engineering-geological, seismic and other conditions, the shelf of Russia is different from others due to a number of features:

- severe ice conditions (large drifting ice fields, ice ridges, floating ice hummocks, etc.);
- shallow waters (depths less than 100 m) leading to significant increase in wave loadings;
- high level of seismicity (on the Far East shelf); and
- difficult engineering and geological conditions.

In designing platforms for the Russian shelf, as a rule, it is necessary to consider a combination of at least three factors from listed above. This is unlike world practice.

When selecting this or that type of platform jack design along with environmental conditions, it is necessary to take into account the impact of the field development general scheme, production method and hydrocarbons transportation technology as well as terms of platform fabrication and transportation on site.

When developing scientifically grounded methodology of design of gravitation-type platforms for use on Russia shelf, i.e., design providing the required reliability and safety level and, as much as possible, based on the lessons learned by the international and Russian specialists in design, construction and operation of platforms, it is necessary to:

- analyze the Russian and foreign regulating documents;
- set up an integral approach to platforms reliability and safety assurance at different stages of their life cycle;
- select correct existing and develop new methods of definition of environment loads;

- develop the concept of consideration of engineering-geological conditions;
- provide consideration of the level, nature and duration of dynamic impacts;
- formulate additional requirements to be imposed on sea engineering surveys; and
- justify the range of design cases for assessment of bearing capacity and stress-deformed state of the system "construction—foundation."

### 2.5 The analysis of external and internal factors and threats for safety of sea platforms

The analysis of threats for off-shore oil and gas production platforms is the first stage of the accidents' risks analysis for the specified objects and provides estimation of their safety level [1, 2]. The threat for SP is the probabilistic characteristic defining a possibility of the impact of affecting factors of specific type, intensity and duration in response to some dangerous (extreme) event that can take place both in the territory of the object and in the external environment. Therefore, the analysis of threats for SP has to be preceded by assessment of dangerous events which can initiate impact of the affecting factors on platforms.

The secondary dangers occur and provoke secondary affecting factors when some object's process modules – SP parts are damaged. The possibility of initiation of these secondary threats will be defined by vulnerability of an object in relation to the primary threats. Thus, the analysis of threats has to be made in an agreement with assessment of vulnerability of the SP parts in relation to the affecting factors acting on them.

The danger to SP is defined by the pattern of random events or processes ($Th$): extreme external natural and technogenic impact, wrong personnel actions and operating conditions of the object technical systems having the potential which can lead to accident. Examples of such events are: seismic activity, extreme wave or ice loads (external dangers), loss of the oil tank containment or of fatigue damages accumulation (internal dangers). The danger of an extreme event is a random variable which, in the simplest case, can be characterized by the probability of occurrence of an event $P(Th)$ during a certain period (1 year) or the during the platform's operational lifetime (**Figure 7**).

Threats for SP are characterized by impacts on an object of the affecting factors of dangerous events. The threat is also a random event (process) $H$, which

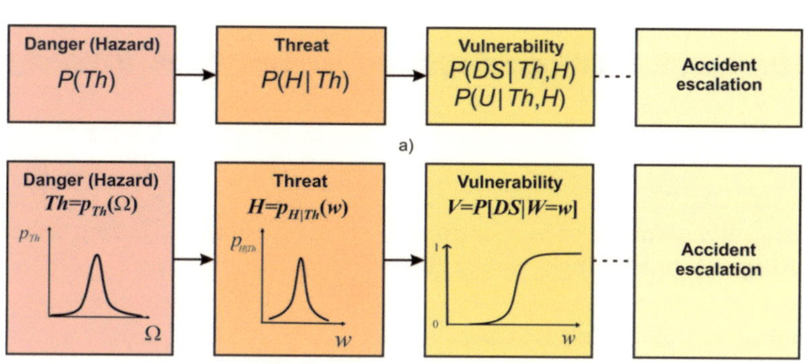


**Figure 7.**
*Presentation of accident occurrence and development as a complex event. (a) Probabilities of the elementary events are described with the help of point estimations, and (b) probabilities of the elementary events are described with the help of probabilistic determination.*

can take place in case of occurrence of a dangerous event and be characterized by conditional probability $P(H|Th)$. For the abovenamed dangers, the events listed below will act as threats:

- in earthquake case, the seismic wave will reach the site where object is located; and
- loss of the oil tank containment will cause the oil leak.

Vulnerability of SP to threat of this type is defined as the conditional probability in case of the affecting factor's impact on an object when the latter one will get a certain damage rate $P(DS_k|\ H)$, where $k$ is an object damage rate.

If it is required to get more accurate description of danger of an extreme event, it should not be characterized by the point estimation of probability of occurrence of a dangerous event $P(Th)$ but by the distribution curve of danger intensity $p_{Th}(\Omega)$ or integral distribution function $P_{Th}(\Omega)$ presented in **Figure 7** (where $\Omega$ characterizes dangerous event intensity). In particular, the seismic hazard can be characterized by distribution of probabilities of potential earthquake intensity degree, while threat from loss of tank containment can be characterized by the distribution of probability of the effective opening area. At the same time, threats will be characterized by family of the conditional distribution functions $p_{H|Th}(w)$ corresponding to different intensity $\Omega$ of a dangerous event. Then the dangers of an earthquake and loss of containment mentioned above will correspond to the threats described by family of probabilistic distributions of amplitudes of vibration accelerations of soil on site of platform location at different earthquake magnitudes and family of probabilistic distributions of volume of leaked oil for different diameters of effective openings.

Vulnerability of an object relative to impact of the affecting factor with intensity $w$ will be characterized by the vulnerability curve $V = P(DS|W = w)$, which defines the conditional probability of sustained damage of level $DS$ with the proviso that a random value intensity $W_{\text{takes a certain value}}$ $(W = w)$.

When making decision on what physical parameter of impact of dangerous process on an object to select for threat intensity evaluation, it is necessary to consider vulnerability of an object relative to action of different components of such impact: for example, in case of seismic impact on the platform, some parts of the equipment and structures are the most sensitive impact from soil vibration accelerations, while the another to vibration amplitudes.

Within that narrative, the accident initiation on SP can be considered as the complex event occurring in case of occurrence of simultaneous random events cascade (**Figure 7a** or **b**):

1. danger: realization of the extreme initiating event $Th = p_{Th}(\Omega)$;
2. threat: impact of affecting factor of dangerous event on SP parts $H = p_{H|Th}(w)$; and
3. vulnerability: damage of SP's parts as a result of impact affecting factors of the initiating extreme event $V = p_{V|H}(w)$.

### 2.6 Damaging and affecting factors

SP operation is associated with production, storage and transportation of considerable volumes of dangerous materials, transformation of considerable volumes of energy, running of hazardous technological processes on the platform as well as with presence in areas of SPs' location of external sources of natural and technogenic nature hazards which are resulting in extreme external impacts on the platform. Depending on the location of danger source (i.e., location of the place where the initiating event starts) outside or inside the platform boundaries, it should be taken into account the external and internal threats damaging and affecting factors. Risks $R(\tau)$ used in expressions (1)–(9) depend on them.

Internal threats for SP are initiated by dangerous process potential of the following [1–3]:

- mass and composition of chemically dangerous substances W which are on the platform; and

- amount of the reserved on the object energy $E$.

Among internal threats to SP are operational loads on parts and components of oil and gas production facility (OGPF), impact of harsh chemical environment, control system failures, etc. The considerable segment of internal threats range for OGPF is caused by human factor action (mistakes at a design stage, construction and operation of the platform, including violation of regulations, etc.).

Among external threats are affecting factors resulting from natural and technogenic events (processes) happening outside SP boundaries. Seismic impacts, hurricane, technogenic accident on the neighboring object, collision with the sea vessel, extreme weather conditions, etc. are between initiating events of the external type. Besides mentioned above, external threats include the events connected with interruptions in work of energy, telecommunication and transportation infrastructures which lead to breakdown of technological processes, damage of platform's control and supply systems and terrorist attacks which also can be classified as an external threat to the platform.

The probabilistic approaches usually are used for description of the initiating events and affecting factors [1–4]. The necessity to use the probabilistic methods is determined by lack of knowledge about comprehensive system "SP—the environment," on the one hand, and by stochastic nature of the processes occurring in a system and environment and by high uncertainty inherent to the examined system (uncertainty of system parameters, materials strength characteristics, external loads, etc. and also the uncertainty explained by limited knowledge of an object) on the other hand.

The threats (affecting factors) H(t) influencing SP (**Figure 8a**), in general, should be considered not only as the separate and determined processes (a) but also as random events (**Figure 8b**) and *stochastic processes* (**Figure 8c**). This is due to the fact that during analysis of the platforms' vulnerability relative to the prevailing threats, an essential role is played by damages' accumulation and fatigue mechanism of ultimate limit states reaching. Such approach necessitates review threats as dynamic task taking into account history of operational loads and dynamic and cyclic impacts of the affecting factors (external loads, influence of extreme temperatures, harsh environment, etc.).

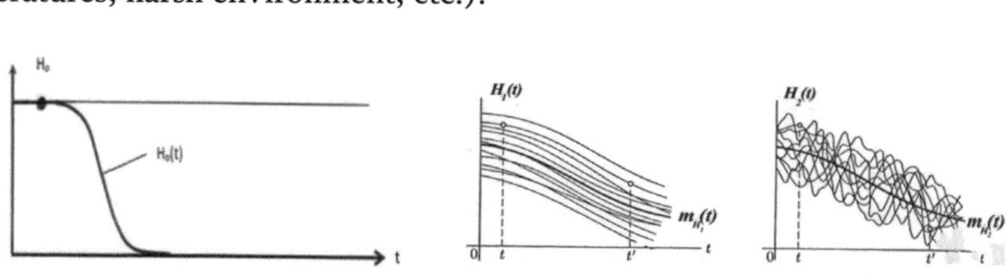


(a) (b) (c)

**Figure 8.**
*Presentation of the threat as a random process.*

In such problem formulation, the definition of threat for SP will be characterized by the random vector-process which is functional of a vector of internal and external force actions $\overline{Q}(t)$, temperature influences $T(t)$, fields of dangerous substances concentration $\bar{c}(t)$, radiations $\overline{\psi}(t)$ and information flows $\bar{I}(t)$.

$$\overline{H}(t) = F\left(\overline{Q}(t), T(t), \bar{c}(t), \overline{\psi}(t), \bar{I}(t)\right\} \tag{10}$$

Physical and chemical bases of the analysis of accidents occurrence and evolution conditions are defined both by work processes in technical SP systems, and by external impacts on these processes.

It is important to note that requirements to detailed level of this object threats' description are defined based on the used destruction mechanisms—external and internal types. The analysis of threats to SP has to be carried out in a manner to provide required data for further calculations of the following:

- stress, stiffness and withstandability (with use of material resistance methods);
- stress and cyclic life and life time (with use of methods of theory of high- and low-cycle fatigue);
- stress and life capability—life time (with use of methods of creep theory and creep-rupture strength theory);
- dynamic strength and life time (with use of methods of crash and fracture dynamics); and
- crack growth resistance (with use of methods of linear and nonlinear fracture dynamics).

If, on the contrary, the fatigue mechanism of destruction is used, the threat cannot be considered as a separate extreme event and has to be characterized by process of on-stream loading.

The quantitative description of development of accidents initiation on SP can be performed on the basis of fundamental mechanisms of disasters physics, chemistry and mechanics. At the same time, the stages of occurrence and development of emergencies can be characterized by various combinations of physical, chemical and the mechanical affecting and damaging factors.

Analysis of the majority of accidents of technogenic and natural-technogenic nature occurred on SP demonstrates that they are determined by three major dangerous factors according to equation (5):

- uncontrolled leak of dangerous substances $W$;
- uncontrolled hazardous energy $E$ release (mechanical and thermal); and
- uncontrolled flows of diagnostic and controlled information of $I$.

If to take into account (**Figure 5**) the classification of accidents on critical infrastructure objects as well as parameters $W$, $E$, $I$ mentioned above, then for classification of oil and gas production facility (OGPF), it is possible to set their critical states' limit areas (**Figure 9**). When talking about critical infrastructure objects, without no doubt, the off-shore oil- and gas production platforms (local—1,

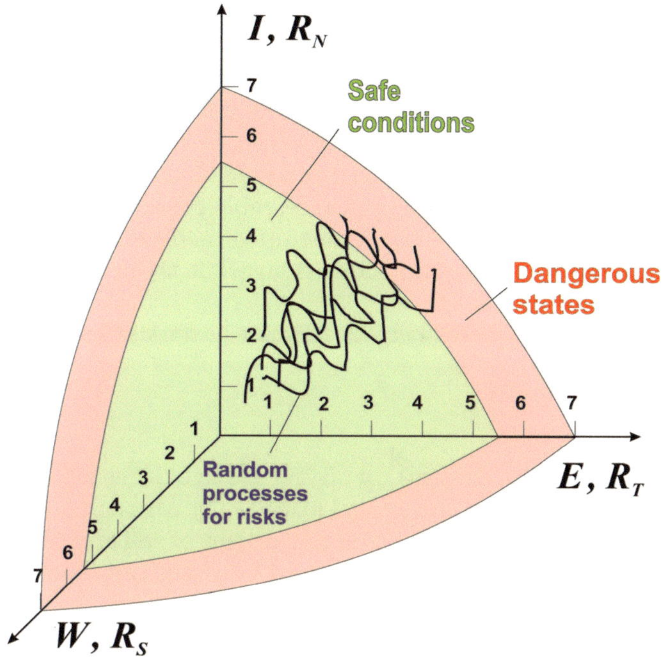


**Figure 9.**
*Areas of dangerous and safe states of the system.*

facility-based—2, domestic—3, regional—4, national (federal)—5, global (transboundary)—6, planetary—7) shall be included in the list of such objects.

Then, radius-vector in space of *W, E, I* will become a quantitative index of dangers to OGPF

$$|T_h| = \sqrt{\overline{W}^2 + \overline{E}^2 + \overline{I}^2}, |T_n| = \overline{W} \cdot \overline{E} \cdot \overline{I} \tag{11}$$

where $\overline{W}, \overline{E}, \overline{I}$ is the hazard class of object for each of accidents classes (from 1 to 7). In the first case, the quantitative value of this hazard will vary from 1.73 to 12.2; and in the second case, it varies from 1 to 343.

The hazards related to external natural processes in the territory of OGPFs location are evaluated in another way and with use of other criteria (earthquakes intensity degree, force of winds, level of floods, extremeness of climatic temperatures, depths of holes, mass of landslides, volume of rainfall, etc.)

The equation (11) can be accepted as unified for different types of dangers: technogenic, natural and natural-technogenic.

In traditional formulation when performing analysis of threats to OGPD initiated by dangerous processes, the first stage of the analysis or problem solving is assessment of losses and risks relevant to accidents on OGPD objects. The solution of the inverse task making it possible to classify the threats to OGPD coming from known consequences of accident occurred on an (**Table 2**) is of interest.

At the solution of such tasks, the intensity of threats is subdivided into the following groups:

Group U1: the threats causing hypothetical accidents which can occur at the options and scenarios of development which are not predicted in advance, with

the greatest possible damages (total destruction of OGPD) and a large number of the victims.
Group U2 group: the threats causing the beyond-design-basis accidents which are followed by permanent damages of the SP critical components with high level of damages and fatalities.
Group U3: the threats causing the design accidents followed by standard outperformance with predictable and acceptable consequences.
Group U4: the threats causing the SP operating mode accidents followed by deviations from normal operation conditions while OGPD is operating in standard mode.
Group U5: the threats when an object operates in standard mode.

## 2.7 Design loads

The loading on offshore ice-resistant oil and gas structures can be split in three groups: permanent, temporary and special loads [1, 2]. Among permanent loads are the loads of the structure weight $P_{s.w.}$ and self-weight of soil and soil pressure on fixed piles. The temporary loads are subdivided into long and short term, namely:

- Long-term load:
    - weight of equipment and drilling rig;
    - weight of liquids, bulk materials and stocks of drill pipes and tubing;
    - weight of warehouse equipment and tools; and
    - weight of drilling cuttings (bore mud, etc.).
- Short-term load:
    - load on drilling rig in and derrick table during drill string trip;
    - snow loads (used for design of bowl type helicopter deck);

| Type of accident | Threat causing the accident | Type of threat |
|---|---|---|
| Hypothetical accidents (Type T1) | Combination of unknown, unlikely and the difficult to predict constructive, technological initiating events and affecting factors of huge intensity, including terrorist attacks. | U1 |
| Beyond-design-basis accidents (Type T2) | The affecting factors, the initiating events and damages development are not known in full. | U2 |
| Design accidents (Type T3) | The affecting factors are known and predictable. | U3 |
| Operating mode accidents (deviations from standard conditions) (Type T4) | The affecting factors are studied and controlled. | U4 |
| Normal (standard) operating conditions (Type T5) | The affecting factors are well understood and controlled. | U5 |

**Table 2.**
*Accident and threat types.*

- due to structural icing;
- wind loads $P_{\text{wind}}$;
- waves $P_{\text{wave}}$ and current $P_{\text{curr}}$ loads;
- loads caused by sheet and hummocked ice $P_{\text{ice}}$;
- docking impact load; and
- helicopter impact load.

The special loads are the seismic ones $P_{seism}$ and those initiated by natural phenomena (structure base subsidence, additional dynamic loads due to impact of ice filed on the structure imbedded in ice); and ice load due to hummocked nature of ice fields (collision of the structure and iceberg). Seismic impacts are taken into account during design of stationary platforms constructed in different regions with seismic magnitude of 7, 8 and 9.

For definition of seismic loads, it is required to have data on seismological parameters of seismic zones: magnitudes, depths of earthquake sources, the epicentral distances, earthquakes frequency, seismicity of the site and spectral characteristics of seismic impacts depending on engineering-geological conditions on construction sites.

Various types of loads on ice-resistant stationary platforms are schematically presented in **Figure 10**.

When calculating the wind and wave loadings, it is expedient to accept load factor for one of loadings equal to 0.9, and for another equal to 1. This assumption is based on more realistic knowledge (from physical point of view) by reference to correlation between these processes. In the case of basic combination, the calculated values of short-term loadings (wind, wave and current) respectively refer to the reliability factor which is equal to 1. For special combinations, these loadings are calculated with factor 0.8, however, at the same time, as well as in the previous case, two possibilities of wind and wave impacts on ice-resistant structures are taken into consideration.

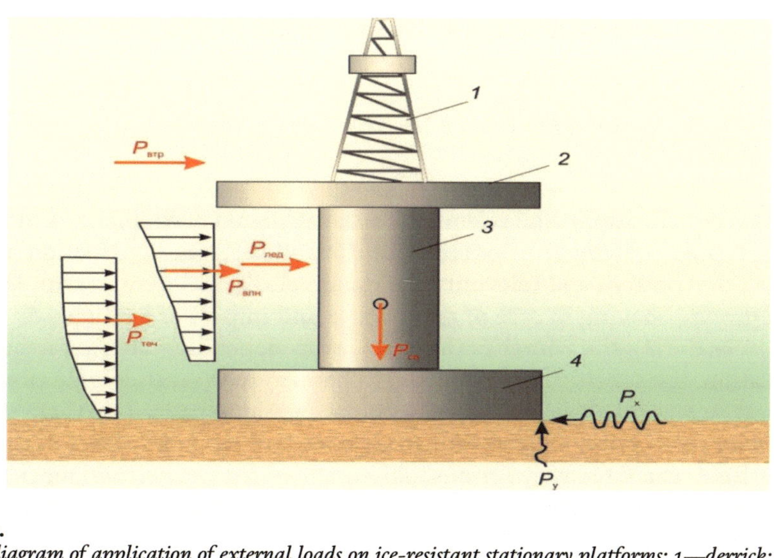


**Figure 10.**
*Symbolic diagram of application of external loads on ice-resistant stationary platforms: 1—derrick; 2—deck; 3—jack structure; and 4—bottom module. For loads, the following symbols are used: $P_{sw}$—gravity force; and $P_x$, $P_y$—horizontal (shear) and vertical (transverse) reactions.*

| Types of calculated loads | Combinations | | | | | |
|---|---|---|---|---|---|---|
| | I | II | III | IV | V | VI |
| Dead loads | 1.0 | 1.0 | 0.9 | 1.0 | 1.0 | 1.0 |
| Long-term live loads | 0.95 | — | 0.8 | 1.0 | 0.95 | 0.95 |
| Short-term live loads: | | | | | | |
| • *ice load* ($h$ = 0.8 m); | — | — | 0.8 | — | — | 1.0 |
| • *wave load* (repeated once in 100 years); | 1.0 | 1.0 | — | — | — | — |
| • *wind load;* | 0.9 | 0.9 | 0.8 | 1.0 | 0.8 | 0.9 |
| • *current load* | 0.9 | 0.9 | 0.8 | — | 0.8 | 0.9 |
| Special loads: | | | | | | |
| • *ice load* ($h$ = 2.5 m); | — | — | — | — | 1.0 | — |
| • *seismic load* | — | — | 1.0 | — | — | — |

**Table 3.**
*Factors of loads combinations.*

As an example of the case when simultaneous impact of the wide spectrum random loadings on ice-resistant structures for sea of Okhotsk conditions can use the approach based on factors of loads combination shown in **Table 3**.

In the given case, it is proposed to analyze the following loads combinations:

I. basic combination of loads during ice-free season;

II. combination of loads during construction and assembling works in ice-free season;

III. special combination allowing for seismic loads;

IV. combination for calculation of maximum efforts in structures of the topside facilities;

V. special combination allowing ice loads occurring during freeze-up period; and

VI. basic loads combination during freeze-up period depending on cycles' number.

In special combinations, the seismic load of calculated earthquake with magnitude 8 is accepted allowing the possible side dynamic effects: liquefaction of soil in the construction bottom and relevant subsidence, additional hydrodynamic loadings from ground shaking in case of open water and impact of ice fields on construction jacks during the winter period. However, depending on the earthquake source location, the specified side effects can happen with considerable time lag with respect to ground shake time, and summation of the caused by them dynamic impacts on a construction with seismic loads does not happen. Impact of the hummocky ice-fields can have very serious consequences for a construction; therefore, such case has to be separated as special loading and be analyzed in other special combination of loads.

In terms of (1)–(5), the total risk $R$ of SP operation as mathematical expectation of incurred damages $U$ should be presented as follows [3]:

$$R = \sum_{i=1}^{n} P(A_i)U_i, (i = 1, \dots, n), \quad (12)$$

where $P(A_i)$ is the probability of causing damage $U_i$ to a technological object and other objects, the population and the environment in case of the emergency scenario No. $I$ and $n$ is the number of possible outcomes of an emergency.

Generally, the probability $P$ of occurrence of analyzed unfavorable event (or its components $P_i$) is defined as the function of function (functionality) depending on sources, corresponding affecting factors and objects subject to damage: person $N$, off-shore technosphere object $T$ and environment $S$; taking into account (3), the probability will be defined by the formula:

$$P = F_p\{P_N(t), P_T(t), P_S(t)\} = \sum_i F_{P_i}(P_{N_i}(t), P_{T_i}(t), P_{S_i}(t)). \quad (13)$$

$P_N(t), P_T(t), P_S(t)$ are the probabilities of occurrence at time $t$ of unfavorable (dangerous) event initiated correspondingly by human factor, technosphere or nature.

The general damage $U$ or its components $U_i$ defined by damages affected by the population $N$, objects of a technosphere $T$ and the environment $S$ as follows:

$$U = F_U\{U_N(t), U_T(t), U_S(t)\} = \sum_i F_{U_i}(U_{N_i}(t), U_{T_i}(t), U_{S_i}(t)). \quad (14)$$

$U_N(t), U_T(t), U_S(t)$ are damages caused by unfavorable (dangerous) events at time $t$ from which suffered population $N$, objects of a technosphere $T$ and the environment $S$ correspondingly.

At the present stage of technical regulation, it is recommended to estimate the quantities of damages $U$ and total risk $R$ from unfavorable events by two indicators: economic—in dollars, rubles (conventional units) and in human losses (fatalities or non-fatal outcomes). Human losses should be estimated by the number of injured or probability of fatalities.

Taking into account expressions of (13) and (14), components of damages and probabilities of accidents can be calculated separately by use of various methods of risk assessment. Also from the expression of the risk (12) presenting the summation of risks of different emergencies, it becomes clear that to define the total risk, the various methods for definition of its components can be used, i.e., the combined approach is applied.

Combined risk analysis is based on the systematic approach that provides review of the system of interest in a formalized manner, i.e., by studying of subsystems' components by considering structural and functional features of this system at the same time.

The damages and losses $U$ caused by technogenic accidents and disasters are defined by three basic components by taking into account expression (4):

$$U = U_T + U_S + U_N, \quad (15)$$

where $U_T$ are damages to off-shore technosphere objects; $U_S$ are environmental damages; and $U_N$ are damages to the population (to the person and society).

Damages $U_T$ are defined by summation of damages from destruction of industrial buildings and constructions of $U_{TII}$ type; damages from destruction of civilian (residential) objects of $U_{TГ}$ type; and damages from destruction of infrastructure of $U_{TИ}$ type (transportation, energy, pipeline, telecommunication systems, etc.):

$$U_T = U_{TП} + U_{TГ} + U_{TИ}. \tag{16}$$

Environmental damages $U_S$ defined by summation of damages to soil $U_{SП}$, aquatic $U_{Sa}$, air $U_{Sв}$ environment, flora $U_{Sp}$ and fauna $U_{Sж}$ are as follows:

$$U_S = U_{SП} + U_{Sа} + U_{Sв} + U_{Sр} + U_{Sж}. \tag{17}$$

Damages to the personnel and population $U_N$ are defined by summation of losses from fatalities $U_{Nб}$ and losses from injuries (permanent injuries and health damages) $U_{Ny}$, which are as follows:

$$U_N = U_{Nб} + U_{Ny}. \tag{18}$$

Damages and losses quantitatively are defined by two types of parameters:

- in physical units—scales (number of damaged objects and injured people, polluted and damaged territories by area); and

- in equivalent economic units (rubles, dollars).

In statistical estimation of the above damages, the summarized information about emergencies from the state reports of departments can be used.

In probabilistic estimation of damages, the data from simulation modeling, data on probable areas covered by the affecting factors, and probabilistic and statistical data on vulnerability of objects, the environment and the population at various emergencies are used.

In the analysis and risk assessment, various aspects of accidents and disasters occurrence and development including various dangerous processes, the factors initiating events, scenarios of development, objects and personnel pattern damage function, etc. can be considered.

The variety of issues to be studied in the analysis process and risk assessment requires application of various methods at various stages of the systems analysis of examined object safety, as well as their integrated application.

Some methods in nature are integral ones; for example, the logical-and-probabilistic method, which includes a graph method, a probabilistic method, a logical reasoning method, event tree analysis and fault tree analysis are probabilistic methods implementing the graph method.

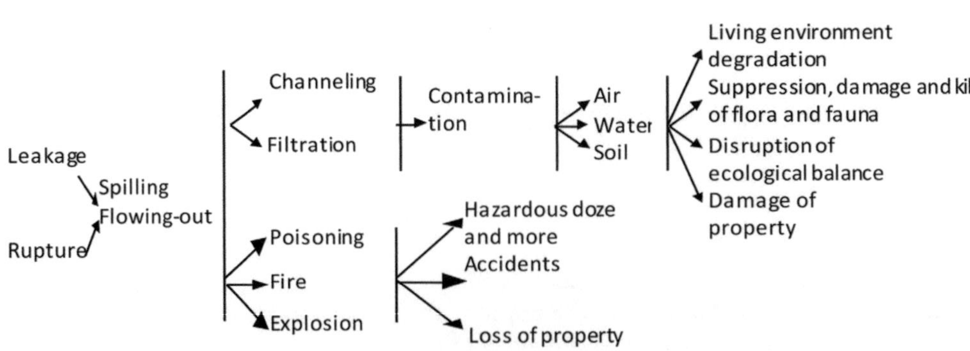


**Figure 11.**
*Basic scenarious of accident development on sea platform (SP).*

The main possible events chains for scenarios of accidents on OGPF are presented in **Figure 11**. The main events (faults) causing accidents are the leak and rupture of technical pipelines. These faults cause development of accidents in various scenarios and corresponding damages. All these possible scenarios and corresponding damages have to be taken into account.

### 2.8 Consideration of ultimate limit states at risk assessment of SP condition

When forming a system of classification of ultimate limit states in routine operating conditions of objects and in case of occurrence of accidents and disasters in comprehensive technical systems, it is required to identify various combinations of states for five groups of situations [1, 2, 5]:

- ultimate limit states for regular service conditions;
- ultimate limit states for abnormal service conditions;
- ultimate limit states for designed accident;
- ultimate limit states for beyond-design-basis accident; and
- ultimate limit states for hypothetical accident.

Ultimate limit stress for normal service conditions have to be in full reflected in design codes of potentially hazardous objects, consider a set of design operating modes and proceed from all previous operating experience of similar objects.

In case of violation of normal (i.e., abnormal) service conditions (at any deviation from planned operating procedure causing the necessity to change operating mode or stop an object without necessity to activate or use all safety systems) the given above types of ultimate limit states can be used, or more extensive and wide. Such expansion is caused by the increase of number of work abnormalities and range of operation parameters changes.

When analyzing a design accident requiring the stop of an object and activation of safety systems, in addition to mentioned above types, it is necessary to consider those types of ultimate limit states which occur at increased mechanical, thermal, electromagnetic and other loads at scheduled stages of accident development.

For beyond-design-basis accidents followed by full activation of safety systems, it is not possible to exclude considerable damages of the most critical components and the equipment in general; in this case, the ultimate limit states include not only standard ones, but also new ultimate limit states that are object specific at broad variation of load conditions at all stages of accidents development.

The hypothetical accidents are most severe, hardly probable and poorly studied, and the worst combination of the affecting factors and that is why it is necessary not only to provide the analysis of the ultimate limit states stated above but also to analyze the states at which significant changes of conditions of working substances and structural and mechanical conditions of engineering materials are possible.

When accidents (explosions, destruction, fires, collisions, collapses, chemically dangerous substances release) are occurring in the load bearing structures, the corresponding ultimate limit states are arising. At different stages of accidents development, these limit states can change both in the direction of scaling up of consequences, and in the direction of localization and full stop of the accident development.

When determining safety of the most important objects, the following types of ultimate limit states have to be considered: plastic deformation and forming; short-duration elastic failure; delayed or fast brittle failure; long-term static fracture; cyclic (low- and multi-cycle) destruction; creep strain accumulation; cyclic strain accumulation; buckling; dangerous vibrations occurrence; coupled units wear; single loading cracks initiation and propagation; cyclic cracks initiation and propagation; corrosion, corrosion and mechanical, cavitation and erosive damages; leakages; and change of structures and a condition of the bearing components.

The ultimate limit states listed above identify methods, structure and criteria of safety analysis by integrated approaches of mechanics, physics and chemistry of disasters.

In the process of design of structure, its components and, at the bottom, the following groups of the ultimate limit states are taken into consideration. The first group with unacceptable plastic strain and damages includes ultimate limit states surpassing of which will cause total unusability of the structure or total (or partial) loss of supporting capacity of the platform substructure. The second group with damages accumulation and development includes the ultimate limit states where surpassing makes impossible the normal operation of the platform substructure.

It should be noted that the above-listed ultimate limit states were taken into account at design of the reinforced concrete support substructure of gravity type for offshore stationary platforms on the sites of the Sakhalin-II project for Piltun-Astokhsky (PA-B) and Lunsky (LUN-A) fields.

The design elements of the platform substructure can be split into criticality categories depending on the external impacts taken into account:

**High criticality design elements**—these are elements whose destruction can cause fatalities, serious damages to constructions and environment contamination.

**Low criticality design elements**—these are elements whose destruction will not cause fatalities, serious damages to constructions and environment contamination.

Between high criticality design elements, the following ones shall be listed:

- design elements of skirt and skirt interface with caisson bottom;
- column walls in areas of their connection with the bottom and top plates of overlapping of a caisson;
- parts of walls and columns overlapping subject to significant loads concentration;
- design elements contacting with ice;
- connection of deck with the column;
- outer walls, floor slabs and caisson bottom;
- internal waterproof walls;
- design elements of supporters of the critical and safety equipment including riser holders; and
- structures which damage and destruction will cause dramatic environment contamination including risers.

Between low criticality design elements, the following ones shall be listed:

- internal structure not involved in provision of general strength; and
- design elements of equipment supporter not identified as elements of critical importance.

### 2.9 Comprehensive assessment of risk, strength, in-service life, reliability and safety

Characterization of initial strength, in-service life, risk and safety of the bearing elements of the sea oil and gas production platform in terms of impact of a complex of loads (including such specific service conditions as collisions with the drifting ice floes, impact of storm and gale-force winds, existence of the corrosive environment, low-temperature embrittlement effects, etc.) is the comprehensive problem considering occurrence of the cyclic dynamic loads corresponding to these conditions and, consequently, nonlinear change in time of the kinetic fields of stresses and deformations in these elements of SP under the impact of irregular loads [1–4]. In this regard in zones of design concentration, the local stresses and deformations have the increased values and the processes of material damage run more intensively leading to appearance of local destructions zones (cracks) eventually developing into macrodestructions (loss of bearing capacity). In such conditions, depending on the nature of loading and the operational environment, various mechanisms of accumulation of damages and destruction are implemented.

For the analysis of operational load of SP (as well as on other objects of energy, transport, oil and gas chemistry) at all stages of the life cycle, curves of the parameters dependent on calculated or real force impact on the bearing elements of the oil and gas production platform (set in the specification or measured during operation) are plotted. Among these parameters are number of loading cycles $N$, time $\tau$, temperature $t$ as well as service forcing $P$, stress $\sigma$ and deformation $e$. The curves of parameters $P$, $t$, $\sigma$ and $e$ as function of time (**Figure 12**) are plotted for all stages and operational phases.

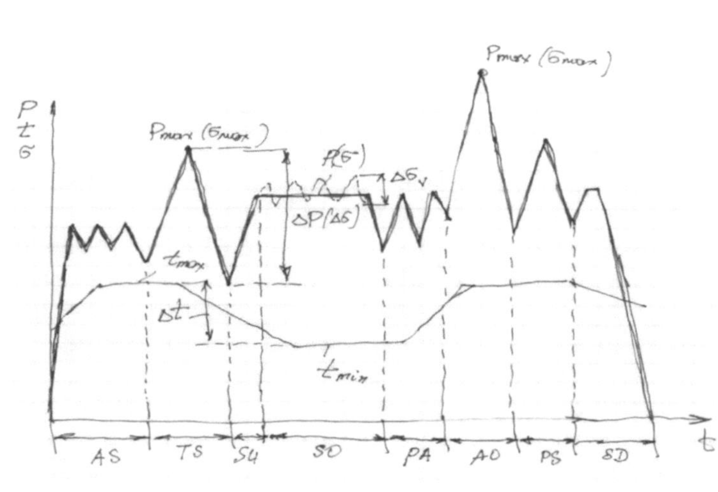


**Figure 12.**
*Diagram of operational loads and their basic parameters.*

These dependences are initial for the analysis of strength, in-service life, risk and safety of elements of engineering designs both for their initial states and for the damaged states. Values P, $t$ and $\tau$, are, as a rule, set by the modes of operation and can be registered by instrumentation and control diagnostic systems or by monitoring equipment.

At the same time, $\sigma$ and $e$ parameters of the general and local stress-deformed states can be obtained with the help of calculation based on the values of parameters $P$, $t$ and $\tau$ or purposely measured by means of full-scale strain gauging and thermometry.

In **Figure 12** where a block of external and internal technological operational loadings are presented, the following standard modes of loading of the SP elements are highlighted: assembling (*AS*), tests (*TS*), start-up (*SU*), stationary the mode with maintenance of set operating parameters (*SO*), basic parameters adjustment (*PA*), accident occurrence (*AO*) (including those after of earthquakes), protection systems actuation (*PS*) and shut-down after planned or fault situation (*SD*).

When analyzing the initial and residual strength, service-life, survivability, risk and safety of the oil and gas production platform, the key phase is decomposition of SP and selection and identification of its potentially dangerous critical components, defining the greatest risks of accidents and disasters occurrence. The critical zones of SP components and critical points in them are identified on the basis of experimental and computational studies of stress-deformed and ultimate limit states. In such zones and points, as a rule, processes of local destructions are initiated followed by tramline destructions. At the same time for further experimental and computational evaluations of initial and residual strength, service-life, survivability, risk and safety, the following characteristics of history of loading (**Figure 12**) are accepted:

- maximum rated load $P_{max}$;
- maximum (minimum) rated temperature $t_{max}(t_{min})$;
- time of standard load conditions $\tau_i$ and total time pf all modes and blocks of modes $\tau_\Sigma$ (life capacity).

With the help of this history of loading set are additional design parameters:

- peak-to-peak range of forcing $\Delta P$ and forcing amplitude;
- peak-to-peak range of temperature variations $\Delta t$; and
- peak-to-peak range of vibration loads $\Delta P_B$ (dual- or multi-frequency) loads.

From the analysis of all $i$ modes according to standard calculation, the most adverse combinations of $P$ and $t$ are identified: $(P_{max}, t_{max})$—for heavy loadings and areas of increased temperatures impacts, and $(P_{min}, t_{min})$—for heavy loadings and low temperatures (including cryogenic). A set of such combinations is defined by taking into account the number and geometrical shape of the designed details or elements and number of critically dangerous zones, sections and points in them.

For quantitative evaluation of static and cyclic strength, as well as in-service life [1, 2, 5], experimental and computational diagram in coordinates of "$\sigma_a$-$N$" (**Figure 13**) is used.

$$\sigma_a = e_a \cdot E = F\{N, r_\sigma, \sigma_T, \sigma_B\}, \tag{19}$$

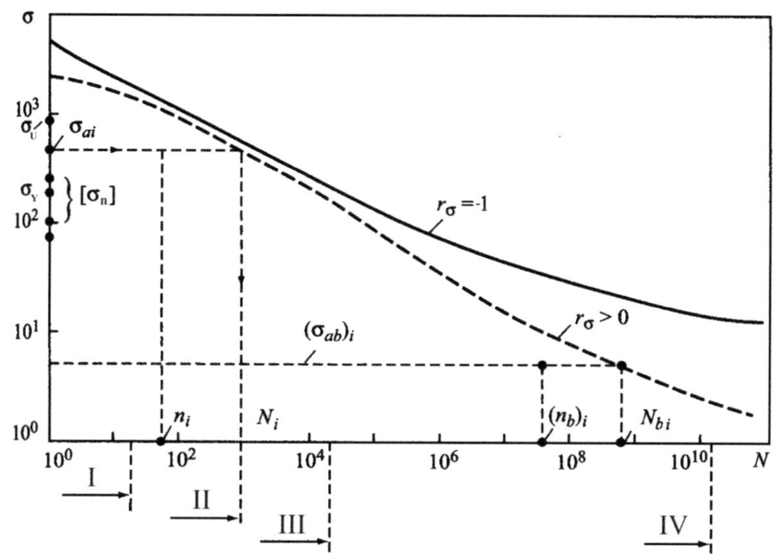


**Figure 13.**
*Diagram that is used for identification of static, cyclic and long-term initial strength and in-service life parameters: I—AS, SU, TS, PS, SD; II—SO; III—PA; and IV—V (vibration $\Delta\sigma_v$—).*

where $\sigma_a$, $e_a$ are strength and deformation amplitudes; $E$ is the elasticity modulus; $N$ is the number of cycles prior to destruction; $r_\sigma$ is the stress ratio; $\sigma_m$, $\sigma_в$ are yield and ultimate stress limits of the structural material.

On **Figure 13**: $\sigma_{ai}$, $(\sigma_{aв})_i$ is the amplitude of basic and vibratory stresses for $i$-mode; $n_i$, $(n_в)_i$ is the number of cycles for basic and vibratory loads; $N_i$, $N_{вi}$ is the number of destructive cycles; $\sigma_m$, $\sigma_в$ are yield and strength limits; $[\sigma_a^*]$, $[N]$ are acceptable tensions amplitudes $[\sigma_a]$ and endurance capability $[N]$ are defined on the basis of traditional calculations with consideration of ultimate factor of safety $n_\sigma$ and marginal life $n_N$

$$[\sigma] = \frac{\sigma}{n_\sigma}, [N] = \frac{N}{n_N}. \tag{20}$$

When making stress assessment, the characteristics $\sigma_B$ and $\sigma_T$ have to be set with taking into account service conditions—impact of loading cycling, temperatures and operating environment.

### 2.10 Criteria of strength, in-service life, safety and protection level (security)

As it was noted above, the solution of fundamental problems of provision of safety, risks and security of critically and strategically important infrastructure facilities is based on the analysis and development of fundamental scientific approaches to issues relevant to strength and in-service life, development of engineering methods of calculations and tests, creation of norms and rules regulating design and fabrication of objects of offshore technosphere, ensuring their functioning within identified limits of the design and beyond-design modes and parameters. Nowadays, the analysis and development of all components of the criterial sequence "Strength → rigidity → consistency → in-service life → reliability → survivability → safety → risk → protection level (security)" became the basic ones, step by step raising requirements imposed on their routine (normal) functioning and ensuring realization of design parameters at all stages of life cycle.

The specified requirements implemented in this knowledge area are imposed on operability of critical structures and expressed by means of the corresponding characteristic parameters of criteria dependences for the above sequence.

A “pyramid” of provision of technosphere objects’ operability according to the main criteria (**Figure 14**) was constructed based on requirements and parameters providing safe operation conditions of these objects.

From **Figure 14**, it is clear that every element located above the other one is supported by the lower elements, i.e., it is laid on it as on foundation. It eventually means that the solution of the task of security, risk and safety provision has to rest upon the solution of problems of “survivability → reliability → in-service life → rigidity → consistency → strength” with passing through traditional stages of their interaction I → VIII. Fundamental results of identification and provision of strength (stage I) were obtained in the beginning of the nineteenth century and it took a long time, while complete analysis of rigidity and resistance (consistency) (stage II) came to the end by the end of this century. In the twentieth century, the theory and practice of provision of “in-service life → reliability → survivability” (stages III, IV, V) were formed. At the end of the last century, the fundamental problem of the analysis and safety and risk provision (stage VI) was formulated for all potentially hazardous civilian and defense objects with transition to management (stage VII) of safety and security according to risks criteria. At these stages, safety and security requirements were formulated like governing, and this provoked development of the new line where consequence “VII → I” becomes the basis for the future technosphere development. At the beginning of this century, the new task (stage VIII) was formulated and this is provision of safety and security of crucial objects based on anti-accidents and anti-disasters of technogenic, natural and anthropogenic character performance.

According to abovementioned and expressions (1)–(9) and **Figure 14**, the proofness of SP is the function of function (functional) Fz of the basic change in time $\tau$ parameters

$$Z_\kappa(\tau) = F_z\,\{R_\sigma\,(\tau), R_{N\tau}(\tau), L_{ld}(\tau), P_{PR}(\tau), S(\tau)\}, \qquad (21)$$

where $Z_\kappa$ is the proofness determined by the ability of an object to resist to accidents occurrence and development of adverse situations in normal and abnormal conditions; $R_\sigma$ is the strength determined by resistance of the bearing object elements to destruction under normal and emergency impacts; $R_{N\tau}$ is the in-service life (endurance capability) determined by time $\tau$ or cycles number $N$ prior to

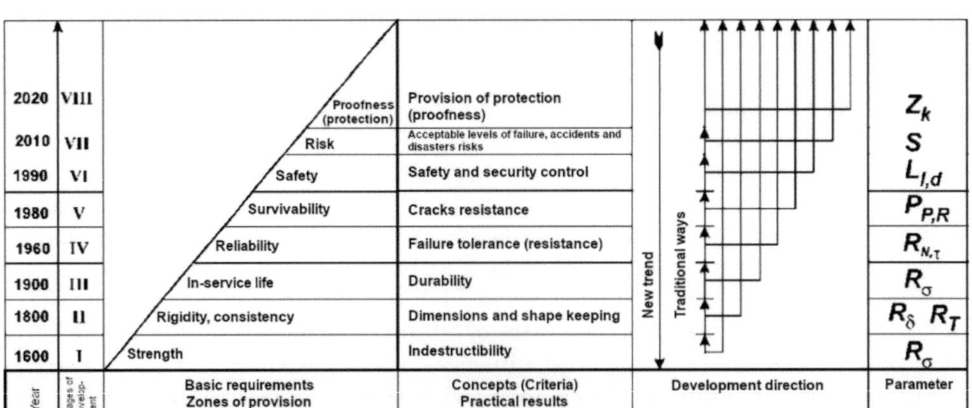


**Figure 14.**
*General structure of provision of technosphere objects operability.*

destructions or instability; $L_{ld}$ is the survivability determined by ability of an object to perform limited functions at damages d and dimensions of defects $l$ that are inadmissible according to norms; $P_{PR}$ is the reliability determined by ability of an object to perform specified functions in the known or defected state at specified loadings $P$ or service-life $R_{N\tau}$; and $S$ is the safety determined by the ability of an object not to pass into a catastrophic state causing significant damages to the person, the technosphere and the environment.

As it was already mentioned, operational conditions of loads of SP are characterized by a significant amount of various factors and parameters; among them are loading conditions and levels of static and dynamic mechanical loads (**Figure 15a**) and impact of corrosive environment, of external factors, etc. These factors taken together and each one individually can cause significant change of nature of behavior of material, its mechanical properties, ability to resist cyclic deformation in comparison with standard design loading specifications (stationary application of cyclic load, room temperature, etc.) at which standard experiments are usually conducted to define the corresponding characteristics. They also may contribute changes in the corresponding patterns of damages accumulation in the material of the equipment components experiencing their influence when in operation.

Cyclic loading waveform of random operating modes as a rule has more sophisticated nature than widely used in experimental practice sinusoidal or triangular waveforms of cyclic loadings.

In some cases, it is obviously possible to schematize and replace actual conditions of loadings by more simple, single-frequency modes. However, generally, the patterns of change of the loadings influencing the structural elements have random nature (**Figure 15b**).

The actual loading modes are schematized (**Figure 15c,e**) in the process of the loading history tracking (**Figure 15e**). Approximation of simulated loading conditions of the equipment as accurate, in respect to reality, as possible for each factor occurring during equipment operation and taking into consideration of impact of these factor on parameters of the characteristic equations and equations describing damages accumulation process is an effective step for adjustment of applied methods for calculations of strength, endurance capability and reliability of the oil and gas production platform components' and hence to identification of really grounded and justified their safe in-service life.

Cyclic strength $\sigma$ and endurance capacity $N$ are defined by the use of the stress-cycle relationship and the equation

$$e_a = \frac{1}{2\cdot(4N)^{m_p}} ln\frac{100}{100-\psi_K} + \frac{1}{(4N)^{m_e}}\cdot\frac{S_K}{E\left(1+\frac{1+2}{1-2}\right)}, \tag{22}$$

where $\psi_K$ is the limit plastic yield of contraction, $S_K$ is the rupture strength of contraction and $E$ is the elasticity modulus defined in the process of standard tests of static tension. Value of index of plastic $m_p$ and elastic $m_e$ components of deformation $e_a$ in the absence of direct data on their values can be determined with the help of material yield stress and ultimate stress values, which are as follows

$$m_p = 0,36 + 2\cdot10^{-4}\sigma_B, m_e = 0,132\cdot lg(S_K/\sigma_{-1}), \tag{23}$$

where value of fatigue limit $\sigma_{-1}$ can be defined as $\sigma_{-1} \cong 0,45\sigma_B$, and rupture strength of contraction $S_K$, dependent on ultimate stress limit $\sigma_B$ and relative narrowing of contraction $\psi_K$, correspondingly comes from relation $S_K = \sigma_B(1+1,4\cdot\psi_K)$. Parameters $\sigma_{в}$, $\psi_K$, $S_K$ in general case are dependent on time $\tau$, operational temperature and full size cross-sections of SP bearing elements.

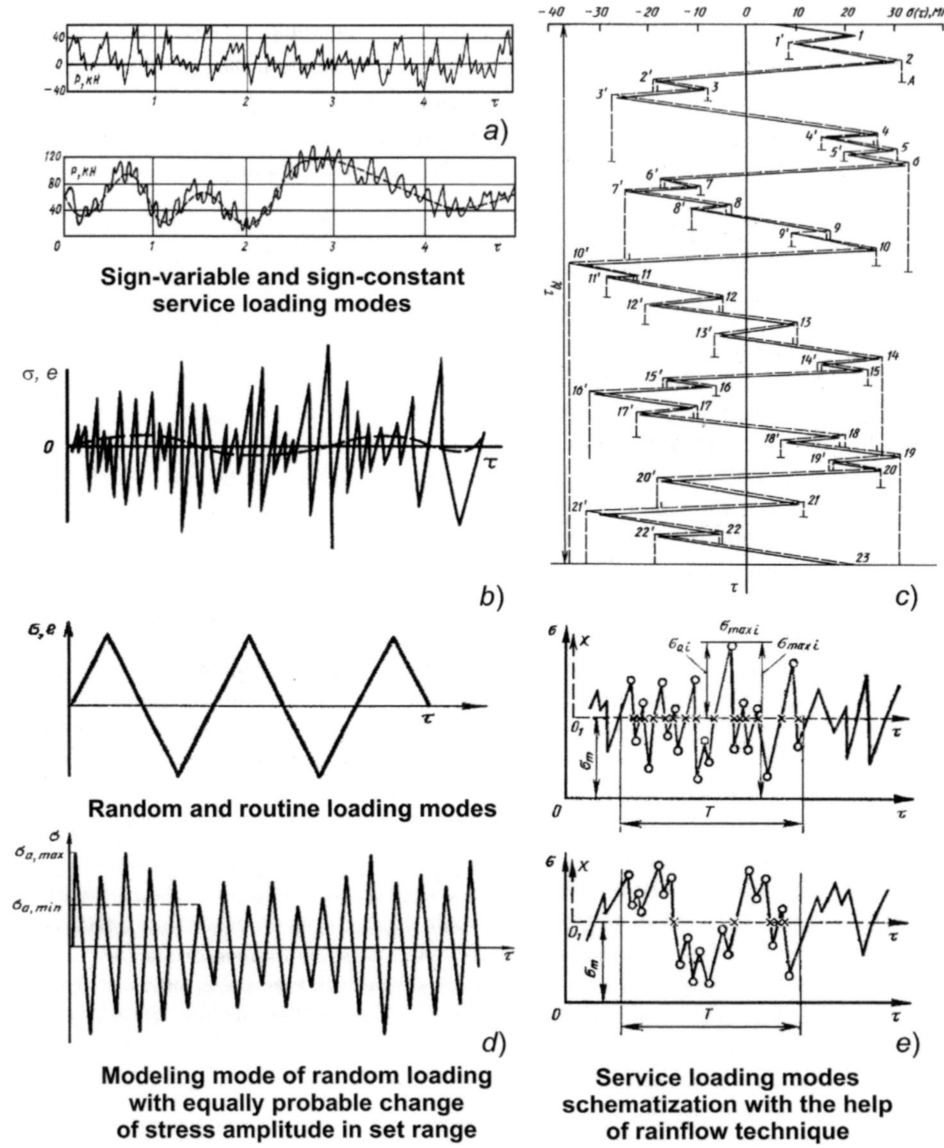


**Figure 15.**
*Methods of schematization of operational modes of loads. a) Sign-variable and sign-constant service loading mode. b) Random and routine loading modes. c) Service loading modes schematization. d) Modeling mode of random loading with equally probable change of stress amplitude in set range. e) Rainflow technique.*

## 2.11 Probabilistic analysis of strength, in-service life and risks

Because SP is functioning in the conditions of the high level of uncertainty concerning external impacts during operation period and bearing capacity level changing due to structures degradation, the criteria in expressions (21)–(23) have to be probabilistic [2–6].

Let function of ultimate limit states for the considered platform element is defined by a ratio of bearing capacity and loading $l$. Generally, function of ultimate limit states

$$g(r,l) = r - l \tag{24}$$

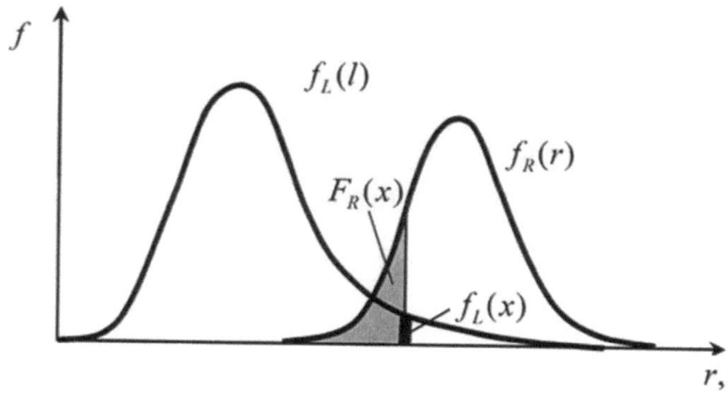


**Figure 16.**
*Probability density functions for bearing capacity and loading. Probability curves for design parameters at assessment of chances of failure.*

is written as $g(\boldsymbol{x})$, where $\boldsymbol{x} = (x_1, x_2, \cdots, x_n)$ is the vector of variables describing the element state. Then the element failure can be generally described as follows (**Figure 16**)

$$\boldsymbol{F} = \{x | g(\boldsymbol{x}) \leq 0\} \tag{25}$$

Conditional probability of failure in case when the element is under load $L = l$ is defined by function $F_R(x)$ (this is due to the fact that $F_R(x) = P(R < l)$). Then, using the theorem of total probability, it is possible to write expression for the probability of element failure as follows:

$$P(\boldsymbol{F}) = \int_{-\infty}^{\infty} F_R(x) f_L(x) dx, \tag{26}$$

where $R$ is the bearing capacity and $x = l$ is the load.

Let us consider the random variable of margin of safety, in-service life and proofness (safety) $M = R - L$ equal to excess of bearing capacity over load. As $R$ and $L$ are random variables, $M$ is also the random variable with mathematical expectation $\mu_M$ and rms deviation $\sigma_M$. They can be calculated from mathematical expectation and rms deviation of values $R$ и $L$:

$$\begin{aligned} \mu_M &= \mu_R - \mu_L \\ \sigma_M &= \sqrt{\sigma_R^2 + \sigma_L^2 + 2\rho_{RL}\sigma_R\sigma_L} \end{aligned} \tag{27}$$

The probability of system failure which is equal to the probability of value $M$ be less or equal to 0.

$$P(\boldsymbol{F}) = P(M \leq 0) = \Phi\left(-\frac{\mu_M}{\sigma_M}\right) = \Phi(-\beta), \tag{28}$$

where $\beta$ is the proofness (safety) index (this variable sometimes is called reliability index) of the element analyzed upon its ultimate limit state $g(\boldsymbol{x})$. Value $\beta$ characterizes the distance of the ultimate limit state surface and can be treated as safety (proofness or security) characteristics of element relative to analyzed failure mechanism.

If the destruction mechanism relative to excess of maximum permissible load is considered, then equation of the surface of ultimate limit states takes the form

$$g_U(\boldsymbol{x}) = R - L, \tag{29}$$

where $R$ is the strength (bearing capacity) of the element and $L$ is the maximal load during the analyzed period.

Safety (proofness or security) upon the criterion of exceed of maximal permissible load will be presented by the expression:

$$\beta_U = \frac{\mu_R - \mu_L}{\sqrt{\sigma_R^2 + \sigma_L^2 + 2\rho_{RL}\sigma_R\sigma_L}}. \tag{30}$$

If to talk about the fatigue mechanism of element destruction, then equation of the surface of ultimate limit states takes the form $g_F(x) = N - n$ where N is the number of cycles prior to destruction at the set level of stresses range and n is the number of cycles to which the element is exposed during use. Then the proofness of element upon criterion of fatigue failure will look as follows:

$$\beta_F = \frac{\mu_N - \mu_K}{\sqrt{\sigma_N^2 + \sigma_K^2 - 2\rho_{NK}\sigma_N\sigma_K}}. \tag{31}$$

Because of hostile environment influence on the OGPF elements and relevant degradation processes in them, the function of element ultimate limit states has to depend on time. In the considered statement, the proofness (safety or security) reserve of a critical element is estimated in the form of $M = R\text{-}L$, where $R$ is the bearing capacity in critical cross-section and $L$ is the loading in the same cross-section. If to consider that both random variables of $R$ and $L$ in real systems can depend on time, then the bearing capacity can change because of degradation of material properties (corrosion, fatigue, etc.); loading, in its turn, can change due to change of service conditions, of external environment, etc. At that their mathematical expectations $\mu_R(t)$ $\mu_L(t)$ and rms deviations $\sigma_R(t)$ $\sigma_L(t)$ will change as well. Then the margins of bearing capacity can be written as follows:

$$M(t) = R(t) - L(t). \tag{32}$$

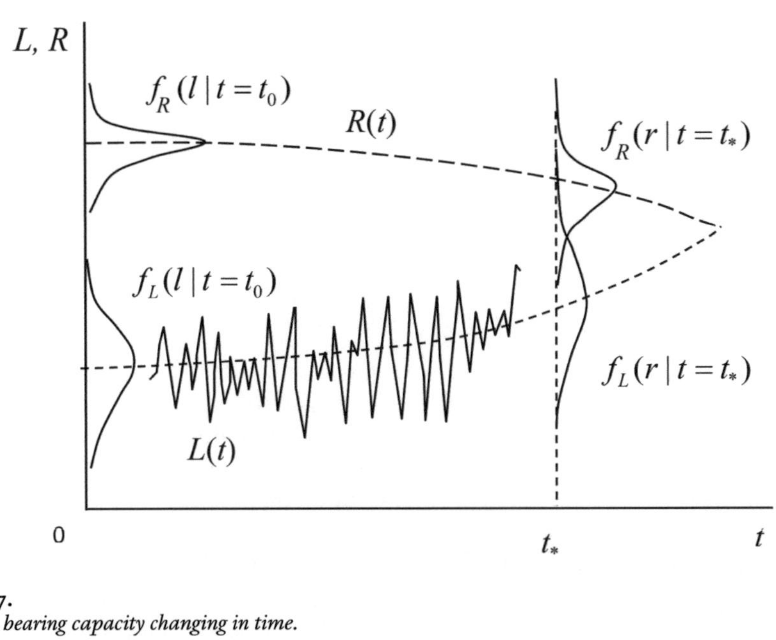


**Figure 17.**
*Load and bearing capacity changing in time.*

In this case, the probability of failure becomes the function of time:

$$P_f(t) = P\{R(t) \leq L(t)\} = P\{g(X(t)) \leq 0\}, \tag{33}$$

where $g(X(t)) = M(t)$, depending on time proofness margins as per bearing capacity. The probability of system failure is

$$P_f(t) = \int_{g(\mathbf{X}(t)) \leq 0} f_{\mathbf{X}(t)}\mathbf{x}(t)d\mathbf{x}(t). \tag{34}$$

The identification of time moment $t_*$ when loading $L(t)$ for the first time will exceed the bearing capacity of an element $R(t)$ is an important task (**Figure 17**).

## 2.12 Engineering justification of strength, in-service life and safety

As it was noted above, continuously raising requirements to regular (normal) and abnormal functioning are imposed on modern SP. In modern conditions of the analysis and provision of safe operation of technosphere objects, the new task about identification and safety and security provision upon criteria of actual $R(\tau)$ and acceptable $[R(\tau)]$ bearing capacities are used in expressions (7) and (9). Within that narrative [1–4, 7], only characteristics of safety with the set levels of risks give justification to acceptance (or rejection) of decisions on permission of new projects realization or permission to operate running offshore objects.

Operational impacts on the SP elements in general (periodically arising ice loadings, service, wind and seismic loads) are characterized by the following parameters, in particular numbers of loading cycles N, time of loading $\tau$ and ambient temperature t. At the same time, $N$ and $\tau$ define in-service life of the examined object, while $t$ defines its cold brittleness. The imperfection of bearing structures is defined by the sizes of cracks of $l$, their shape and location. Sizes $l$ are initial for determination of objects survivability. Characteristic of flexibility, rigidity, stability $\lambda$ of the bearing component of the analyzed element depends on a shape and dimensions of cross section, length and type of supporting. It defines his stability.

External routine and abnormal impacts (including accidents and catastrophic) generate in the analyzed element design stress level $\sigma$ and deformations $e$; they depend on the applied loads (mechanical, temperature, aero hydrodynamic, seismic, etc.), a way of their application, the sizes and shapes of cross-sections. If these impacts increase, then at some point in the bearing elements, the ultimate state limits (critical) are achieved, and these elements are destructed, losing stability and getting inadmissible deformations. Stresses and deformations at this moment achieve extreme (critical) values $\sigma_f$, $e_f$.

Values of characteristics $\sigma_f$, $e_f$ according to **Figures 12–14** depend on values $N$, $\tau$, $l$, $t$, $\lambda$. Based on these dependences and upon experimental and laboratory studies of the construction materials, the following are plotted:

- fatigue curves (live curve) for stresses "$\sigma$–$N$" and deformations "$e$–$N$";
- stress rupture curves for stresses "$\sigma$–$\tau$" and deformations "$e$–$\tau$";
- crack resistance curve (survivability) for stresses "$\sigma$–$l$" and deformations "$e$–$l$";
- temperature resistance curve (cold- and heat resistance) in coordinates of stresses "$\sigma$–$t$" and deformations "$e$–$t$"; and

- stability curves (general or local) in coordinates of stresses "$\sigma$–$\lambda$" and deformations "$e$–$\lambda$."

At relatively low levels of external routine impacts when occurring deformations are elastic, the calculations relevant to stresses and deformations have identical results. At the increased abnormal and stress impacts when occurred are general and local plastic deformations, the calculations made with respect to stresses $\sigma$ and deformations $e$ are divergent—the values of stresses $\sigma_f$ happen to be insensitive to $N$, $\tau$, $l$, $\lambda$ variation. This fact predetermines the importance of transition from the traditional determined calculations in terms of stresses $\sigma_f$ to probabilistic calculations in terms of deformations $e_f$ [2, 5, 7–9].

In case of the integral analysis of strength, in-service life and safety, the deformation curve in true coordinates (the true stress $\sigma$ and true deformations $e$) is presented as follows

$$\sigma = \sigma_T\ (e/e_T)^m, \tag{35}$$

$$m = lg(S_k/\sigma_T)/lg(e_k/e_T), \tag{36}$$

where $\sigma_T$ is the yield stress; $m$ is the work hardening exponent ($0 \le m \le 0.3$); $S_k$ is the tension strength; $e_T = \sigma_{T/E}$; $e_k = S_{k/E}$; and $E$ is the elasticity modulus.

The strength-duration curves $\sigma^t_{B\tau}$ and ductility property $\psi^t_{k\tau}$ for time $\tau$ are the basic ones in case of long-term loading at increased temperature

$$\sigma^t_{B\tau} = \sigma^t_B\ (\tau_o/\tau)^m{}_B, \psi^t_{k\tau} = \psi^t_k(\tau_o/\tau)^{m_\psi} \tag{37}$$

where $\tau_0$ is the time of short-time tests ($\tau_0 \approx 0.05$ h); and $m_B$, $m_\psi$ are the material characteristics depending on temperature $t$ and yield stress $\sigma^t_T$ ($0 \le m_в \le 0.08$, $0 \le m_\psi \le 0.15$). Then, it is possible to obtain the cyclic stress curve "$\sigma^*$ - $N$" as per parameter $\tau$.

In estimating the effect of temperatures $t$, different from room temperature $t_0$ = 20°C (both in the range of low climatic temperatures 20°C $\ge t \ge -60$°C, including cryogenic range -60°C $\ge t \ge -270$°C and elevated 20°C $\le t \le$ 350°C and high temperatures 350°C $\le t \le$ 1000°C), standard tests are carried out in thermocryocameras. In the absence of such tests' results, the estimated dependences of mechanical properties on temperature of $t$ °C or $T$ °K ($T = t + 273$) T are plotted

$$\{\sigma^t_m, \sigma^t_{\text{B}}\} = \{\sigma_m, \sigma_{\text{B}}\} \cdot \{\beta_m, \beta_{\text{B}}\}\left(\frac{1}{\text{T}} - \frac{1}{\text{T}_o}\right) \tag{38}$$

where $T$ is the temperature in Kelvin degrees ($T = t^o + 273$); and $\beta_T$ and $\beta_B$ are designed material characteristics dependent on $\sigma_T$. Limiting yielding is calculated via $\psi_\kappa$, $\sigma_T$ and $\sigma_B$ at room temperature.

For dynamically loaded components of the SP, the values of $\beta_T$ decrease from 120 to 50 with $\sigma_T$ changing from 300 to 700 MPa, and at increased deformation velocities $\dot{e} = de/d\tau$ ($10^0$ c$^{-1} \le \dot{e} \le 10^3$ c$^{-1}$), there is increase of yield stress and ultimate stress limit defined experimentally or calculated with the help of polynomial equation

$$\sigma^{\dot{\text{e}}}_{\text{B}} = \sigma_{\text{B}}(\dot{\text{e}}/\dot{\text{e}}_0)^{m_{\text{eB}}}, \sigma^{\dot{\text{e}}}_{\text{T}} = \sigma_{\text{T}}(\dot{\text{e}}/\dot{\text{e}}_0)^{m_T} \tag{39}$$

Dynamic plasticity performance calculation is done via $\psi_\kappa$, $\sigma_T$ and $\sigma_B$, with the help of the same relations that are used for temperature effects description. Eqs. (37)–(39) provide possibility to calculate work-hardening index $m$ in Eq. (36).

The entire system of experimentally defined ($E$,$\sigma_T$, $\sigma_B$, $\psi_\kappa$) and designed ($m$, $S_\kappa$, $m_p$, $m_e$, $m_B$, $m_\psi$, $\beta_T$, $\beta_B$, $n_B^{\dot{e}}$, $n_T^{\dot{e}}$) characteristics is identified with regard to results of mechanical tests of smooth standard samples.

The real bearing SP components have various zones of concentration and various sizes of cross-sections. Performance of the mechanical tests for assessment of sensitivity to a factor of tension concentration (in elastic and inelastic areas) and size factor represents essential methodical difficulties and is time-consuming.

For big group of constructive metal materials due to use of the modified analytical decisions (of Neuber type), it is possible to receive correlation of tension concentration factor $K_\sigma$ and deformations $K_e$ in elastoplastic domain with theoretical concentration factor $\alpha_\sigma$ in elastic domain, taking into account the relative level of the effective stress $\sigma/\sigma_T$ and work-hardening index $m$

$$\{K_\sigma, K_e\} = F_k\{\alpha_\sigma, \sigma/\sigma_T, m\}. \tag{40}$$

For existing offshore structures $1 \le \alpha_\sigma \le 5$, $1 \le K_\sigma \le \alpha_\sigma$, $\alpha_\sigma < K_e \le \alpha_\sigma^2$.

For experimental evaluation of size facto impact (sizes $F$ of transverse cross-section) on mechanical properties of large-size SP components a set of polynomial equations is recommended:

$$\sigma_T^F = \sigma_T(F_0/F)^{m_{TF}}, \sigma_B^F = \sigma_B(F_0/F)^{m_{BF}}, \psi_k^F = \psi_k(F_0/F)^{m_{\psi F}}, \tag{41}$$

where $m_{TF}$, $m_{BF}$, $m_{\psi F}$ –characteristics not separate steels, but their groups (as per the stress level and doping level ($m_{TF} \approx m_{BF}$ = 0.013, $m_{\psi F}$ = 0.024–0.04).

For assessment of survivability characteristics based on crack resistance criteria in presence in the SP bearing structures of cracks like defects, the standard, unified and special tests with variation of cracks sizes $l$, cross-sections $F$ and loads technique $Q$ shall be conducted. The critical value of the stress intensity factor within the frameworks of the linear fracture mechanics is generally viewed as fundamental characteristic of crack resistance at cyclic loading

$$K_{Ic} = \sigma_c\sqrt{\pi l} \cdot F\{l, F, Q\}, \tag{42}$$

where $\sigma_c$ is the failure stress for the sample with limitation $\sigma_c \le (0.9–1.0)\sigma_T$.

At the same time, by numerous experiments, it was shown that at change of $l$, $F$, $Q$, and temperature $t$, time $\tau$, deformation velocity and stress voluminosity $I_\sigma$, $D_e$ the basic characteristic $K_{Ic}$ changes (in the same manner as change other basic design characteristics $\sigma_T$, $\sigma_B$, $\psi_\kappa$.).

As the first assumption in technical practice use is made of minimal values of $K_{Ic}$ depending on the temperature $t$ as this not always is counted as safety factor. The most acceptable in comprehensive assessment of strength, in-service life and safety of the SP components is the use of the minimum values defined on cylindrical samples with a circular crack with further calculation of $K_{Ic}$ value as per basic characteristics $\sigma_T$, $m$, $e_\kappa$ with regard to changes caused by variation of parameters $l$, $F$, $Q$, $t$, $\tau$, $I_\sigma$, $D_e$. In more general case when conditions of linear mechanics of destruction are not satisfied and there are considerable deformations of plasticity and creep, instead of the standard characteristics $K_{Ic}$ (or critical integral $J_c$ and critical cracks opening $\delta_c$), the deformation criterion of $K_{Iec}$ is developed and implemented, where $K_{Iec}$ is the critical factor of deformations intensity [5–7]. Factually this factor plays the same role as deformations concentration factor $K_e$ in (41) upon condition of similarity of $\alpha_\sigma$ and $K_I$. At the same time, the modified analytical solution with regard to (4.14) type gives dependence

$$\overline{K}_{Ie} = \overline{K}_I^{p_{ke}}, \tag{43}$$

where $\overline{K}_I = \frac{K_I}{\sigma_m}$; $\overline{\sigma}_n = \overline{\sigma}_n/\sigma_{\text{T}}$; $\sigma_m$ is the yield stress; $P_{ke} = \frac{2-n(1-m)(1-\overline{\sigma}_n)}{1+m}$ is the generalized parameter depending on work-hardening index $m$ and relative level of rated stresses; $m$ is the work-hardening index for deformation curve; and $n$ is the characteristic of structural material type $n \approx 0.5$.

The value of stress intensity factor in terms of operation at stress $\sigma_n$ with regard t0 (4.14) equals to

$$\overline{K}_I = \overline{\sigma}_n \sqrt{\pi l} \cdot F\{l, F, Q\}. \tag{44}$$

Expressions (41) and (44) make it possible to get conditions of local destruction—crack formation (41) and its development according to (43).

In presence of cracks and use of local criterion obtained is expression to plot the fracture diagram connecting increment of the crack length $\Delta l$ with the rated stress value and designed parameters of mechanical properties

$$\Delta l = \frac{1}{2\pi}\left(\frac{\overline{K}_{Ie}}{\overline{e}_f}\right). \tag{45}$$

where $\overline{e}_f = \frac{1}{e_{\text{T}}}(e_k)$.

If loading process is cyclic, the value $\Delta l$ is equivalent to crack increment in preplanned cycle $\Delta l = \frac{dl}{dN}$, and the main parameter of loading appears to be the peak-to-peak range of deformations intensity factor $\Delta K_{Ie}^{(N)}$ in this very loading cycle $N$ with a variable work-hardening index $m = m(N)$. The value $m = m(N)$ depends on cyclic properties of materials that can be as follows:

- cyclically stable—$m(N)$ does not change depending on number of half-cycles of $N$;
- cyclically hardening—$m(N)$ increases with growth of $N$; and
- cyclically softening—$m(N)$ decreases with growth of $N$. Then

$$\frac{dl}{dN} = \frac{1}{2\pi}\left(\frac{\Delta\overline{K}_{Ie}^{(k)}}{\overline{e}_f}\right)^2 = \frac{1}{2\pi\overline{e}_f^2}\left(\Delta\overline{K}_{Ie}^{(k)}\right)^2 = C_e\left(\Delta\overline{K}_I\right)^{m^e}. \tag{46}$$

Expression (46) with regard to expressions (43), (44) is similar to known Paris-Erdogan equation when $C$ and $m_k$ are material constants; however, in expression (46), the values $C$ and $m$ are variables and are calculated. Mechanical tests for identification of $\overline{K}_{Ie}, \overline{K}_{Iec}, dl/dN$ within the frames of nonlinear destruction mechanics are more comprehensive than those in linear destruction mechanics when identified are values of $K_{Ic}$ and $dl/dN$. In non-routine events, emergency and catastrophic situations in nonlinear setting of the problem analyzed are the following essential effects of redistribution of the local plastic deformations and creep deformations depending on $m$, $t$, $\tau$, $N$, $F$, $I_\sigma$, $D_e$ in case of probabilistic approach. Noted complexity is overcome within deformation destruction criteria at setting of the general problems of strength, in-service life, reliability, survivability, risks, safety and SP equipment protection.

Theoretical and practical solutions of the considered problems of strength, in-service life, reliability, crack resistance were already performed for such high-risk objects as nuclear reactors, hydraulic and thermal power stations, aircraft, main pipelines and unique engineering constructions.

The ground for the analysis and risk management directed to quantitative evaluation of critical and acceptable risks is based on the matrix of risks. Qualitative and quantitative risk assessment is based on the standard matrices of criticality determined by probabilities of adverse events occurrence (destructions, failures, etc.) and consequences of these events. However, within risk matrixes, the mechanisms of material and the bearing SP components degradation relevant to the erosion and corrosion processes are considered.

The listed above approaches, methods, criteria, design schemes and calculation dependences give the chance to carry out assessment of SP technical condition and risks monitoring.

## 3. Development of methods of calculations and justification of strength, in-service life and safety

### 3.1 Techniques of provision and enhancement of strength, in-service life and safety

Taking into account a possibility of reaching in time of the ultimate limit states in the wide range of loading parameters, further it is required to define the following groups of situations occurring during SP functioning as presented in **Table 2**.

Each class of situations corresponds to diminution of safety level of the analyzed objects while diminution of safety level can be estimated on expressions (1)–(9) as per values of risks $R_i^{э}(t)$ of objects operation on a specified time interval of operation. Quantitative values of risks $R_i^{э}(t)$ are calculated as product of the probability of occurrence of each of the specified situation $i$ - $P_i^{э}(t)$ by economic losses values as per analyzed situation $U_i^{э}(t)$. At the same time, the condition of safety provision takes the following form $n_R = R_c(t)/R_t^{э}(t)$ where $R_c(t)$ is critical (inadmissible, unacceptable risk), $R_t^{э}(t)$ is designed risk for the moment of operation $t$ for mode $i$ and $n_R$ is the safety margin as per risks.

According to **Table 2**, the last three abovementioned groups of the situations (T5, T4, T3) occurring during objects functioning can be referred to a kind of the risks which are monotonously increasing up to critical values. Such risks, mainly, are caused by the controlled processes of damages and degradations of physical-mechanical properties of material relevant to its aging. The first two groups (T2, T1) correspond to the occurrence of the most dangerous situations with extreme impact parameters (earthquakes, tsunami, acts of terrorism and military actions). These cases require use of the most difficult calculations, tests, modeling, diagnostics, monitoring and protection. In this case, classic methods of a material consumption justification, constructability and efficiency are insufficient. In such statement, the approaches presented in clauses 2.9–2.12 have to be implemented.

### 3.2 Risk-based inspections

In case of use of foreign and domestic safety standards for risk analysis, the approaches given in [1, 2, 10, 11] can be rather efficient:

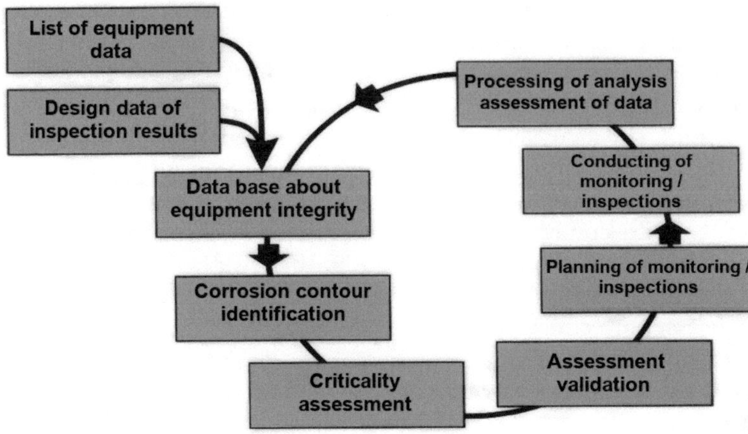


**Figure 18.**
*Basic diagram of implementation risk based inspections technique.*

- flow chart and fault-tree construction techniques (**Figure 11**);
- probabilistic modeling technique (**Figures** 7 and 8); and
- risk-based inspection (RBI) technique developed by Shell Global Solutions International company for residual life evaluation and planning of the objects' high-pressure equipment health monitoring frequency with consideration of risks-analysis (**Figure 18**). Inspections and tests planning is performed upon analysis of data about current technical condition of specified equipment item.

In the approach (**Figure 19**) presented above by analogy with **Figure 4**, the classes and categories of criticality, consequences of damages from accidents and accidents can be assessed in a similar way to **Figure 4**.

The risks analysis technique is based on information about scenarios of dangerous situations and probabilities of their occurrence received a priory. It is possible for SP for which design and operation experiences are accumulated already. In engineering design performed according to clauses 2.9–2.12, the inspections frequency can be obtained upon calculations as per expressions (18)–(41).

| | | RBI StF | IPF DR | RCM ETBF | Class of criticality of risks and reliability management | | | | |
|---|---|---|---|---|---|---|---|---|---|
| Probability class | H | High | 0-0.5 year | | No need in additional monitoring of technological process | Not applicable for fracture mechanism not related to ageing | | | |
| | M | Medium | 0.5-4 year | | | | | | |
| | L | Low | 4-20 year | | | L | M | MH | H |
| | In | Insign. | >20 year | | | | L | M | MH |
| Consequences category | | Economical (USD) | | | No / Minor loss <10 Thd. | Minor loss 10-100 Thd. | Local loss 0.1-1 mln | Severe damage 1-10 mln. | Widespread damage >10 mln. |
| | | For safety and health | | | No / minor injuries | Moderate injuries | Serious injuries | Isolated cases of injury followed by fatality | Multiple fatalities |
| | | For environment | | | No / minor consequences | Moderate consequences | Local consequences | Serious consequences | Severe consequences |
| Class of consequences | | | | | Minor | Low | Medium | High | Extreme |

**Figure 19.**
*Criticality and risks matrix.*

### 3.3 Monitoring and seismic protection of offshore platforms

One essentially important question in the problem of protection of objects of offshore and land infrastructures is provision of SP seismic stability; this can be achieved with the help of developed scientific bases of design of self-lubricating, and self-adjusting sliding supports with reverse motion used as seismic-insulators for bridges, industrial and civil constructions. These works are also used for oil and gas offshore platforms on the continental shelf of the Russian Federation on the Sakhalin Island [1, 2, 17, 18].

It was proposed offshore structures protection against earthquakes to use the friction pendulum bearings (FPB) as the seismic-insulators [1, 12–14]. A calculation method for the service life of a FPB and the method of assessment of friction coefficient were experimentally developed [17, 18].

The real possibility of pendulum sliding supports use as efficient mean for absorption of energy from external force appeared in the last 30–40 years thanks to development of new technologies (in particular in connection with development of space research works in the USSR and the USA) and to introduction of new tribotechnical materials (such as the antifriction self-lubricant weaved fibrous materials).

In the SP pendulum bearings used are pendulum characteristics, providing increase of the natural oscillations (vibrations) period of the isolated structure in a manner to avoid the maximal forces occurring at an earthquake. During an earthquake, the articulation slide block in the bearing moves (slides) along a stainless steel concave surface, forcing a support to move within small pendulum displacements. The schematic view of the bearing is presented in **Figure 20**. The plate with a spherical concave surface is mounted on the top as viewed from the deck; this is done to arrange convenient operation. At such location of a concave plate, the grease does not get on the slide face. The lower plate of the case is mounted on the jack structure.

If forces occurring during an earthquake do not exceed the level of friction forces, then the structure supported by the bearing corresponds to the standard structure lying on the jack and has its own oscillation (vibration) period without insulator. As soon as the level of friction forces is exceeded, the structure starts oscillate with designed period; at that the dynamic response and damping are defined by bearing properties.

The hemispherical design of the articulation slide block allows getting relatively uniform distribution of pressure under the slide block and this reduces the movement judder and prevents occurrence of high local pressure in the bearing.

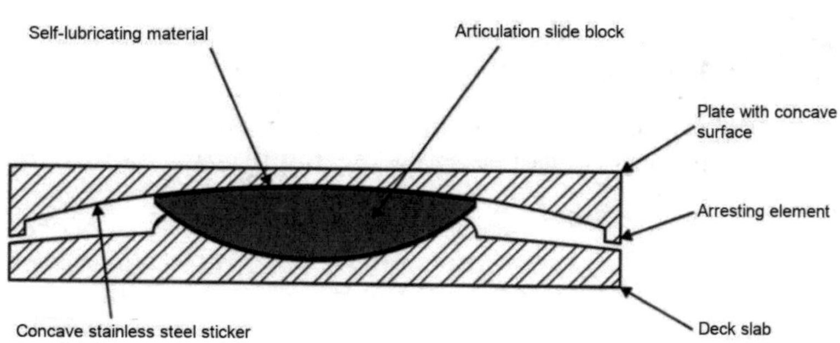


**Figure 20.**
*Bearing structure diagram.*

| Platform component | a | b |
|---|---|---|
| Deck | 0.24 | 0.73 |
| Deck, level (+)27 m | 0.31 | 0.65 |
| Deck, level (+)38 m | 0.25 | 0.74 |
| Deck, level (+)47 m | 0.31 | 0.84 |
| Flare unit | 2.00 | 4.37 |
| Drilling module | 0.61 | 1.22 |
| Crane on the North side | 0.82 | 1.74 |
| Crane on the South side | 1.46 | 2.27 |

**Table 4.**
*Average side accelerations ∂($m/sec^2$) of the oil and gas platform components when pendulum bearings are used (a) and without such bearings (b).*

As the displacements caused by an earthquake initially occur in bearings that are seismic-insulators, the side loadings and vibration motions transferred to a construction drop significantly.

In **Table 4**, the mean peak accelerations are presented, influencing, at designed earthquake, on the components of the oil and gas platform Lun-A for cases when friction pendulum bearings are in use and are not in use.

Accelerations drop is at 1.5–3 times that leads to significant reduction of wear of bearings and the antifriction self-lubricant film.

Development of oil and gas fields, as a rule, is carried out in the seismically active areas (their activity reaches magnitude 8–9 on 1–9 scale), and this is one of the main difficulties to be overcome in the process of such developments execution.

Sea platforms "Lun–A" and "PA-B" of the Sakhalin-II project are installed on the shelf of the Sakhalin Island in 2007. The weight of the gravity based structure is: for the"Lun–A" platform—103 thousand tons and for the "PA-B" platform—106 thousand tons. The weight of the topsides of "Lun–A" is 28 thousand tons and of "PA-B"—34 thousand tons. Service life of sea platforms "Lun–A" and "PA-B" is 30 years. Their design shall provide operation of equipment without damages and failures and resist loads occurred in the process of earthquake with probable repeatability once in 200 years and keep running without serious damages after impact upon such seldom earthquake that may occur once in 3000 years.

For the first time in world practice on "Sakhalin-II" project were installed frictional pendulum sliding supports (**Figure 21**) to provide seismic insulation between sea platform concrete gravity based structure and topside.

Such FPB previously were used for construction of highways, bridges and airports never before they were used in sea platforms.

Four bearings—seismic insulators installed in the catwalk of four concrete supports provide damping of extreme horizontal loadings due to isolation of the topside buildings from the most destructive pushes and due to reduction of loads on the topside buildings caused by impact of daily temperature changes, pressure of ice and waves.

### 3.4 Comprehensive on-line diagnostics, monitoring and the automated protection

Comprehensive on-line diagnostics and monitoring of sophisticated constructive components of SP equipment as per strength criteria, in-service life and crack resistance takes on greater and greater importance in the course of studies and works with regard to technogenic safety [1, 2, 4, 5, 7, 15, 16]. So far, the solution of these tasks is difficult because of absence of enough nomenclature and number of means for multi-parameter and multi-factor diagnostics of the damaged SP elements with taking into account scenarios of accidents. When looking for methods and diagnostic means and monitoring performance, it is necessary to apply the system concept providing umbrella approach for: the preliminary analysis of the stress-deformed states by analytical and numerical methods; identification of the most loaded and dangerous zones; nondestructive testing and diagnostics at all stages of equipment life cycle; and development of a system of diagnostic data collection and exchange between design offices, manufacturers and operators.

Only based on this understanding, it is possible to provide high system reliability, sufficient depth and validity of diagnosing.

### 3.5 New offshore subsea technology solution for shelf development

Along with expert evaluation of above-water and above-ice technologies, the feasibility studies and assessment of basic features of subsea systems, including issues of energy security, were carried out. This analysis is made by the community of the specialized sea organizations: RNTs "Kurchatov institute" and Institute of machine science named after A.A. Blagonravov RAS (Moscow) with participation of the National laboratory Sandia (USA).

As a solution acceptable from the economical and technical point of view of above task is related to the transition to the system of underwater and under-ice technology of exploration, production, treatment and transportation of hydrocarbons (oil and liquefied natural gas—LNG (**Figure 22**) that so far is not available. Higher price of such underwater and under-ice system is compensated by the reduction of the subsequent costs required to provide safety and physical protection. Estimates show that the possible losses caused by technogenic accidents of above-water natural threats and terrorist impacts on the objects of a underwater technologies complex is 10 times less, than from impact of similar risk factors for traditional above-water technologies. The appraisals done by the specialized organizations show the technical capability of Russia to develop for the Arctic shelf the underwater and under-ice atomic technologies (**Figure 22**).

Calculations done with taking into account information from clause 2 make it possible to obtain the risks values for both traditional (on-land and above-water sea) technologies and for new (underwater) technologies. The following risks'

a) b)

**Figure 21.**
*Use of frictional pendulum bearings (sliding supports) on sea oil platforms installed in the top part of four concrete jacks. a) PA-B Platform; b) Lun-A Platform.*

**Figure 22.**
*Scheme of under-water and under-ice technological complex: 1—ice coverage on the sea surface; 2—underwater LNG carrier or tanker; 3—subsea oil and LNG tank; 4—subsea natural gas liquefaction and oil unit; 5—Field processing unit; 6—subsea power generation unit; 7—subsea condensate storage facility; 8—subsea condensate carrier; 9—LNG terminal; and 10—subsea well.*

| Types of risks | $R$ | $R_*$ | $R(t)$ | $R_*(t)$ | $R_{US}(t)$ | $R_{US^*}(t)$ |
|---|---|---|---|---|---|---|
| 1. Risks for on-land infrastructures $R^H$ | 48.3 | 48.9 | 71.5 | 72.4 | 59.9 | 60.6 |
| 2. Risks of above-water sea transportation $R^M$ | 4.1 | 4.13 | 6.1 | 6.2 | 5.08 | 5.12 |
| 3. Risks of on-land and above water technologies $R^{HM}$ | 52.4 | 53.0 | 77.6 | 78.6 | 64.0 | 65.7 |
| 4. Risks of terrorist attacks on on-land infrastructures $R_T^H$ | 6.0 | 6.1 | 55.1 | 56.1 | 46.2 | 46.4 |
| 5. Risks of terrorist attacks in case of above-water sea transportation $R_T^M$ | 0.8 | 2.15 | 3.1 | 8.29 | 2.46 | 6.4 |
| 6. Risks of terrorist attacks on land infrastructures and on sea transportation $R_T^{HM}$ | 6.8 | 8.25 | 58.2 | 64.4 | 48.7 | 52.8 |

**Table 5.**
*Risks of the traditional LNG technologies (million dollars per year).*

characteristics are given in **Table 5**: $R$—risks estimates currently for on-land, above-water and under-water technologies without loss of LNG and oil at accidents and disasters; $R^*$—the same risks, but taking into account losses; $R(t)$—expected risks in 10 years at growth of LNG production, transportation and consumption rate in 4% without LNG loss; $R^*(t)$—the same risks taking into account LNG loss; $R_{US}(t)$—expected risks for LNG and oil transportation in the USA without LNG losses; and $R_{US^*}(t)$ —expected risks for the USA taking into account loss of LNG and oil.

The estimated cost efficiency of new underwater technologies (liquefaction and transportation) increases in comparison with the traditional (on-land and above-water) technologies. Risks of new technologies at an initial stage are $(8.6–10.1)\cdot 10^6$ of dollars/year; and for traditional ones $(59.2–61.3)\cdot 10^6$ dollars/year.

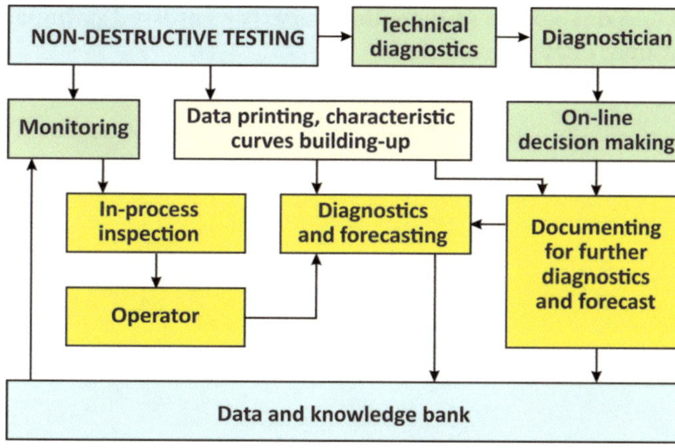


**Figure 23.**
*Diagnostics and monitoring procedure chart.*

These risks have to be considered at stage of economic assessment of all newly created technologies related to shelf developments.

The person (operator) or an automatic system, when conducting diagnostics and monitoring (**Figure 23**), closely follow change of parameters and use their abilities to identify and forecast the processes and the phenomena., most actively joins in control processes. The software provides comprehensive processing of the obtained information and active assistance to the operator by performing additional data processing and presenting upon operator's request necessary information recorded in the computing system memory.

A set of the principles, methods and means of defects finding and detection or, in another words, arranging of diagnostic assurance of crack resistance of equipment elements during production and in service, forms the basis for accidents prevention, actuation of the automated protection and safety enhancement.

Such approach was implemented during Sakhalin-II projects execution for protection of SP from earthquakes and tsunami.

## 4. Conclusion

Issues of development of the world and Russian continental shelf for exploration, drilling, production, treatment, storage and transportation of hydrocarbons become more and more important socioeconomically and in scientific and technical aspects. Unique sea platforms for a temperate and Arctic climate, undoubtedly, fall into group of critically and strategically important objects of infrastructures of life activity and life support. The fundamental studies and applied research works in the field of provision of strength, in-service life, survivability and risks play key role in comprehensive solution of issues relevant to the sea platforms safety, security and protection from accidents and disasters.

Their implementation is focused on scientific justification of classification of continental shelf technosphere objects, classification of routine and abnormal situations, development of methods and systems of diagnostic, monitoring and protection.

In the future-oriented technologies for safe continental shelf development, the results of advanced scientific theoretical and experimental developments in such industries as nuclear, airspace and transport will be used. The specified research works and development have both clearly expressed national and general international character.

## Author details

Gennady Yu. Shmal[1], Vladimir A. Nadein[2]*, Nikolay A. Makhutov[3], Pavel A. Truskov[4] and Viktor I. Osipov[5]

1 Academy of Mining Sciences, The Union of Oil and Gas Producers of Russia, Russia

2 Member of the Society of Petroleum Engineers, LLC, Oil and Gas Safety—Energodiagnostika, Russia

3 Engineering Sciences, RAS, Mechanical Engineering Research Institute of the Russian Academy of Sciences (IMASH RAN), The Pipeline Transport Institute (PTI, LLC), Russia

4 Engineering Sciences, The "Sakhalin Energy Investment Company Ltd.", Russia

5 Geological Faculty of the MSU, Department of Geo-Engineering and Protection of Geological Environment, RAS, Geological and Mineralogical Sciences, Russia

*Address all correspondence to: vladimir_nadein@ogsed.ru

## References

[1] Safety and Security of Russia. Basis of Safety of Continental Shelf Development. Moscow: MGOF Znanie; 2013. 768p

[2] Safety and Security of Russia. Justification of Strength and Safety of the Continental Shelf Objects. Moscow: MGOF Znanie; 2015. 668p

[3] Safety and Security of Russia. Fundamental and Applied Problems of Integrated Safety and Security. Moscow: MGOF Znanie; 2017. 992p

[4] Kostoqryzov A, editor. Probalistic Modelinq in System Engineering. London: IntechOpen; 2018. 278p

[5] Makhutov NA. Strength and Safety. Fundamental and Applied Studies. Novosibirsk: Nauka; 2008. 528p

[6] Safety and Security of Russia. Risks Analysis and Safety and Securities Issues. Parts I-IV. Bases of the Analysis and Safety Regulation. Moscow: MGOF Znanie; 2006. Part I—p. 639; Part II—p. 752; Part III—p. 800; Part IV—p. 857

[7] Makhutov NA. Safety and Risks. Systems Studies and Development. Novosibirsk: Nauka; 2017. 714p

[8] Strength Calculation Norms for Equipment and Piping of Nuclear Power Units (PNAE G-7-002-86). Gosatomnadzor of the USSR. Moscow: Energoatomizdat; 1989. 525p

[9] Alekseev Yu N, Afanasiev VP, Litonov OE, Mansurov MN, Panov VV, Truskov PA. Ice-Engineering Aspects of Oil and Gas Offshore Field Developments. St-Petersburg: Hydrometeoizdat; 2001. 356 p

[10] Risk Based Inspection Methodology. API Recommended practice 581. 3rd edition. April 2016. Addendum 1, April 2019

[11] Guidelines on Risk Based Inspection. OP 04-30260. Netherlands: SGS; 2009

[12] Lee DE. The base isolation of Koeberg nuclear power station 14 years after installation. In: Proceedings of the Post-SMiRT Conference Seminar on Isolation, Energy Dissipation and Control of Vibrations of Structures; Capri, Italy; 1993

[13] Clarke CSJ, Buchanan R, Efthimiou M, Shaw C. Structural platform solution for seismic arctic environments—Sakhalin II offshore facilities. In: 2005 Offshore Technology Conference – 17378; 2005. p. 21

[14] Drozdov YN, Pavlov VG, Puchkov VN. Friction and Wear in Extreme Conditions. Moscow: Mashinostroenie; 1986. 223p

[15] Drozdov YN, Nadein VA, Puchkov VN. Tribological characteristics of friction, pendulum type seismic-insulators. Friction and Wear. 2007;**28** (2):119-127 (in Russian)

[16] Makhutov NA, Gadenin MM. In: Kyuev VV, editor. Technical Diagnostics of Residual Life and Safety. Study Guides. Moscow: Spektr; 2011. 187p. (Diagnostics of safety)

[17] Nadein VA, Drozdov Yu. N, Puchkov VN, Puchkov MV. Characteristics of Mendulum Sliding Bearings—Seismic Insulators Vestnik Mashinostroeniya, 2. Moscow. 2007. p. 47–53 (in Russian)

[18] Safety and Security of Russia. Safety of Means for Storage and Transportation of Energy. Moscow: MGOF Znanie; 2019. 928p

Chapter 5

# Probabilistic Modeling, Estimation and Control for CALS Organization-Technical-Economic Systems

*Igor Sinitsyn and Anatoly Shalamov*


## Abstract

Theoretical propositions of new probabilistic methodology of analysis, modeling, estimation and control in stochastic organizational-technical-economic systems (OTES) based on stochastic CALS informational technologies are considered. Stochastic integrated logistic support (ILS) of OTES modeling life cycle (LC), stochastic optimal of current state estimation in stochastic media defined by internal and external noises (including specially organized OTES-NS (noise support) and stochastic OTES optimal control) according to social-technical-economic-support criteria in real time by informational-analytical tools (IAT) of global type are presented. OTES-CALS are nonlinear and continuous-discrete. So we use approximate methods of normal approximation of probabilistic densities both for modeling and estimation. Spectrum of possibilities may be broaden by solving problems of OTES-CALS integration for existing markets of finances, goods and services. Analytical modeling, analysis, parametric optimization and optimal stochastic processes regulation in limits of illustrate some technologies and IAT given plans.


**Keywords:** continuous acquisition logic support (CALS), estimation control, planning and management technologies, modeling and analysis technologies, organizational-technical-economic systems (OTES), stochastic systems (StS)

## 1. Introduction

Probabilistic foundations of one of the modern directions in the field of after sale product service—integrated logistic support (ILS) are systematically treated. Stochastic continuous acquisition logic support (CALS) is the basis of ILS functioning in the presence of noises and stochastic factors in organizational-technical-economic systems (OTES). While spelling this chapter we firstly tried to explain reader the new approaches for creation informational technologies (IT) of modeling, optimal data processing in stochastic systems (StS) for high-quality manufacturing products (MP). Secondly, we consider optimization problems for complex of enterprises being part of virtual enterprise (VE).[1] In broad sense VE

[1] Virtual enterprise is such an enterprise that consolidates geographically separated economical subjects and interact in the process mutual production using chiefly electronical communicational means.

presents geographically distributed OTES whose consolidated budget at fixed time period is divided between two basic structure types of VE. First structure is responsible for MP creation and production with given functional and exploitational-technical qualities.[2] Second structure is responsible for professional quality and staff life quality (professional skills, medical services, etc.). In this case OTES criteria for complex optimal OTES control[3] are defined by the social-technical-economic efficiency indicators. Such indicators depends on the resources costs at required quality of basic processes in both structures during life cycle (LC).

Besides standard ILS problems solving by such OTES systems modern IT provides deep OTES integration in general structures of local and global markets of finances, goods and services (FGS).

New approaches for OTES control are based on the probabilistic methodology for analytical modeling of stochastic processes coming from stochastic nature of internal and external noises. Special attention is paid to stochastic noises generated by injurious OTES-NS (noise supplier).

In the modern ILS models and strategies ERP (enterprise resource planning) and MRP2 (manufacturing resource planning) only statically deterministic mathematics is used for solving planning problems. Unlike the existing methodology the suggested stochastic methodology firstly takes into account stochastic optimal planning processes dynamics and secondly performs current operative control using modern methods of stochastic analysis, modeling and estimation (filtering, for casting, identification, etc.) and control methods and technologies [1–4]. It gives opportunity to raise the level and the quality of OTES control by means of informational-analytical tools (IAT). There tools are being global control VE net based on CALS principles and technologies. Stochastic imitational models and complex imitational models give the opportunity to estimate the accuracy of analytical models and solve problems of optimal data processing and control in high dimensional and fast OTES-CALS.

Stochastic CALS methodology was firstly developed in [5] for modeling and analysis. Let us consider the development the stochastic estimation and control problems. We hope that these approaches will be useful for probabilistic systems engineering [6].

## 2. CALS technologies and OTES

According to contemporary notions in broad sense ILS being the CALS basis represents the system of scientific, design-project, organization-technical, manufactural and informational-management technologies, means and fractional measures during LC of high-quality MP for obtaining maximal required available level of quality and minimal product technical exploitational costs.

Contemporary ILS standards [7–12] being CALS vanguard methodology not in the right measure answer necessary purposes. CALS standard, have as debatable achievements and the following essential shortcomings:

- informational-technical-economic models being not dynamical;
- integrated data base (DB) for analysis of logistic support (ALS) is super plus on one hand and on the other hand does not contain information necessary for

---

[2] Functional product quality means how technical MP answers the functional purpose, exploitation, maintenance and repair. Exploitational-technical quality of MP is defined as adjustment to control, support and restoration during MP lifetime.

[3] OTES control is being business process management (BPM).

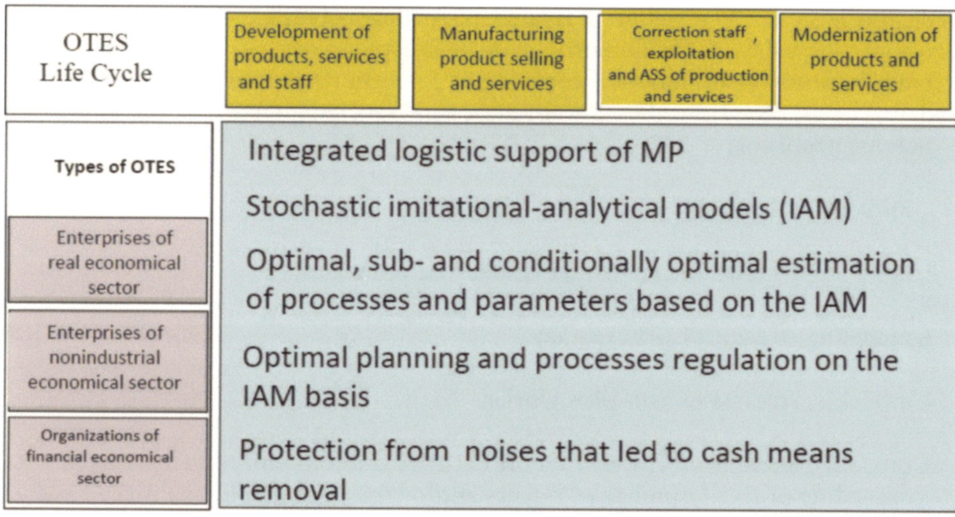


**Figure 1.**
*Integrated modeling, estimation and control technologies for cost life cycle control in after sale support (ASS) systems.*

complex through cost life cycle (CLC) estimation according to modern decision support algorithms;

- computational algorithms for various LC stages are simplified and do not permit forecasting with necessary accuracy and perform optimal control at conditions of internal and external noises and stochastic factors.

So ILS standards do not provide the whole realization of advantages for modern and perspective IT including staff structure in the field of stochastic modeling and control of two interconnected spheres: techno-sphere (technics and technologies) and social ones. These systems form the new the system class: OTES-CALS systems. Such systems destined for the production and realization of various services including staff structure, engineering and other categorical works providing exploitation, after sales MP support and repair, staff, medical, economical and financial support of all processes. New developed approach is based on the new stochastic modeling and control IT (**Figure 1**). These technologies are based on generalized social-technical-economic efficiency indicators for LC processes in comparison with usual CALS standards.

Research and control object in OTES are processes total LC of homogeneous sets of MP and resources. Special attention is paid to staff as object of professional training, improvement and medical service.

## 3. Probabilistic modeling and analysis

### 3.1 Basic elements of OTES stochastic modeling

According to [13, 14] we introduce composite elements (CE) as OTES with the following elements: (1) basic technical means (TM) and TM being part of serving equipment; (2) staff. For creation unique stochastic model of interoperable OTES processes it necessary to define the data set forecasting CLC indicators at given period of exploitation. This set of indicators includes: (1) coefficient of CE performance at planning for given period of exploitation; (2) level of professional and medical readiness; (3) investment readiness. Analogously the technical resources corresponding values of professional, staff health and investment resources are defined. Informational-analytical tools (IAT) perform probabilistic analytical modeling of OTES technical means for after sales servicing (ASS).

Solving problems:

1. initial data forming;

2. modeling technical means and exploitation processes;

3. modeling streams of plan works;

4. modeling streams of non-plan works;

5. modeling streams of TM written off CP after random fault when we have exceeding of fixed number of repairs and/or reaching given resource;

6. modeling spare parts (SP) accumulation processes on stores;

7. modeling of SP delivery processes from stores into exploitation system;

8. cost modeling of after sales CP supply at given period of TM exploitation according payments items;

9. forming total cost model of TM after sale processes for whole CP list at given period of exploitation and providing given level of TM park.

In addition it is possible to give analogous list of problems for modeling staff and medical services.

### 3.2 Structural schemes of state change streams

State graph of TM, graph of equipment infrastructure and state staff graph are the basis of stochastic OTES-CALS model. For example, let us consider (**Figure 2**) basic CP state graph. This graph is constructed in accordance with basic LC processes for TM and OTES infrastructure for each CP. Being CP in any states that corresponds the definite costs of various resources and the total production value. This production value must be the object of monitoring and statistical data processing for estimation of probability characteristics (means, probabilistic moments, distributions, etc.). For each CP being the part of aggregate with the help special technologies are sequentially aggregated for final product (FP).

Main modeling stage of usage and service processes consists in probabilistic forecast of main indicators final values: sum of production costs and technical readiness level for OTES-CALS technical means maintenance at given time period. Thus, graph (**Figure 2**) must be supplemented additional graph for calculating integrals cost values at this time period according to [14] recommendations.

Vertex of basic graph maps CP current LC states in two level operational capability:

1. CP is on the stock in amount $X_1(t)$;

2. CP service able exploiting in aggregate in amount of $X_2(t)$

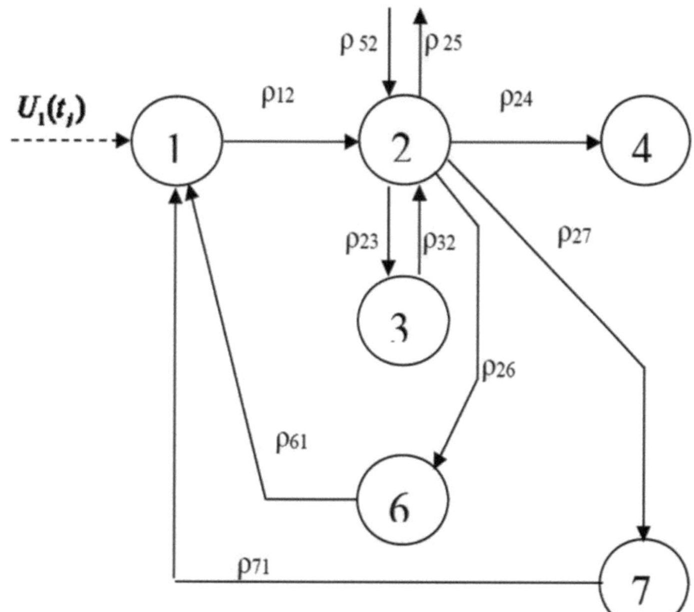


**Figure 2.**
*Basic graph of TM state, professional level and staff health state.*

3. CP is at user repair (first level) in amount $X_3(t)$;

4. CP is written off part and utilized in amount $X_4(t)$;

5. CP is at routine maintenance in amount $X_5(t)$;

6. CP is in factory repair (second level) at supplier in amount $X_6(t)$;

7. CP is in capital repairs at supplier in amount $X_7(t)$.

Graphs edges describe CP transition direction at states changes. Parameters $\mathrm{p_{ij}}(\mathrm{t})$ are transition intensities from one state to another. Values $\mathrm{U_1}(\mathrm{t_j})$ reflect the discrete filling process of the store by SP at time moments $t_j, j = 1, 2, \ldots$ for providing technical readiness of TM. In general case value and time moment of replenishment are random. These factors must be taken into consideration in LS for OTES-CALS model development. Basic graph of TM state, professional level of staff and health state are given on **Figure 2**.

The developed methodology is unique as for modeling TM and OTES-CALS dynamical staff potentials. Therefore, for constructing unique forecasting costs of staff potential the stochastic model it is necessary to apply developed integrated approach for description and modeling professional level personal health state [14, 15].

### 3.3 Stochastic processes and equations for OTES-CALS

Let us consider basic elements of stochastic OTES-CALS modeling and analysis using **Figure 3** for after sales maintenance support system.

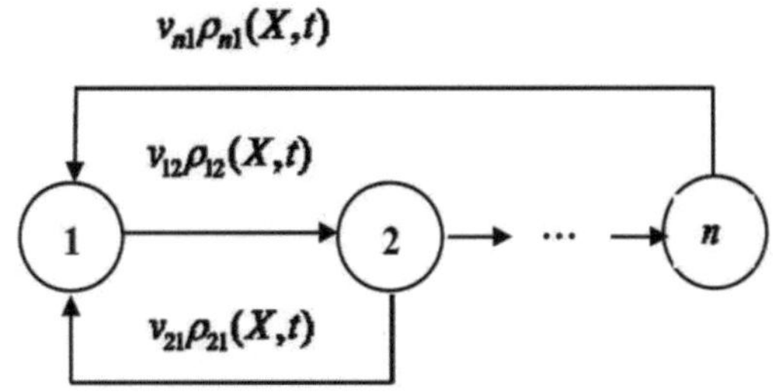


**Figure 3.**
*General state graph for after sales maintenance support system.*

Vertex of graph (**Figure 3**) corresponds 1, 2, ..., $n$ states, where of the same type resource be. Let current amount of resources be $X_1(t), \ldots, X_n(t)$, $X(t) = [X(t_1), \ldots, X(t_n)]^T$. Graphs edges corresponds transition of resources from state to state $h(k, h = 1, 2, \ldots, n,\ k \neq n)$ at random time moments forming Poisson streams transition events properly. Find some of resources states corresponding in queuing system for repeatedly recovery. General capacity is defined by the number of channels and being essentially nonlinear intensity[4] function depending on the amount of input resource units. This fact is mapped by $\rho_{kh}(X,t)$. In general case this nonlinear function has vector argument.

Stochastic equations and corresponding algorithms of analytical modeling for mathematical expectation $m = m(t)$, covariance matrix $\theta = \theta(t)$ and matrix of covariance functions $K(t_1, t_2)$ are as follows [5]:

$$dX = \varphi(X,t)dt + \int_{R_0^q} c(X,v,t)P(dv,dt) = \varphi(X,t)dt + \sum_{k,\,h=0}^{n} \int_{R_0^q} S_{kh}^T(v_{kh},X,t)P(dv_{kh},dt),\ X(t_0) = X_0, \tag{1}$$

$$\begin{aligned} &\dot{m} = \mathrm{M}[\varphi(X,t) + S\rho],\quad m(t_0) = m_0, \\ &\dot{\theta} = \mathrm{M}\{[\varphi(X,t) + S\rho]X^{0T} + X^0\left[\varphi^T(X,t) + \rho^T S^T\right] + S\,\mathrm{diag}(\rho)S^T\},\quad \theta(t_0) = \theta_0, \\ &\frac{\partial K(t_1,t_2)}{\partial t_2} = \mathrm{M}\{X_1^0\left[\varphi^T(X_2,t_2) + \rho^T(X_2,t_2)S^T\right], K(t_1,t_1) = \theta(t_1). \end{aligned} \tag{2}$$

Hence M is symbol of mathematical expectation; $P$ is symbol of probabilistic measure; $X$ and $X^0$ are noncentered and centered state vectors; $\varphi(X_t, t)$ is in general vector nonlinear function reflecting current value of OTES-CALS efficiency criterion; $S(v)$ structure matrix of Poisson streams of resources (production) with values $v$ according to state graph; $S_u(v) = \left[s_{uk1}(v_{uk}) \ldots s_{ukn_g}(v_{uk})\right]$ is $u$-row of matrix $S(v)$; $\rho$ is the intensity vector.

Forming the structure matrix $S = [S_{kh}]$ for OTES-CALS (**Figure 2**) is shown in **Table 1**.

*Column 1.* Transition direction in system graph. Total amount of rows is equal to amount transitions—$m$; amount of columns is equal to states amounts—$n$ (graph vertex); two servicing columns: first and $n+2$.

*Column 2—n + 1.* Values of intensities of transition—$\rho_{kh}$.

So columns from 2 till $n+1$ presents $m \times n$ matrix $S$ and column under number $(n+2)$ is $m$ dimensional vector with intensity vector $\rho$.

[4] Stream of random events intensity is being mean number of events per time unit.

| Transitions 1 | 2 | 3 | 4 | 5 | ... | $n$+1 | $n$+2 |
|---|---|---|---|---|---|---|---|
| k → h | $s_{kh1}$ | $s_{kh2}$ | $s_{kh3}$ | $s_{kh\eta}$ | ... | $s_{khn}$ | $\rho_{kh}$ |
| 1 → 2 | $-v_{12}$ | $v_{12}$ | 0 | 0 | ... | 0 | $\rho_{12}$ |
| 1 → 3 | $-v_{13}$ | 0 | $v_{13}$ | 0 | ... | 0 | $\rho_{13}$ |
| ... | ... | ... | ... | ... | ... | ... | ... |
| 1 → η | $-v_{1\eta}$ | 0 | 0 | $v_{1\eta}$ | ... | 0 | $\rho_{1\eta}$ |
| ... | ... | ... | ... | ... | ... | ... | ... |
| 2 → 1 | $v_{21}$ | $-v_{21}$ | 0 | 0 | ... | 0 | $\rho_{21}$ |
| 2 → 3 | 0 | $v_{23}$ | $-v_{23}$ | 0 | ... | 0 | $\rho_{23}$ |
| ... | ... | ... | ... | ... | ... | ... | ... |

**Table 1.**
*Forming the structure matrix S.*

Equation (2) are approximate and valid at conditions of normal (Gaussian) approximation method. For raising the accuracy of analytical modeling and analysis methods of probabilistic distributions (moments, semi-invariants, coefficients of orthogonal expansions of densities) [2, 3] may be used.

The developed ordinary differential equations with initial conditions may be used for basic risk problems of systems engineering [6].

### 3.4 Modeling and analysis of aircraft vehicles park life cycle

Following [16] let us consider informational-analytical tools (IAT) for aircraft vehicle park modeling and control by technical-economic efficiency criteria after sales maintenance products (ASMP).

IAT modular includes:

- normative data base (DB) of passported aggregates;
- operative DB;
- forecasting processes block for ASMP;
- optimization block of delivery programs (annual application);
- catalog of codified items supply.

#### *3.4.1 Normative DB (NDB)*

According to contract supplier creates NDB according to standards DEFSTAN 00-600, S1000D, S2000 M and specifications S1000D, S2000 M. For IAT acceleration there are designed emulated DB (DBE) in the form of additional tables. Information from NDB automatically comes into DBE. These data characterize:

- interrepair resource of final MP (FMP) and CP;
- mean duration of capital repair (CR) of FMP and CP;

- mean capacity of CR of FMP;
- tone between scheduled services of FMP and CP;
- mean duration of regulation works (RW) of FMP and CP;
- mean capacity of RW for FMP;
- fix resource (lifetime) of FMP and CP;
- fixed resource till first and next repair of FMP;
- CP price;
- MP mean time to first failure;
- mean capacity of MP operative repair (OR);
- mean duration of OR;
- mean capacity of overhaul repair (OHR);
- mean duration of OHR;
- sign of availability;
- CP maximum number of repairs;
- sign of repairing indication after fixed resource.

*3.4.2 Data base monitoring*

According to contract software tools as operative DB (ODB) is filling in accordance with corresponding instruction of annual planning.

For current application the following information is formed:

- planning time period;
- actual amount of FMP;
- amount of no repaired of FMP;
- amount of repaired FMP;
- amount of FMP being in capital repair (CR);
- required coefficient of good condition;
- CP mean annual planning lifetime;
- CP mean time to first failure;
- CP frequency of overhaul (OHR);

- mean capacity of regulation works (RW);
- mean cost of the first technical service;
- CP surplus in store.

### *3.4.3 Forecasting block of ASS processes*

Corresponding software tools are described by mathematical model of park state changes amount of FMP and CP during fixed period of exploitation, material-technical technical service and types of repairs.

Output forecasting characteristics are the following:

- coefficient of technical readiness along the park for each CP (0...1);
- delivery cost of SP along CP parks;
- coefficient of technical readiness for FMP park;
- delivery cost of SP along FMP park;
- cost of ASS for LC cost estimation.

### *3.4.4 Optimization block for annual application*

This software tool is destined for automatic work with delivery list SP for forming SP optimal program for any fixed time period of exploitation (including annual application) providing given level of technical readiness of FMP park at conditions of minimal purchasing price.

Structure of initial parameters includes:

- calculating reliability or exploitation for each CP;
- repairing system for each component (period, duration of works, etc.);
- service prices (repair and delivery);
- annual budget for CP repair and SP delivery;
- level of price escalation for planning time period;
- regular amount of FMP;
- CP real mean lifetime;
- amount of CP and FMP being in repair;
- planning time period (from 1 year to lifetime);
- planning annual lifetime FMP and CP;
- required level of technical readiness coefficient.

Among the cost parameters there are specified the following ones: annual budget, price of one CP, price of one CP repair. For each characteristic, annual escalation price coefficient is fixed.

Replenishment and repair programs are optimal if SP superplus on store are minimal. It is equivalent to cost supply minimum.

The received in computer experiments optimal programs are calculated for grant support of fixed level of readiness during exploitation time period at conditions of minimal sufficient budget which is determined in optimization process. In the case of the optimization delivery program at conditions of wittingly restricted budget the developed programs provide readiness level. This level is being maximally close to given value.

Let us demonstrate designed "Optimization SASS, Version 2.0" IAT for prognosis and optimization in various regimes. **Figures 4–6** corresponds search regime SP delivery program and repair capacity in years of planning time period for support fixed level of repair characteristic—0,75 (**Figures 4** and **5**) at minimally sufficient budget. As the result of optimal search program (**Table 2**) the following programs were obtained:

- for SP delivery (0, 0, 1, 3, 4, 5, 6, 6, 6) (in things);
- for CP repair capacity (21, 23, 23, 20, 20, 20, 20, 21, 22) (in years$^{-1}$).

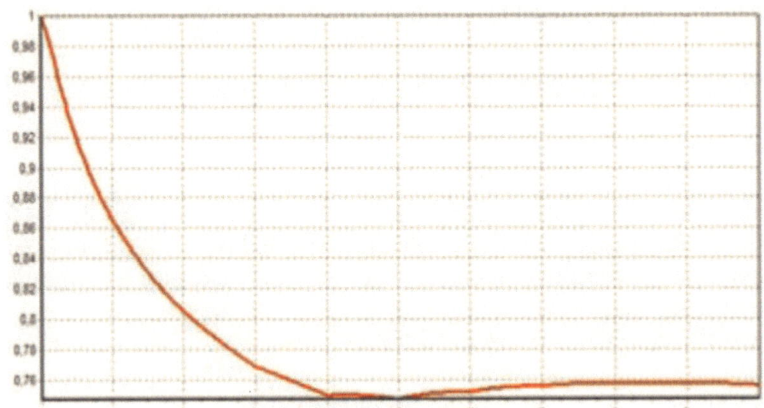

**Figure 4.**
*Mean coefficient of serviceability (minimal adequate budget).*

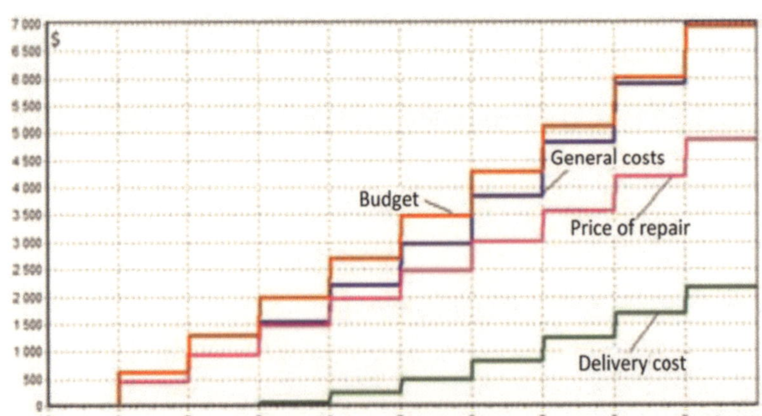


**Figure 5.**
*Dynamics of costs and appropriations (minimal adequate budget).*

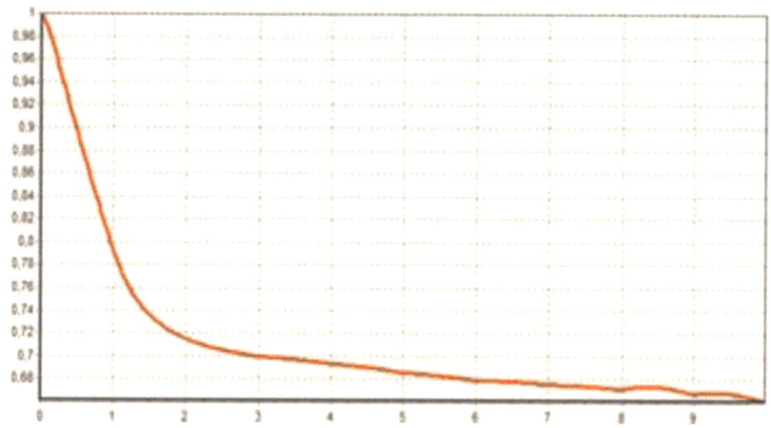

**Figure 6.**
*Mean coefficient of serviceability (wittingly restricted budget).*

| Year | 1 | 2 | 3 | 4 | 5 | 6 | 7 | 8 | 9 | The sum |
|---|---|---|---|---|---|---|---|---|---|---|
| Deliveries | 0 | 0 | 1 | 3 | 4 | 5 | 6 | 6 | 6 | |
| Price of CP | 52,5 | 55,125 | 57,8813 | 60.7753 | 63,8141 | 67,0048 | 70,355 | 73,8728 | 77,5664 | |
| Delivery cost | 0 | 0 | 67,8813 | 182,3259 | 255,2563 | 325,0239 | 422,1301 | 443,2386 | 465,3985 | 2211,25263 |
| Repair capacity | 21 | 23 | 23 | 20 | 20 | 20 | 20 | 21 | 22 | |
| Price of repair | 441 | 507,15 | 532,5075 | 486,2025 | 510,5126 | 536,0383 | 562,8402 | 620.5313 | 682,5844 | 4929,36675 |
| General costs | 441 | 507,15 | 590,3888 | 668,5284 | 765,7689 | 871,0622 | 984,9703 | 1063,7679 | 1147,9829 | 7140,6194 |
| Budget | 630 | 661,5 | 694,575 | 729,3038 | 765,7689 | 804,0574 | 844,2603 | 886,4733 | 930,7969 | 6946,7355 |

**Table 2.**
*Optimization results (restricted budget wittingly).*

General costs for delivery and repair are equal to $7150 at total budget $6947. Dynamics of costs and appropriations are given on **Figure 5**. Evidently, maximal costs are being repair costs. The repair costs are economically sound at given price ratio. Repair coefficient full during 4 years of exploitation is explained by the fact that the maintenance of repair coefficient is unprofitable during period when its value exceeds given level. It is obvious that CP repair begins with outset of exploitation period but SP purchasing after 4 years of exploitation.

**Figures 6** and 7 corresponds programs of search of delivery and repair at conditions of financial restrictions. In this case general delivery and costs do not exceed annual budget and also whole planning time period. But it is impossible to deduct repair coefficient at 0,75 level (**Figure 6**). Optimal programs at the level of repair coefficient equal to 0,66 for the end of time period are given on **Table 3**. In this case budget restrictions are valid. Therefore, we have the following programs:

- for SP delivery (0, 0, 1, 1, 2, 3, 2, 9, 14) (in things);
- for CP repair capacity (21, 23, 23, 19, 18, 18, 17, 2, 0) ($\text{years}^{-1}$).

General costs are approximately equal 5860 $ at total budget 5788 $. Optimal dynamics of delivery and repair costs are given on **Figure 7**.

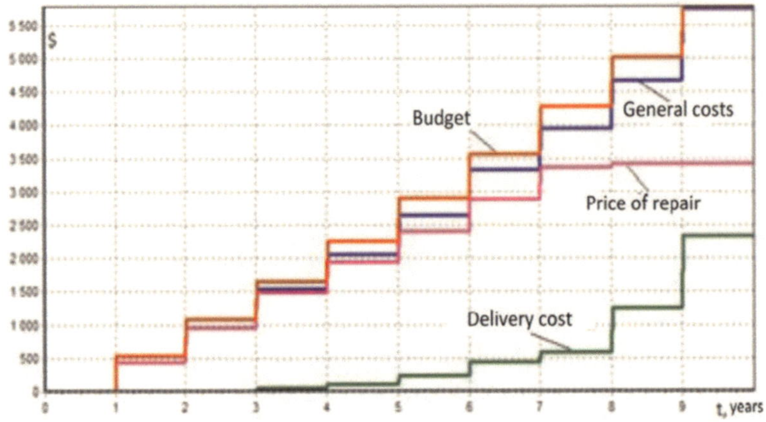


**Figure 7.**
*Program of delivery and repair (restricted budget wittingly).*

| Year | 1 | 2 | 3 | 4 | 5 | 6 | 7 | 8 | 9 | The sum |
|---|---|---|---|---|---|---|---|---|---|---|
| Deliveries | 0 | 0 | 1 | 1 | 2 | 3 | 2 | 9 | 14 | |
| Price of CP | 52,5 | 55,125 | 57,8813 | 60.7753 | 63,8141 | 67,0048 | 70,355 | 73,8728 | 77,5664 | |
| Delivery cost | 0 | 0 | 57,8813 | 60.7753 | 127,6282 | 201,0143 | 140,71 | 664.8549 | 1085,9298 | 2388,79381 |
| Repair capacity | 21 | 23 | 23 | 19 | 18 | 18 | 17 | 2 | 0 | |
| Price of repair | 441 | 507,15 | 532,5075 | 461,8924 | 459,4614 | 482,4344 | 478,4141 | 59,0982 | 0 | 3471,95803 |
| General costs | 441 | 507,15 | 590,3888 | 522,6677 | 587,0895 | 683,4488 | 619,1242 | 723,9532 | 1085,9298 | 5860,7518 |
| Budget | 525 | 551,25 | 578,8125 | 607,7531 | 638,1408 | 670,0478 | 703,5502 | 738,7277 | 775,6641 | 5788,9463 |

**Table 3.**
*Optimization results (wittingly restricted budget).*

## 4. Probabilistic optimal estimation and control

### 4.1 Optimal estimation

Nowadays such IT as filtering, extrapolation, identification, etc., are widely used in technical applications of complex systems functioning in stochastic media. These IT are based on statistical data analysis, modeling and estimation and gives only statistical estimates [2–4]. For OTES-CALS we have latent specially supported noises. Following [17–19, 22] let consider optimal filtering IT for special class of OTES using example 1.

*Example 1.* Let us consider typical OTES as system of after sales servicing (SASS). This system provides maintenance of technical readiness given level for MP park at quality conditions. Corresponding costs are fixed in bookkeeper documentation. Let us name it by "real" bookkeeper. In case of some types of noises it is possible to speak about "additional" bookkeeper. This bookkeeper is generated by fictional SASS for reserved taking out of the surplus of finances. On **Figure 8** the corresponding scheme is given where:

- SASS graph is given in upper part of **Figure 9**, number 1, 2, 3 being the following states: on store, in exploitation, in repair;
- "additional" graph of SASS generating noises $\zeta_i$ being fictitious analogs of $X_i(t)$ is given by dashed line in lower part of **Figure 8**;

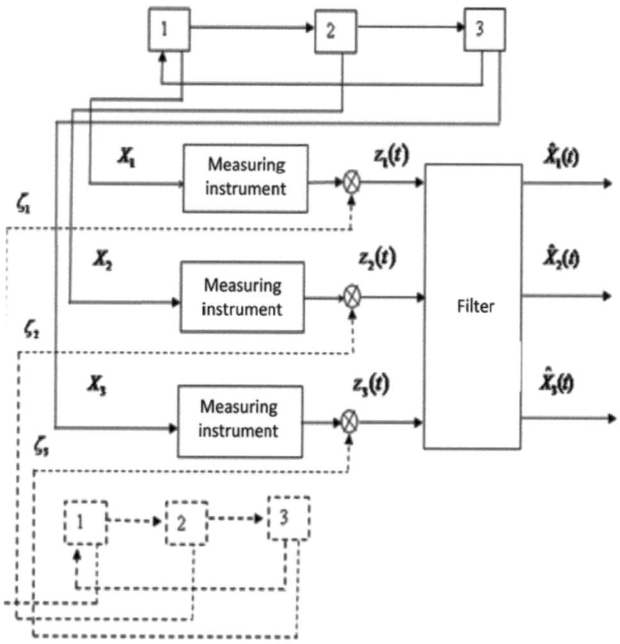


**Figure 8.**
*Structure of stochastic SASS.*

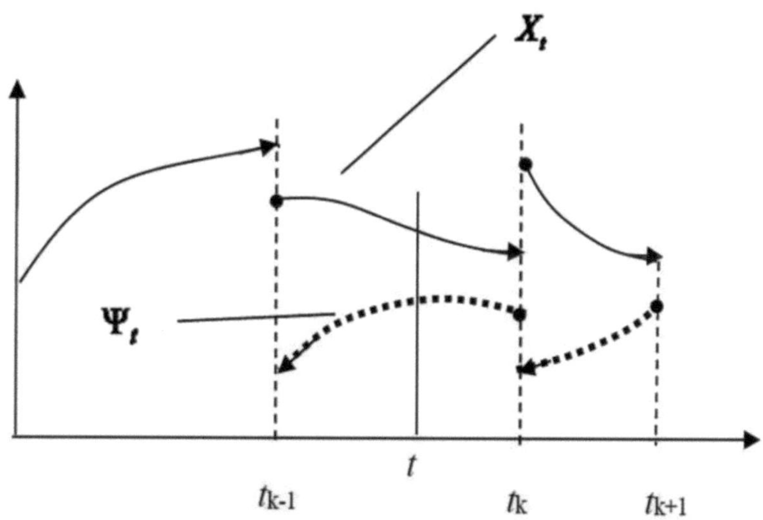


**Figure 9.**
*Continuous discrete self-conjugated processes $X_t$ and $\Psi_t$.*

- in the middle part of **Figure 8** measuring observation devices are given; $z_i(t)$ measure parameters and processes on background noise $\zeta_i$; $\hat{X}_i(t)$ being optimal estimates of real processes $X_i(t)$.

As it is known from [2–4] estimation technologies are based on: (1) model of OTES; (2) model OTES-OS (observation system); (3) model OTES-NS (noise support); (4) criteria and estimation methods; (5) filters (estimators) models.

It is required to develop it for the useful processes $X_i(t)$ filtration from its mixture with processes $\zeta_i(t)$ of system noise.

For solving this problem the linear Kalman filter is used [3, 4]. So we get the following result. Let complex stochastic models OTES, OTES-OS and OTES-NS are described by the following linear differential Equations:

$$\dot{X}_t = \overline{a}X_t + a_1G_t + a_2\zeta_t + a_0 + \chi_x V_\Omega, \tag{3}$$

$$\dot{G}_t = q(D_t)X_t + b_2\zeta_t + \chi_g V_\Omega \tag{4}$$

$$\dot{D}_t = bX_t + \overline{b}_1 D_t + b_0 + \chi_d V_\Omega \tag{5}$$

$$\dot{\zeta}_t = \overline{c}_2\zeta_t + c_0 + \chi_\zeta V_\Omega \tag{6}$$

Hence $X_t, G_t, D_t, \zeta_t$ are OTES, OTES-OS and OTES-NS; $q(D_t)$ is amplification factor of measurement block depending on number of resources in OTES-OS; $V_\Omega(t) = \left[V_x^T(t)\ V_g^T(t)\ V_\zeta^T(t)\ V_u^T(t)\right]^T$ is composite noise vector of white noises; $\chi_x, \chi_g, \chi_d, \chi_\zeta$ are matrices of corresponding dimensions. Then equation for optimal linear Kalman filter at $q(D_t) = q_t$ will be

$$\dot{\hat{X}}_t = \overline{a}\hat{X}_t + a_1G_t + a_2\zeta_t + a_0 + R_t q_t^T {\upsilon_g}^{-1}\left[Z_t - \left(q_t\hat{X}_t + b_2\zeta_t\right)\right] \tag{7}$$

where $\upsilon_g$ is the matrix of white noise intensities of internal noises OTES-OS and external noises from OTES-NS; $\zeta_t$ is the noise in the form Poisson process in OTES-NS; $R_t$ is the solution of the following Riccati Equation:

$$\dot{R}_t = \overline{a}R_t + R_t\overline{a}^T + \upsilon_x - R_t q_t^T {\upsilon_g}^{-1} q_t R_t \tag{8}$$

where $\upsilon_x$ is vector internal OTES noises.

### 4.2 Optimal control

Modern OTES are class of large scale systems of microeconomics of special type corresponding to five technological structures. These systems satisfy modern standards but need further effective control systems based on stochastic system theory during the whole LC of OTES elements.

Effective control of OTES needs optimal technologies for solution of the following problems:

- planning stocks of various recourses on the basis of forecasting external and internal demand;
- planning necessary manufacturing capacity and product delivery system in accordance with expected demand;
- distribution of finished products;
- planning loading capacities for repair and service with long work time, etc.

In general, case functional organization and control for the definite processes in OTES we need to solve the following problems:

- design mathematical models of various variants of logistical systems;
- work out methods of optimal complex planning of such processes as delivery, production, marketing and transporting of finished products, etc.

At modern design practice logistical systems and in whole OTES of given destination and for functional control known standards MRP/ERP, DEFSTAN 00-600, MIL STD 1388 [7–12] are used. These standards are typical for relatively stable economics. Such approaches are rested upon deterministic consumer demands models and corresponding local optimization methods implemented only in isolated nodes of logistic and production chains but global OTES optimization. For this reason MRP/ERP approaches cannot permit optimal complex planning LC processes at given period of time with calculating boundaries of proper values. This problem is very important for customer corresponding to profit finances. Deterministic approaches are useful only at conditions of economics with stable state of markets. At stochastic conditions, it is necessary to use corresponding approaches and analysis and synthesis OTES informational technologies based on stochastic models and methods that permit to calculate optimal control strategy.

Control of OTES being the integrated net of enterprises includes: (1) processes planning in accordance with goals and criteria; (2) effective operational (situational) control. The first concerns program control, the second—state regulation. Program control principles are as following: orientation finite goal expressed by goal graph and corresponding efficiency indicators, working restrictions, though planning and order-continuous principle [20, 21].

Basic stages of program control are the following: (1) optimization of resources distribution between goals and frequency of financing; (2) priorities of program separate goals and achievement means; (3) adaptation of program goal to changing external conditions. For one type of production, program control includes planning and adaptive distribution general budget between OTES participants (net nodes) based on stochastic estimation of processes and parameters and adaptation to external processes (noises).

For the stochastic OTES control actions are as follows:

- streams intensity transition between nodes;
- parameters of probabilistic distributions values of resources parties or products which involve the intransitions;
- frequency and size of discrete supply at replenishment (works volume) at calendar services;
- mean capacity service personal.

Analysis of these control categories shows that first part of control functions is continuous. The second part is discrete time functions (supply plan, plan of resources service and products at calendar plan, etc.). So it is necessary to consider OTES as mixed continuous-discrete stochastic system (StS) and apply probabilistic stochastic methods and IT of analysis, modeling, estimation and control.

Let us consider the basis of OTES filtering.

### *4.2.1 Problem statement*

At first let us consider deterministic multidimensional dynamical system described by the following nonlinear differential and difference equations [1, 5]:

$$\dot{X}_t = \xi(X_t, \pi_t, t), \tag{9}$$

$$X_k = B(X_{k-0}, U_k), X_k = X(t_k) \tag{10}$$

Here $X_t$ is $n_x \times 1$ phase vector; $X_{k-0}$ is the value of $X_t$ precede $t_k$; $\pi_t$ is the control vector in continuous time; $\xi, B$ are continuously differentiable $n_x \times 1$ vector functions; $t_0$ and $t_f$ is initial and terminal time moments; $U_k$ is $n_u \times 1$ dimensional control vector at time moments $t_0 < t_1, t_2, \ldots, t_{N_k} \le t \le t_f$; $X_0$ is initial condition. We choose the following functional $J$ which includes discrete and continuous components representing expenditure functioning and control:

$$J = \sum_{k=1}^{N_k} \Gamma(X_{k-0}, U_k) + \int_{t_0}^{t_f} K(X_t, \pi_t, t)dt + H(X_f) \tag{11}$$

where $\Gamma, K, H$ are known scalar differentiable functions. It is required to define optimal functions of continuous and discrete control $\pi_t$ and $U_k$ jointly supply minimum for functional $J$: $J^* = \underbrace{\min}_{\pi_t,\, U_k} \left\{ \sum_{k=1}^{N_k} \Gamma(X_{k-0}, U_k) + \int_{t_0}^{t_f} K(X_t, \pi_t, t)dt + H(X_f) \right\}$.
Hence for optimal control functions we have: $(\pi_t^*, U_k^*) = \arg\underbrace{\min}_{\pi_t,\, U_k} J$.

### 4.2.2 General solution

Let us find general solution by variational method [1]. For this purpose we compose mixed (from continuous and discrete functions) Lagrange functional (Lagrangian):

$$L = \int_{t_0}^{t_f} \left[ K(X_t, \pi_t, t) + \left(\Psi_t^T \cdot X_t\right)' \right] dt + \sum_{k=1}^{N_k} \Gamma(X_{k-0}, U_k) + \mathrm{H}(X_f). \tag{12}$$

where variable $\Psi_t$ is vector indefinite Lagrange multiplier. Vectors variables $\Psi_t$ and $X_t$ have discontinuity of the first kind at $t = t_k$. At these times $X_t$ is continuous on the right and being continuous on the left from the theory two-point boundary-value problem. We get $\Psi_t$ (**Figure 9**) by integration of corresponding equations from $t_f$ to $t_0$. So taking into account (9), (10) and considering integrand as generalized function of the following form:

$$\begin{aligned} \frac{d}{dt}\left(\Psi_t^T X_t\right) &= \dot{\Psi}_t^T X_t + \Psi_t^T \dot{X}_t + \left[\left(\Psi_{k+0}^T X_k - \Psi_k^T X_{k-0}\right)\right] \delta(t - t_k) = \\ &= \dot{\Psi}_t^T X_t + \Psi_t^T \xi(X_t, \pi_t, t) + \left\{\Psi_{k+0}^T [B(X_{k-0}, U_k)] - \Psi_k^T X_{k-0}\right) \delta(t - t_k). \end{aligned} \tag{13}$$

After substitution (13) into (12) and using $\delta$-function property we get new expression for functional $L$:

$$\begin{aligned} L = \int_{t_0}^{t_f} \Xi_t dt + \sum_{k=1}^{N_k} \Theta_k + \mathrm{H}(X_f) &= \int_{t_0}^{t_f} \left[ K(X_t, \pi_t, t) + \dot{\Psi}_t^T X_t + \Psi_t^T \xi(X_t, \pi_t, t) \right] dt + \\ &+ \sum_{k=1}^{N_k} \left\{ \Gamma(X_{k-0}, U_k) + \Psi_{k+0}^T [B(X_{k-0}, U_k)] - \Psi_k^T X_{k-0} \right) \right\} + H(X_f), \end{aligned} \tag{14}$$

where

$$\Xi_t = K(X_t, \pi_t, t) + \dot{\Psi}_t^T X_t + \Psi_t^T \xi(X_t, \pi_t, t) \tag{15}$$

$$\Theta_k = \Gamma(X_{k-0}, U_k) + \Psi_{k+0}^T [B(X_{k-0}, U_k)] - \Psi_k^T X_{k-0} \tag{16}$$

From Lagrangian stationary conditions we have the following algorithms $\Psi_t$, $\Psi_{k+0}$ from and $\pi_t^*$, $U_k^*$:

$$\frac{\partial L}{\partial X_t} = \frac{\partial \Xi_t}{\partial X_t} = \frac{\partial K(X_t, \pi_t, t)}{\partial X_t} + \frac{\partial \xi^T(X_t, \pi_t, t)}{\partial X_t} \Psi_t + \dot{\Psi}_t = 0 \tag{17}$$

$$\frac{\partial L}{\partial \pi_t} = \frac{\partial \Xi_t}{\partial \pi_t} = \frac{\partial K(X_t, \pi_t, t)}{\partial \pi_t} + \frac{\partial \xi^T(X_t, \pi_t, t)}{\partial \pi_t} \Psi_t = 0 \tag{18}$$

$$\frac{\partial L}{\partial U_k} = \frac{\partial \Theta_k}{\partial U_k} = \frac{\partial \Gamma(X_{k-0}, U_k)}{\partial U_k} + \frac{\partial [B^T(X_{k-0}, U_k)]}{\partial U_k} \Psi_{k+0} = 0 \tag{19}$$

$$\frac{\partial L}{\partial X_{k-0}} = \frac{\partial \Theta_k}{\partial X_{k-0}} = \frac{\partial \Gamma(X_{k-0}, U_k)}{\partial X_{k-0}} + \frac{\partial [B^T(X_{k-0}, U_k)]}{\partial X_{k-0}} \Psi_{k+0} - \Psi_k = 0 \tag{20}$$

$$\Psi(t_f) = \frac{\partial H}{\partial X_f} \tag{21}$$

Relations (17)–(21) are necessary optimal control conditions in given continuous-discrete problem. Thus two-point boundary value is described by closed set of Eqs. (9)–(14) and Eqs. (17)–(21). So from (17), (20) we get equations for $\Psi_t$, $\Psi_{k+0}$ conjugated with $X_t$, $X_{k-0}$. From (18), (19) we get implicit forms for $\pi_t^*$ and $U_k^*$.

#### *4.2.3 Solution of linear-quadratic problem*

Consider linear continuously-discrete system [1, 5]:

$$\dot{X}_t = \xi_x X_t + \xi_\pi \pi_t, \tag{22}$$

$$X_k = B_x X_{k-0} + B_u U_k, \tag{23}$$

where $\xi_t$, $\xi_\pi$ are matrix coefficients of, $n_x \times n_x$ and $n_x \times n_\pi$ dimensions changing at $t = t_k$; $B_x$, $B_u$ are matrices coefficients of $n_x \times n_x$ and $n_x \times n_u$ dimensions.
It is given quadratic efficiency criterion:

$$L = \int_{t_0}^{t_f} \Xi_t dt + \sum_{k=1}^{N_k} \Theta_k + \frac{1}{2} X_{t_f}^T H X_{t_f} = \int_{t_0}^{t_f} \left[ \frac{1}{2} \left( X_t^T K_t X_t + \pi_t^T K_\pi \pi_t \right) + \dot{\Psi}_t^T X_t + \Psi_t^T (\xi_x X_t + \xi_\pi \pi_t) \right] dt +$$
$$+ \sum_{k=1}^{N_k} \left[ \frac{1}{2} \left( X_{k-0}^T \Gamma_x X_{k-0} + U_k^T \Gamma_u U_k \right) + \Psi_{k+0}^T (B_x X_{k-0} + B_u U_k) - \Psi_k^T X_{k-0} \right] + \frac{1}{2} X_{t_f}^T H X_{t_f}, \tag{24}$$

where

$$\Xi_t = \frac{1}{2} \left( X_t^T K_x X_t + \pi_t^T K_\pi \pi_t \right) + \dot{\Psi}_t^T X_t + \Psi_t^T (\xi_x X_t + \xi_\pi \pi_t) \tag{25}$$

$$\Theta_k = \frac{1}{2}\left[X_{k-0}^T \Gamma_x X_{k-0} + U_k^T \Gamma_u U_k\right) + \Psi_{k+0}^T (B_x X_{k-0} + B_u U_k) - \Psi_k^T X_{k-0}\right] - \Psi_k^T X_{k-0}, \tag{26}$$

$K_x$, $\Gamma_x$, $H_{t_f}$ and $K_\pi$, $\Gamma_u$ being positive semidefinite and positive defined matrices of corresponding dimension. It is required to find optimal algorithm control for linear system described by Eqs. (22)–(24). Algorithm of optimal design based state feedback control gives the followings equations:

$$\dot{\Psi}_t = -\xi_x^T \Psi_t - K_x X_t \tag{27}$$

$$\Psi_{t_f} = H X_{t_f} \tag{28}$$

$$\pi_t = -K_\pi^{-1} \xi_\pi^T \Psi_t \tag{29}$$

$$\Psi_k = B_x^T \Psi_{k+0} + \Gamma_x X_{k-0} \tag{30}$$

$$U_k = -\Gamma_u^{-1} B_u^T \Psi_{k+0} \tag{31}$$

Algorithm includes: (1) integration in inverse time with initial condition (28) of vector differential Eq. (27) and difference Eq. (30) with data storage in each step; (2) formulae (29), (31) for calculating controls with usage of stored $\Psi_t$, $\Psi_{k+0}$. Note that during Eq. (27) integration at time moments $t = t_k$, $k = N_f, N_{f-1}, \ldots, 1$ step-wise changes $\Psi_t$ occur according to Eq. (30).

For reducing two-point boundary problem to ordinary we apply known approach and perform linear change of variables in Eqs. (22)–(31):

$$\Psi_t = \tilde{R}_t X_t \tag{32}$$

$$\Psi_k = \tilde{R}_k X_{k-0} \tag{33}$$

$$\Psi_{k+0} = \tilde{R}_{k+0} X_k \tag{34}$$

where $\tilde{R}_t$, $\tilde{R}_{k+0}$ are values of coefficients matrices. These variables are the solutions of continuous and discrete Riccati. These equations are integrated in inverse time. So we get optimal solutions in interconnected continuous and discrete parts of OTES-CALS in the following forms:

$$\pi_t^* = -K_\pi^{-1} \xi_\pi^T \tilde{R}_t X_t, \tag{35}$$

$$U_k^* = -\Gamma_u^{-1} B_u^T \tilde{R}_{k+0} \left(\mathrm{I} + B_u \Gamma_u^{-1} B_u^T \tilde{R}_{k+0}\right)^{-1} B_x X_{k-0}. \tag{36}$$

Expansions (35) and (36) define on-line regulator on the basis of known values of phase current vector $X_t$, $X_{k-0}$ For linear system with quadratic criterion described by the following equations:

$$\dot{X}_t = \xi_x X_t + \xi_\pi \pi_t + \xi_{0t} \tag{37}$$

$$X_k = B_x X_{k-0} + B_u U_k, \tag{38}$$

the optimal control $\pi_t^*$ is expressed by

$$\pi^* = -K_\pi^{-1} \xi_\pi^T \left( \tilde{R}_t X_t + \frac{1}{2} g_t \right) \tag{39}$$

$$\dot{g}_t = \left(\tilde{R}_t \xi_\pi K_\pi^{-1} \xi_\pi^T - \xi_x^T\right) g_t - 2\tilde{R}_t \xi_{0t}, g_{t_f} = 0 \tag{40}$$

Here $\xi_{0t}$ is constant term; formula for $U_k^*$ remains similar.

*Example 2.* For illustration let us consider SASS (**Figure 10**) for the technological supply process by serviceable CP. On **Figure 10** graphs nodes corresponds CP states:

- CP being in usage stage with amount $X_{t1}9$;
- CP being in repair with amount $X_{t2}$ (after usage);
- CP being draft out with amount $X_{t3}$.

It is evidently $X_{t3} = N_0 - (X_{t1} + X_{t2})$; $\rho_{12}$, $\rho_{13}$ being intensity parameters of ordinary CP Poisson streams entering for repair and draft of; $\pi_t$ being repair productivity. Time of CP filling up is equal to $T$.

It is required to determine optimal parameter $\pi_t^*$ of restoration and optimal volumes $U_k^* = U^*(t_k)$, $t_k = kT$ which gives minimum to quadratic functional being sum of expenditure costs:

$$L = \frac{\alpha}{2}\sum_{k=1}^{N} U_k^2 + \frac{\beta}{2}\int_{t_0}^{t_f} \pi^*(t)dt + \frac{\delta}{2} M\left[\left(m_1(t_f) - \gamma\right)^2\right] \tag{41}$$

Here $\alpha, \beta, \gamma, \delta$ are parameters of functional; $m_1(t_f)$ are mathematical expectation of CP remainder at $t_f$ ready for use; $\gamma = K_{TT} N_0$ mean number of aggregates ready for use; $K_{TT} \in [0, 1]$ being coefficient of technical readiness; $N_0$ originate amount of CP.

Is possible to show that at $\varphi(X, t) = 0$ mathematical expectation of CP amount and at step-vise $X_{1t}$ is defined by the following equations:

$$\dot{m}_{1t} = -\rho_{12} m_{1t} + \pi_t m_{2t},\quad m_{1t}(0) = N_0,\quad \dot{m}_{2t} = -\rho_{12} m_{1t} - \pi_t m_{2t},\quad m_{2t}(0) = 0, \tag{42}$$

$$m_{1,k} = m_{1,k=0} + U_{1k}. \tag{43}$$

These mathematical expectations are continuous and discrete variables. This problem being nonlinear because control function $\pi_t$ enters into the right hand of equations in the form of composition with function $m_{2t}$ depending upon control. So it is necessary to use general problem statement Eqs. (9) and ((1) and expressions (17)–(21).

Let us denote.

$$\xi^T(X_t, \pi_t, t) = \left[-\rho_{12} m_{1t} + \pi_t m_{2t} \quad \rho_{12} m_{1t} - \pi_t m_{2t}\right]. \tag{44}$$

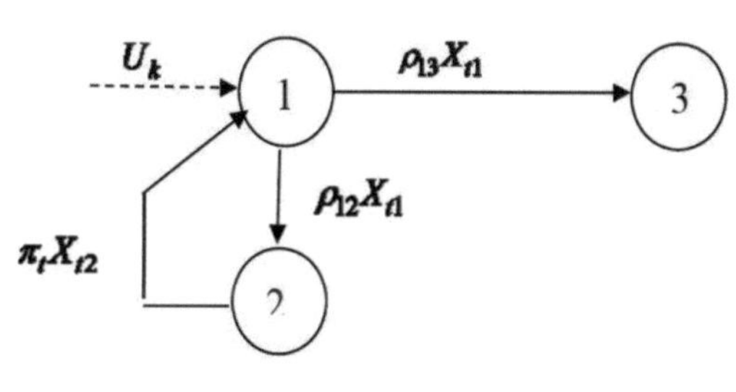


**Figure 10.**
*System state graph.*

For conjugated functions from Eqs. (17), (21) and (41) we get:

$$\dot{\psi}_{1t} = \rho_{12}\psi_{1t} - \pi_t\psi_{2t},\ \psi_{1t}(t_f) = \delta(m_{1t}(t_f) - \gamma),\ \ \dot{\psi}_{2t} = -\rho_{12}\psi_{1t} + \pi_t\psi_{2t},\ \psi_{2t}(t_f) = 0. \quad (45)$$

Than from Eq. (19) follows $\psi_{1t,k+0} = \psi_{1t,k}, \psi_{2t,k+0} = \psi_{2t,k}$. So the conjugated variables $\psi_{1t}, \psi_{2t}$ are continuous functions.

From Eqs. (18) and (19) we have the following expressions for optimal continuous and discrete controls

$$\pi^*(t) = \frac{1}{\beta} m_{2t}(\psi_2 - \psi_1), \quad (46)$$

$$U_k^*(t) = -\frac{1}{\alpha}\psi_{1,k+0}. \quad (47)$$

Taking into account that firstly $\pi^*(t)$ implicitly incoming into right hand of Eq. (42) and $U_k^*$ secondly $U_k^*$ and $\pi^*(t)$ implicitly connected between each other over $\psi_{1,k+0}$ such numerical methods as gradient method may be used [1, 23]. According to gradient method next $(i+1)$ iteration of $\pi^*(t)$ is calculated by $\pi^{i+1} = \pi^i - r\Delta^i$ where $\Delta^i = \beta\pi^i + m_{2t}(\psi_2 - \psi_1)$. Hence the recurrent iteration is calculated by the following expression:

$$\pi^{i+1} = \pi^i - r[\beta\pi^i + m_{2t}(\psi_2 - \psi_1)], \quad (48)$$

where $r$ is chose from convergence and exactness condition.

Numerical results for $m_{1t}$ and $\pi^*(t)$ are given o **Figure 11**, Values of jumps $m_{1t}$ at $t_k = kT, k = 1, \dots, f$ corresponds to optimal values of deliveries $U_k^*$. Values of parameters are: $\rho_{12} = 0,7$ $\rho_{13} = 0,3$; $N_0 = 100$; $T = 0,5$; $t_f = 2$; $2\alpha = 0,5$; $2\beta = 0,1$; $2\delta_0 = 100$; $K_{TT} = 0,8$.

Two main conclusions follows from **Figure 11**:

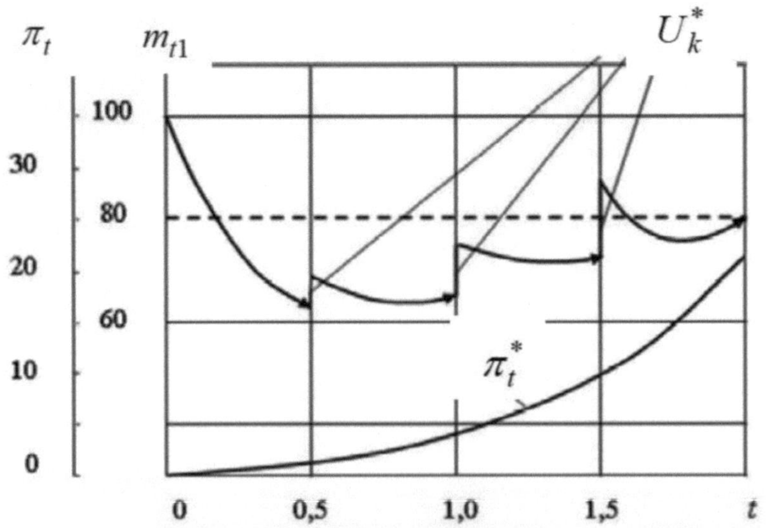


**Figure 11.**
*Volumes $U_k^*$ of delivery SP optimized jointly with repair capacity $\pi_t^*$.*

- optimal nonstationary restoring politics $U_k^*$ and restoring channel capacity goes to $\gamma = 80$ at the end $[t_0, t_f]$;
- values $U_k^*$ are conjugated with $\pi_t^*$;
- by variation $\alpha$, $\beta$ we choose the cost parameters $\alpha + \beta = 1$ type of control depending on cost ratio.

Thus the described restoring politics for given level gives CP owners the opportunity to form separate repair net based on CP order (on supplier side).

*Peculiarities of optimal control stochastic continuous-discrete systems with state feedback.*

For the linear equations:

$$\dot{X}_t = \xi_x X_t + \xi_\pi \pi_t + \xi_{0t} + V_t, \tag{49}$$

$$X_k = B_x X_{k-0} + B_u U_k, \tag{50}$$

where $V_t$ is internal noise being white noise (in strict sense) with known probabilistic characteristics acting in continuous channel; $U_k$ is known discrete function depending on control using formulae for linear stochastic regulator synthesis for system (49) and (50) optimal control and separation theorem we come to the following basic algorithm. It includes two steps:

- optimal deterministic regulator design;
- calculation of optimal estimates $\hat{X}_t$,$\hat{X}_{k-0}$ of (49) and (50) phase vector which is observed in mixture with white noise and substitution into regulator formulae. Exact solution exists only for linear stochastic systems. Using method of normal approximation or statistical linearization [2–5] relatively to state vector it is possible to get the simple approximate algorithm.

### 4.3 Optimal planning and control

As it was already mentioned in Subsection 4.1 OTES-CALS includes complex through along LC on-line planning of processes with goals and objectives and given criteria. Program—object planning is the separate part of applied control theory of LS processes for complex high-technology products which ensure solving LS integration tasks enterprises-participants. We introduce virtual enterprise (VE) as a system developing according with given goals, objectives and programs. On-line realization of plans and programs occur in presence on one side internal noises due to control stochastic and on the other hand by external noises from third party and organizations.

Following [2–4, 23] let us consider optimal regulator for operative control. Within given framework program/plan for OTES-CALS as VE functioning in stochastic media using social-technical-economic effectiveness criteria. We use probability filtering theory based on Kalman and Pugachev filters [2–4]. Optimal stochastic regulator (**Figure 12**) is designed on the basis of the partition theorem. So at first it is necessary to design optimal regulator and then filter for reducing noises.

Using Kalman filtering theory [2–4] for linear continuous-discrete OTES-CALS we get the following equations for stochastic optimal continuous-discrete regulator:

*Filter equations*

$$\dot{\hat{X}}_t = \bar{a}\hat{X}_t + a_2\zeta_t + a_0 + R_t q_t^T v_z^{-1}\left[Z_t - \left(q_t\hat{X}_t + b_2\zeta_t\right)\right] + \xi_\pi \pi_t^* \tag{51}$$

$$\hat{X}_k = B_x\hat{X}_{k-0} + B_x U_k^* \tag{52}$$

$$\dot{R}_t = \bar{a}R_t + R_t\bar{a}^T + v_x - R_t q_t^T v_z^{-1} q_t R_t \tag{53}$$

where $Z_t = \dot{G}_t$, $v_x$ is matrix of in OTES-CALS internal noises intensities; $v_z$ is matrix of OTES-OS and OTES-NS intensifies of noises;
$R_0 = \mathrm{M}\left[\left(X_0 - \hat{X}_0\right)\left(X_0 - \hat{X}_0\right)^T\right]$ is initial conditions for Eq. (53). Direct time integration of Riccati Eq. (54) is used.

*Regulator equations*:

$$\pi_t^* = -K_\pi^{-1}\xi_\pi^T\left(\tilde{R}_t\hat{X}_t + \frac{1}{2}g_t\right), \tag{54}$$

$$\dot{g}_t = \left(\tilde{R}_t\xi_\pi K_\pi^{-1}\xi_\pi^T - \xi_x^T\right)g_t - 2\tilde{R}_{tt}a_2\zeta_t,\ g_{t_f} = 0 \tag{55}$$

$$U_k^* = -\Gamma_u^{-1}B_u^T\tilde{R}_{k+0}\left(\mathrm{I} + B_u\Gamma_u^{-1}B_u^T\tilde{R}_{k+0}\right)^{-1}B_x\hat{X}_{k-0}, \tag{56}$$

*Auxiliary equations*:

$$\tilde{R}_t = -\tilde{R}_t\xi_x - \xi_x^T\tilde{R}_t + \tilde{R}_t\xi_\pi K_\pi^{-1}\xi_\pi^T\tilde{R}_t - K_x,\ \tilde{R}_{t_f} = H_{t_f} \tag{57}$$

$$\tilde{R}_k = B_x^T\tilde{R}_{k+0}\left(\mathrm{I} + B_u\Gamma_u^{-1}B_u^T\tilde{R}_{k+0}\right)^{-1}B_x + K_x \tag{58}$$

Inverse time integration of Riccati Eqs. (57) and (58) is needed Eq. (51). The continuous-discrete Kalman filter equations are inter connected with regulator equations (**Figure 12**).

At last we get equations describing OTES-CALS dynamics with optimal continuous-discrete regulator insuring minimal deviation from given plan during given time interval $[t_0, t_f]$

$$\dot{X}_t = \bar{a}X_t + a_2\zeta_t - \xi_\pi K_\pi^{-1}\xi_\pi^T\left(\tilde{R}_t\hat{X}_t + \frac{1}{2}g_t\right), \tag{59}$$

$$X_k = B_x X_{k-0} - B_u\Gamma_u^{-1}B_u^T\tilde{R}_{k+0}\left(\mathrm{I} + B_u\Gamma_u^{-1}B_u^T\tilde{R}_{k+0}\right)^{-1}B_x\hat{X}_{k-0} \tag{60}$$

where $\zeta_t$ is external noise from OTES-NS also acting on OTES-OS.

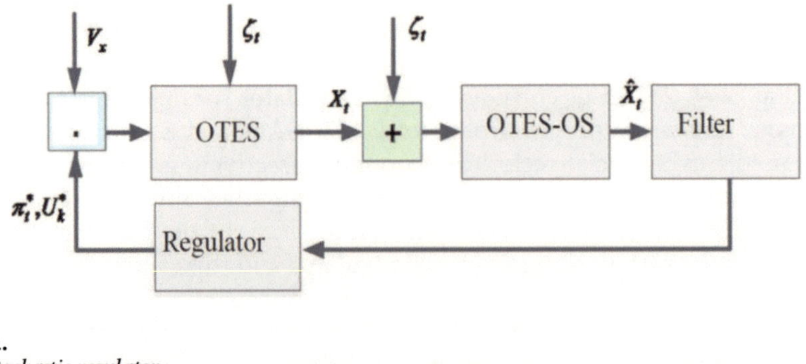


**Figure 12.**
*Optimal stochastic regulator.*

So the design of OTES-CALS includes two stages:

- solution of connected Eqs. (57) and (58) in inverse time with time fixation of coming current data massive;

- solution of differential Eqs. (51)–(53) and (55) in direct time using earlier fixed data and with substitution Eqs. (54) and (56).

## 5. Conclusion

The suggested probabilistic methodology for OTES-CALS allows to solve:

- problems of systems analysis, risk prognosis substations of forestall measures stability of extraction of latent effects on the basis of stochastic analytical modeling for applied LC problems;

- problems of optimal estimation and control on the basis probabilistic estimation and control methods.

Such systems are industrial, energetical, transport systems, financial and economic systems, insurance companies, etc.

Optimization being realized using social-technical-economic criteria. This permits to optimize project budgets for providing given quality MP and OTES-CALS staff potential.

## Acknowledgements

The authors would like to thank Federal Research Center "Computer Science and Control of Russian Academy of Sciences" for supporting the work presented in this chapter.

Authors much obliged to Mrs. Irina Sinitsyna and Mrs. Helen Fedotova for translation and manuscript preparation.

## Author details

Igor Sinitsyn* and Anatoly Shalamov
Federal Research Center "Computer Science and Control of Russian Academy of Sciences", Moscow, Russia

*Address all correspondence to: sinitsin@dol.ru

## References


[1] Pugachev VS, Sinitsyn IN. Lectures on Functional Analysis and Applications. Singapore: World Scientific; 1999. 730 p

[2] Pugachev VS, Sinitsyn IN. Theory of Stochastic Systems. 2nd ed. Moscow: Torus Press; 2001. p. 1000. (in Russian)

[3] Pugachev VS, Sinitsyn IN. Stochastic Systems. Theory and Applications. Singapore: World Scientific; 2001. 908 p

[4] Sinitsyn IN. Kalman and Pugachev Filters. 2nd ed. Logos: Moscow; 2007. p. 772. (in Russian)

[5] Sinitsyn IN, Shalamov AS. Lectures on Theory of Integrated Logistic Support Systems. Moscow: Torus Press; 2012. p. 624. (in Russian)

[6] Kostogryzov A. Probabilistic Modeling in System Engineering. Intech Open; 2018

[7] DefStan 00-600. Integrated Logistic Support. Requirements for MOD Projects

[8] JSP-886. The defence logistic support chain manual. Volume 7: Integrated logistic support

[9] S1000D. International specification for technical publications utilizing a Common Source Data Base

[10] S2000M. International specification for Material management

[11] S3000L. International procedure specification for Logistics Support Analysis

[12] S4000M. International procedural handbook for developing scheduled maintenance program for military aircrafts

[13] Sinitsyn IN, Shalamov AS. Basic technologies for cost management life cycle in highly available organizational-technical-economical systems. Part 1. Principles and approaches. Highly Available Systems. 2014;**10**(4):18-47. (in Russian)

[14] Sinitsyn IN, Shalamov AS. Basic technologies for cost management life cycle in highly available organizational-technical-economical systems. Part 3. Methodical support of stochastic cost management modeling processes. Highly Available Systems. 2015;**11**(1): 23-60. (in Russian)

[15] Lelykova OI. Russian academy of civil service. Health and professional longstanding of government employee. Available from: http://www.psi.lib.ru/statyi/sbornik/soszd.htm (in Russian)

[16] Sinitsyn IN, Shalamov AS, Belousov VV, Gumnilova TS. Basic technologies for cost management life cycle in highly available organizational-technical-economical systems. Part 5. Test example of cost management life cycle aerotransport OTES park. Highly Available Systems. 2015;**11**(1):85-101. (in Russian)

[17] Sinitsyn IN, Shalamov AS, Belousov VV. Software tools for analytical modeling of high dimensional processes in after sale production support systems. Highly Available Systems. 2016;**12**(1): 45-49. (in Russian)

[18] Sinitsyn IN, Shalamov AS. Basic technologies for cost management life cycle in highly available organizational-technical-economical systems. Part 4. Stochastic modeling of processes on market of finances, goods and services. Highly Available Systems. 2015;**11**(4): 61-84. (in Russian)

[19] Sinitsyn IN, Shalamov AS. Basic technologies for cost management life cycle in highly available organizational-technical-economical systems. Part 6. Stochastic methods of dynamical processes microeconomical modeling. Highly available systems. 2015;**11**(2): 3-12. (in Russian)

[20] Sinitsyn IN, Shalamov AS, Titov YP. Methodological problems of stochastic modeling, filtering and optimization. Highly Available Systems. 2016;**10**(2): 38-94. (in Russian)

[21] Sinitsyn IN, Shalamov AS. Models and basic technologies in highly available mass OTES. Highly available systems. 2018;**14**(1):43-58. (in Russian)

[22] Shalamov AS. Optimal control in differential-difference system with random time quantum. Automation and Remote Control. 1990;**8**:46-51. (in Russian)

[23] Csaki F. Modern Control Theories. Nonlinear, Optimal and Adaptive Systems. Academiai Kiado: Budapest; 1972. p. 1096

Chapter 6

# Combined Calculated, Experimental and Determinated and Probable Justifications for Strength of Trunk Crude Oil Pipelines

*Dmitry Neganov and Nikolay Makhutov*

## Abstract

Within the long-term Russian and foreign practice, deterministic methods of basic strength calculations have been developed and are being developed at the design stage of long-distance pipelines. Occurring operational damages, failures, accidents, and catastrophes show there are no direct substantiations for the prevention of such emergencies in the framework of existing calculations. In order to respond to these situations, the following are developed: additional precise deterministic, static, and probabilistic calculations with linear and nonlinear criteria of deformation and fracture mechanics, complex diagnostics of the state of the pipeline using in-line pigs, and laboratory, model, bench, and field tests of pipelines with technological and operational defects. The results of systematic scientific research and applied developments are presented.

**Keywords:** pipeline, strength, calculation

## 1. Statement of the problem

The trunk pipelines for transporting liquid and gaseous hydrocarbons are one of the main types of transport infrastructure in the world.

The most important pumped media are crude oil, gas, broad hydrocarbon fractions, and petroleum products. The total length of trunk pipelines in the world is more than 1.5 million km; in Russia it is about 230 thousand km, and the length of oil pipelines in the world is 170 thousand km; in Russia it is about 70 thousand km.

The length of the largest individual oil pipeline systems is Canada-USA 4700–5300 km with pipe diameters of 450–1220 mm, China-Kazakhstan 2200 km with a diameter of 813 mm, Azerbaijan-Georgia-Turkey 1768 km with a diameter of 1067 mm, Tanzania-Zambia 17,210 km with a diameter of 200–300 mm, and Italy-Germany 1000 km with a diameter of 660 mm.

In Russia, the largest oil pipelines are the Eastern Siberia-Pacific Ocean, 4740 km with a diameter of 1020–1200 mm, and Druzhba, 5500 km with a diameter of 520–1020 mm; eight trunk oil pipelines have a length of more than 1000 km.

**Figure 1.**
*Oil trunk pipelines (Russia).*

The operating pressures in the main oil pipelines range from 2 to 10 MPa (**Figure 1**).

The trunk oil pipelines are operated in a very wide range of climatic conditions (from −70°C to +60°C) and natural hazards (seismic, landslides, geological faults), with ground, underground, and underwater laying.

Despite the large, more than century-old experience of research, testing, construction, and operation of oil trunk pipelines in the world and in Russia, there were large-scale accidents and disasters. These accidents were accompanied by the release of large amounts of oil (up to 100–600 thousand barrels) into the environment (land, water) with great environmental damage, fires, death and injury to people, and pollution of hundreds of hectares of land. Economic damages from such accidents are estimated at $ 10–100 million. The total number of accidents on oil pipelines in the world over the past 20 years is more than 2000, and the number of large oil leaks is more than 4500. For every million tons of pumped oil, 3–5 tons fall into leaks.

In general, the accident rate on the trunk oil pipelines is reduced. However, at present it is at the level of 0.1–0.3 per 1000 km per year (**Figure 2**).

These data indicate the need for further research and practical development to reduce accidents and improve the safety of trunk pipelines.

In recent years, four basic approaches to determining the strength, resource, and safety of oil pipelines have emerged:

- Deterministic
- Statistical
- Probabilistic
- Combined

## 2. Solving the problems of strength by basic and calibration methods

### 2.1 Basic deterministic calculations

The system of domestic and foreign trunk oil pipelines that took shape in the second half of the twentieth and the beginning of the twenty-first century is characterized by multistage creation and development of integrated approaches to justifying their strength [1–4]. These approaches were initially formed on the basis of the fundamental theories of thin-walled shells, classical theories of strength; they made it possible to form the main computational methods for the selection of computational schemes and computational cases and the assessment of static strength, taking into account the types of stress and limit states.

The basic strength condition was then recorded in the simplest form:

$$\sigma^{э}_{\max} \leq \sigma_{on}, \tag{1}$$

where $\sigma^{э}_{\max}$ is maximum operating voltage stress and is hazardous stress.

For a thin-walled pipe with a diameter $D$ with wall thickness ($\delta << D$), ring stresses are maximal:

(**Figure 3**).

$$\sigma^{э}_{\max} = \frac{P^{э}_{\max} \cdot D}{2\delta}, \tag{2}$$

where $P^{э}_{\max}$ is maximum operating pressure (**Figure 3**).

**Figure 2.**
*Accidents on oil pipelines (Venezuela, China, Russia).*

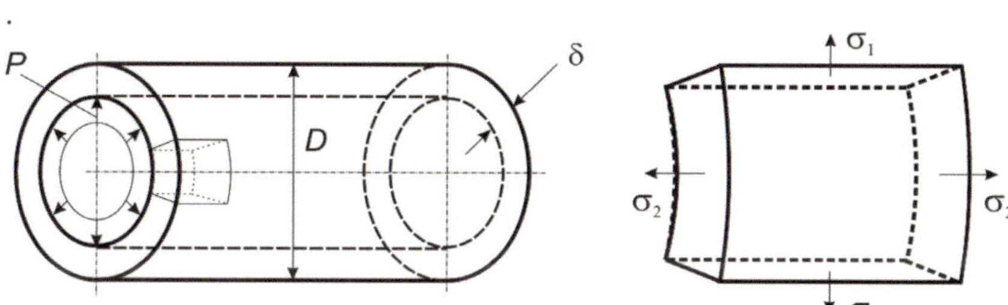


**Figure 3.**
*The main design scheme.*

Since in engineering calculations of static strength according to (1) and (2), a whole set of design, technological, and operational methods remained unclear, permissible stresses $[\sigma]$ with corresponding safety margins were entered into the calculation:

$$[\sigma]=\frac{\sigma_{on}}{n_{в}}=\min\left\{\frac{\sigma_{т}}{n_{т}},\frac{\sigma_{в}}{n_{в}}\right\}. \tag{3}$$

Dangerous stresses were applicable ones corresponding to:

- Ultimate strength $\sigma_{в}$, which excluded the occurrence of fracture (the first significant limiting state).

- Yield strength $\sigma_{т}$ (or conditional yield strength $\sigma_{0,2}$), which excluded the formation of unacceptable plastic deformations (the second most significant limiting state). For modern pipeline systems transporting petroleum and petroleum products, a number of main life cycle stages, as measured up to 30–60 years, are introduced into the strength analysis:

- Feasibility study of the project

- Outline and detailed design

- Construction and testing of pipeline systems

- Operation of pipeline systems with diagnostic and repair and rehabilitation works

- The withdrawal of sections of pipelines or pipeline systems from operation

- For each of these stages and for the entire life cycle, to date, in our country and abroad, certain approaches and methods to substantiate strength have been formed.

These methods are divided into two main groups:

- Basic strength calculations

- Verification calculations of strength for the used construction material

The formation of methods of basic and calibration calculations is currently associated with the stages of the life cycle. At the same time, an important role is always played by scientific studies to substantiate the strength criteria, the choice of design schemes and design cases, followed by the introduction of safety margins. This is a scientific basis for solving applied problems of strength—the development of strength standards with their design characteristics (**Figure 4**).

For all pipe steels $\sigma_{т} \leq \sigma_{в}$ and to fulfill condition (3), the margins must satisfy.

$$1 \leq n_{т} \leq n_{в}. \tag{4}$$

The development of pipeline transport during the decades of the twentieth to twenty-first centuries [5, 6] was accompanied by a gradual increase in the

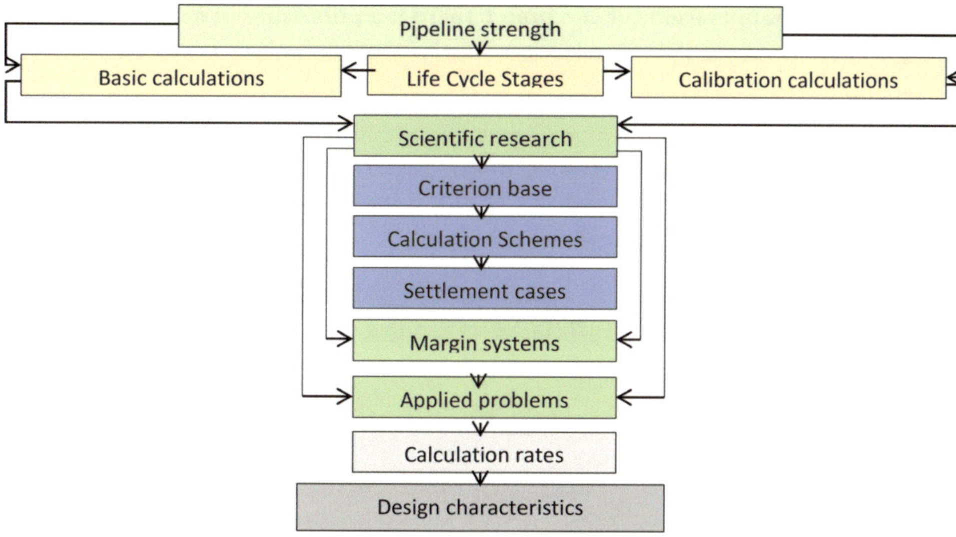


**Figure 4.**
*Generalized scheme of research and rating of strength.*

mechanical properties $\sigma_{\text{т}}, \sigma_{\text{в}}$ of pipe steels ($200 \leq \sigma_{\text{т}} \leq 800$, $350 \leq \sigma_{\text{в}} \leq 950$) and a decrease in strength margins ($2.8 \geq n_{\text{т}} \geq 1.5$, $4.0 \geq n_{\text{в}} \geq 1.8$).

Expressions (1)–(4) were and remain central to foreign strength standards [2–4]. In Russian practice [1, 4, 5], expressions (1) and (2) were retained, but the strength margins $n_{\sigma}$ in (3) were presented in a differentiated form:

$$n_{\sigma} = \{n_{\text{т}}, n_{\text{в}}\} = \frac{n \cdot K_1 \cdot K_{\text{н}}}{m}, \tag{5}$$

where $n$, $K_1$, and $K_{\text{н}}$ are the reliability factors for load, material, and purpose ($n, K_1, K_{\text{н}} \geq 1$) and $m$ is the operating condition ratio ($m \leq 1$).

In view of **Figure 4** and expressions (1)–(5) in the feasibility study of the project, two main parameters are defined and assigned, $p^{\text{э}}_{\max}$ and $D$, ensuring the specified pipeline performance.

At the stage of preliminary and detailed design, the main calculation is reduced to the calculated determination of the minimum wall thickness of the pipeline according to the given $p^{\text{э}}_{\max}$ and $D$, taking into account economically and technologically reasonable choice of pipe steel with characteristics $\sigma_{\text{т}}, \sigma_{\text{в}}$ (according to technical conditions or standards):

$$\delta_{\min} \geq \frac{p^{\text{э}}_{\max} \cdot D}{2[\sigma]}. \tag{6}$$

Since the values $p^{\text{э}}_{\max}$ at each of the calculated sections of the pipeline depend on their height position, which determines the hydrostatic part of the pressure, the calculated values $\delta_{\min}$ are variable in length.

At the stage of construction and pre-hydraulic pressure tests $p^{\text{г}} \geq p^{\text{э}}_{\max}$, calibration calculations with an assessment of the maximum hydraulic tests are carried out in accordance with (2), their comparison with permissible values in accordance with (3) and (5) and confirmation of the absence of destruction or the formation of unacceptable plastic deformations

$$\sigma^{\text{г}}_{\max} = \frac{p^{\text{г}} \cdot D}{2\delta_{\min}} \leq \frac{\sigma_{on}}{n^{\text{г}}_{\sigma}}. \tag{7}$$

At the stage of operation for the time $\tau$ there is a possibility of accumulation of damage, and a decrease in wall thickness $\delta_{\min}(\tau)$ due to corrosion, erosion, as well as a change in mechanical properties $\sigma_{т}(\tau), \sigma_{в}(\tau)$.

These parameters are determined according to periodic in-line inspection, as well as according to mechanical testing of samples from damaged sections of pipelines. The verification calculation of the strength for this stage is reduced to the assessment of the strength margin $n_{\sigma}(\tau)$.

$$n_{\sigma}(\tau) = \frac{\sigma_{on}(\tau)}{\sigma^{э}(\tau)}; \qquad (8)$$

$$\sigma_{э}(\tau) = \frac{p^{э}_{\max} \cdot D}{\delta_{\min}(\tau)}; \qquad (9)$$

$$\sigma_{on}(\tau) = \{\sigma_{т}(\tau), \sigma_{в}(\tau)\}. \qquad (10)$$

If for the analyzed stage $\tau$ margin по (8) is not less than in in (3) and (5), then the operation of the pipeline can be continued.

To calculate the estimated time $\tau_p$ of the next in-line inspection, it is necessary to have data on monitoring $\sigma_{т}(\tau_p), \sigma_{в}(\tau_p)$ and $\delta_{\min}(\tau_p)$ on the basis of previous in-line inspections operations.

If such initial information is absent, then the construction of calculated curves is possible:

$$\{\sigma_{т}(\tau), \sigma_{в}(\tau)\} = F\{(\sigma_{т}, \sigma_{в}) \cdot \tau^{m_{\tau}}\}, \qquad (11)$$

where $m_{\tau}$—If such initial information is absent, then the construction of calculated curves is possible [6, 7] ($0 \leq m_{\tau} \leq 0.03$ for $\tau$ in hours).

## 2.2 Statistical strength analysis

In actual practice, in the manufacture and testing of pipes, the construction of pipeline sections and the operation of pipeline systems, all specified parameters of expressions (1)–(11), are statistically variable, despite the determination of the main calculations in the design of pipelines throughout the system of design expressions.

The statistical analysis of the calculated parameters in the framework of the basic calculations of the strength of pipelines is aimed at establishing:

- Minimum (min), average (m), and maximum (max) parameter values
- Comparability with the values adopted in the project documentation
- Deviations of the calculated parameters to the dangerous and safe side in comparison with the statistically determined.

On this basis, two decisions are made about the possibility or impossibility of further operations of pipelines.

In the first case, the main requirement for the strength of pipelines must be met; in the second case, the strength is considered not ensured if the maximum operating stresses exceed the allowable.

$$\sigma^{э}_{\max\ c} \le [\sigma]; \sigma^{э}_{\max\ c} > [\sigma]. \tag{12}$$

For the scientific substantiation of the need and possibility of extending the operation of pipelines in cases of failure to meet the strength conditions, it is possible to reduce the operating pressure to level.

$$p^{э}_{c} < p^{э}, \quad \text{when } \sigma^{э}_{c} < \sigma^{э} \le [\sigma] \tag{13}$$

Simultaneous fulfillment of conditions (1)–(3) requires the mandatory calculation of the strength of the pipeline—its pipes and sections, where the realization of expression (12) is detected.

This calculation should include:

- Maximum values of operating pressure $p^{э}_{\max}$
- Maximum values of the diameter $D_{\max}$ of the pipe or section of the pipeline
- Minimum wall thickness $\delta_{\min}$
- The minimum values of the characteristics of mechanical properties $\sigma_{\text{Т}\min}, \sigma_{\text{В}\min}$.

In this case, you can write

$$\sigma^{э}_{c} = F_{\sigma}\{p^{э}_{\max}, D_{\max}, \delta_{\min}\} \le [\sigma] = \min\left\{\frac{\sigma_{\text{В}\min}}{n_{\text{В}}}, \frac{\sigma_{\text{Т}\min}}{n_{\text{Т}}}\right\} \tag{14}$$

The strength condition according to expression (14) should be checked according to the statistical analysis and when conditions (12) are fulfilled.

At the same time, both for condition (12) and for conditions (13) and (14), it is advisable to give an assessment of the strength according to (1):

- With average values of all parameters $p^{э}_{m}$, $D_m$, $\delta_m$, and $\sigma_{\text{Т}\ m}, \sigma_{\text{В}\ m}$

$$\sigma^{э}_{c} = F_{\sigma}\{p^{э}_{m}, D_m, \delta_m\} \le [\sigma] = \min\left\{\frac{\sigma_{\text{В}\ m}}{n_{\text{В}}}, \frac{\sigma_{\text{Т}\ m}}{n_{\text{Т}}}\right\} \tag{15}$$

- At extreme (extreme) values $p^{э}_{\max}$, $\delta_{\max}$, $D_{\min}$, and $\sigma_{\text{Т}\ \max}, \sigma_{\text{В}\ \max}$

$$\sigma^{э}_{\max} = F_{\sigma}\{p^{э}_{\max}, D_{\min}, \delta_{\max}\} \le [\sigma] = \min\left\{\frac{\sigma_{\text{В}\ \max}}{n_{\text{В}}}, \frac{\sigma_{\text{Т}\ \max}}{n_{\text{Т}}}\right\}. \tag{16}$$

Thus, according to (12)–(16), the calculated (average, minimum, maximum) values of $p^{э}$, $D$, $\delta$, and $\sigma_{\text{Т}}$, $\sigma_{\text{В}}$ are due to a whole range of design, technological, and operational factors.

### *2.2.1 Operating pressure*

The statistical nature of operating pressures $p^{э}$ satisfies inequalities.

$$p^{э}_{\min} \le p^{э}_{m} \le p^{э}_{\max} \tag{17}$$

due to changes in the actual pressures in a given pipe or in a given section of the pipeline due to

- Actuation systems to maintain the specified working pressure at pumping stations

$$p^{э}_{н\ min} \le p^{э}_{н\ m} \le p^{э}_{н\ max} \tag{18}$$

- Deterministic design and actual operational differences of hydrostatic pressures $\Delta p_г$ from changes in the profile of the heights of laying pipelines

$$\Delta p^{э}_{г\ min} \le \Delta p^{э}_{г\ m} \le \Delta p^{э}_{г\ max} \tag{19}$$

- Deterministic design pressure changes $\Delta p_{гс}$
- Deterministic design pressure changes due to changes in hydraulic resistance to the movement of oil and oil products (due to changes in flow areas, viscosity, and temperature of the transported working fluid)

$$\Delta p_{гс\ min} \le \Delta p_{гс\ m} \le \Delta p_{гс\ max} \tag{20}$$

- Deterministic design and actual operating pressure changes due to external effects on the pipeline (seismic, temperature, vibration, aero-hydrodynamic)

$$\Delta p_{в\ min} \le \Delta p_{в\ m} \le \Delta p_{в\ max} \tag{21}$$

In the basic calculations using expressions (1)–(11), for deterministic and statistical estimates of the static strength of pipelines, pressure components should be included when the pipelines are operating at maximum design conditions:

$$p^{э}_{р} = p^{э} + \sum \Delta p^{э}. \tag{22}$$

Deeper in scope, cyclic pressure changes due to software changes in pipeline operation modes (start-up, shutdowns, performance change—throughput) are subject to accounting for cyclic strength and durability calibration calculations. Statistical information on the change in pressure is obtained from the registration data at pumping stations.

### *2.2.2 Diameter of pipelines in operation*

The diameter D, which is included in expressions (1), (4)–(6), and the pipeline, is characterized by the scattering of its actual values. It is due to pipe manufacturing technology and is reflected in the maximum and minimum technological tolerances on the diameter $\Delta D_т$:

$$\{D^{т}_{max}, D^{т}_{min}\} = D_{п} \pm \Delta D_{т}, \tag{23}$$

where $D_п$ is design diameter.

Values $\Delta D_т$ in either direction may be the same or different.

The diameters of $D^{э}$ pipes in various parts of pipelines that are fixed during operation during inspections and diagnostics of pipelines may differ from the diameters $D_т$ after the manufacture of pipes:

$$\{D^{э}_{max}, D^{э}_{min}\} = \{D^{т}_{max}, D^{т}_{min}\} \pm \Delta D^{э}. \tag{24}$$

Values $\Delta D^э$, as a rule, have a positive value due to the possible deformation under the action of test or operating modes with increased pressure.

The second factor of change in diameters $D^э$ can be ovalization of the cross-section during transportation, construction, and operation (usually while maintaining the length of the perimeter of the pipeline):

$$D^э = F_D\{D^э_{min}, D^э_{max}\}, \tag{25}$$

where $D^э_{min}, D^э_{min}$ are minimum and maximum diameter in the zone of ovalization.

For ovalized sections, the calculated determination of stresses according to (1) should take into account their increase.

The calculated justification of static strength in the framework of the basic calculations according to (1)–(21) should be mainly oriented:

- To the maximum values in $D_{max}$ in (23) and (25)
- At maximum operating stress $\sigma^э_{max}$

*2.2.3 Pipeline wall thickness in operation*

Pipeline wall thickness $\delta$ has the most significant effect on operating stresses $\sigma^э$ and strength conditions.

The statistical variation of values, as well as diameters D, is due to:

- Rolling sheet technology, which is a blank for pipes:

$$\{\delta^т_{min}, \delta^т_{max}\} = \delta_п \pm \Delta\delta_т, \tag{26}$$

where $\delta_п$ is design wall thickness and $\Delta\delta_т$ is technological tolerance for thickness.

The change in wall thickness during pipe rolling can be neglected, taking into account the main bending deformations.

When testing pipes at the manufacturing stage and during construction, it is possible to change the wall thickness downwards:

- Due to plastic deformations from high-pressure tests $p_и$

$$\Delta\delta_и = F_\delta\{p_и, \Delta D^э, \mu\}, \tag{27}$$

where $\mu$ is Poisson's ratio ($0.3 \le \mu \le 0.5$) and $\Delta D^э$ is a possible increase in the diameter $D^э$ of the pipeline, defined by (25).

- Due to corrosion and erosion damage

$$\Delta\delta_{кэ} = F_\delta\{p^э, \tau^э, c_{кэ}\}, \tag{28}$$

where $\tau^э$ is operation time and $c_{кэ}$ is the rate of corrosion erosion damage.

In the basic calculations of static strength according to the basic calculations of static strength according to the expressions (1)–(28), the minimum value of the wall thickness should be used:

$$\delta^э_{min} = \delta_п - \Delta\delta_т - \Delta\delta_и - \Delta\delta_{кэ}. \tag{29}$$

Expression (29) under condition (14) will mean the maximum increase in operating stresses $\sigma^э$.

### 2.2.4 Characteristics of mechanical properties

Mechanical properties with strength characteristics $(\sigma_в, \sigma_т)$, $(R_1, R_2)$, as well as $p^э, D, \delta$ are stochastic. In order to ensure and justify the static strength of pipelines ,their minimum values should be entered into the calculation. The statistical variation in the characteristics of the mechanical properties of pipe steels is determined by a set of technological factors:

- Chemical composition and structural structure (grain size $d$)
- Modes of thermal and thermomechanical $(t_т)$ processing
- The level of preliminary plastic deformations $e_n$ during sheet rolling, rolling of tube blanks, and testing of pipes, sections, and sections of pipelines
- Temporary factors of aging and degradation in time $\tau$

$$\{\sigma_в^э, \sigma_т^э\} = F_\sigma\{d_3, t_п, e_n, t^э\}. \tag{30}$$

For widely used pipe steels, the increase in strength characteristics $\sigma_T, \sigma_B$ is usually combined with a decrease in ductility.

In the main calculations of the strength of pipelines, it is recommended to use statistical data on the reduction of strength characteristics:

$$\{\sigma_в^э, \sigma_т^э\} = \min\{\sigma_в, \sigma_т\}. \tag{31}$$

### 2.2.5 Reflection of statistical factors of strength in margin

Use in domestic and foreign basic regulatory calculations of the strength of the system of strength margins $n_\sigma$ (when calculating the permissible stresses $[\sigma]$) and reliability coefficients $K_1, K_2, m$, and $n$ (when calculating the limiting states and resistances) makes it possible to obtain a connection between them in the form of expression (5).

All coefficients of expression (5) in a deterministic form, taking into account the statistics of design parameters for expressions (12)–(31), reflect the general variation of design, technological, and operational strength factors.

The strength margins $n_\sigma$ of (5) in the deterministic basic and calibration statistical calculations take into account the level of the necessary reduction of operating stress $\sigma^э$ compared to dangerous stresses $(\sigma_{max}^э < \sigma_{оп})$:

$$n_\sigma = \frac{\sigma_{оп}}{\sigma_{max}^э}. \tag{32}$$

At the same time, dangerous stresses $\sigma_{оп}$ are understood not only as deterministic but also as statistical limits of strength $\sigma_в$ (to exclude one-time static damage) and plasticity $\sigma_т$ (to exclude one-time static damage) and plasticity (to prevent the formation of unacceptable plastic deformations):

$$\sigma_{оп} = \min\{\sigma_в, \sigma_т\}. \tag{33}$$

The statistical information about the values $\sigma^э$ is obtained on the basis of the analysis of the stress–strain state by statistical parameters $p^э, D, \delta$ on the basis of the entire system of expressions (1)–(33).

In combination with the statistical data on the hazard values of the criterial characteristics $\sigma_{оп}$ in the form of tensile strengths $\sigma_в$ and yield strengths $\sigma_т$ (or design resistances $R_1$, $R_2$), a scatter can be obtained $n_\sigma(n_в, n_т)$:

$$n_{\sigma\min} = \left\{\frac{\sigma_{В\ \min}}{\sigma^э_{\max}}, \frac{\sigma_{Т\ \min}}{\sigma^э_{\max}}\right\}; n_{\sigma\ \max} = \left\{\frac{\sigma_{В\ \max}}{\sigma^э_{\min}}, \frac{\sigma_{Т\ \max}}{\sigma^э_{\min}}\right\}. \tag{34}$$

To make decisions about the admissibility of safety margins, $n_\sigma$ should be estimated, and their average values

$$n_{\sigma\ ср} = \left\{\frac{\sigma_{В\ m}}{\sigma^э_m}, \frac{\sigma_{Т\ m}}{\sigma^э_m}\right\}. \tag{35}$$

The strength of the pipeline, determined by the allowable stresses in the statistical interpretation, can be considered as secured if the normatively specified margin $n_{\sigma н}(n_{вн}, n_{тн})$ satisfies the inequality

$$n_{\sigma\ \max} \geq n_{\sigma\ m} \geq n_{\sigma\ \min} \geq n_{\sigma Н}. \tag{36}$$

According to these statistics, it is possible to quantify statistical variations of the coefficients $m$, $n$, $K_1$, and $K_н$. On this basis, you can make a conclusion about the strength of the pipeline, if combinations are performed:

$$\{m_{\min}, K_{\text{I}\max}, K_{н\ \max}, n_{\min}\} > \{m, K_1, K_н, n\}_н. \tag{37}$$

Failure to comply with conditions (36) and (37) requires making decisions about conducting refined basic and calibration calculations by deterministic and statistical methods.

## 2.3 Basics of probabilistic strength calculations

The accumulation of statistical information in the form of histograms of the main design parameters of strength makes it possible to proceed to a probabilistic analysis in the form of a distribution of strength. They are reflected in regulatory calculations for limiting states and limiting resistances [5] through the safety factors for the material, load, working conditions and purpose, and load in expression (25):

The essence of this analysis [1, 6, 8] is:

- to obtain the probability density functions $p$ of external and internal effects (the number of pressure $p^э$) and the corresponding design stresses $\sigma^э$ and design resistances (yield strength $\sigma_T$ and strength $\sigma_в$) with the subsequent determination of the probability of failure $P_p$ in areas, where areas with extremely low probabilities ($P_p \leq 10^{-7}$);

- to construct probability functions $P(\sigma^э)$ and $P\{\sigma_T, \sigma_в\}$ with the definition of the relationship between strength margins $\{n_T, n_в\}$ and given probabilities $P(\sigma^э)$, $P\{\sigma_T, \sigma_в\}$, corresponding to the volume of the initial statistical and probabilistic information.

There is a simple relationship between probability $P$ and the amount of initial statistical information:

$$n = \frac{i - 0,5}{P}, \tag{38}$$

where $i$ is the sequence number of the measured value and $n$ is the total number of measurements.

With a commonly used sample of 20 measurements, the value is $P = 2.5 \cdot 10^{-2}$ (or 2.5%).

To estimate the values of $P$ at the level of $10^{-2}$ (or 1%) it is necessary to make already 50 measurements, and for the probability of $10^{-4}$ – 5000.

In statistical and probabilistic studies of the mechanical properties of structural steels, the volume of samples n is in the range of 20–22.000 [6, 10]. According to the histogram of the limit distribution functions $\sigma_Т, \sigma_В$, the functional $F$ is obtained for the strength margins $n_Т, n_В$:

$$\{n_T, n_в\} = F\{P(\sigma^э), P(\sigma_T, \sigma_в)\} \tag{39}$$

The number of laboratory samples of steel 17G1S, cut from pipes in the initial state and after 40 years of operation is 28.

To solve probabilistic problems of strength in terms of expression (39) in the zone of small probabilities of destruction $P_p$, a large amount of statistical information is needed with samples measured in the hundreds and thousands, which is practically impossible in many real cases. In this connection, it is more promising to use expressions (38) and (39), which allow $\{n_T, n_в\}$ estimating reserves for a given probability $P$ of calculated characteristics, corresponding to the availability of experiments on the distribution functions, $P(\sigma^э)$ and $P\{\sigma_T, \sigma_в\}$, with the choices of tens and hundreds.

**Figure 5** shows the scheme for the implementation of a probabilistic analysis of reserves: along the ordinate axis, the probabilities $P(\sigma^э)$ and $P(\sigma_в)$ on a scale corresponding to the normal distribution law. Then by the median values $\sigma^э_m$ и $\sigma_{вm}$ for the probability $P$ = 50% and for other values of $P$ ($P$ < 50%).

$$(n_{\sigma в})_m = \frac{(\sigma_в)_m}{(\sigma^э)_m}; (n_{\sigma в})_p = \frac{(\sigma_в)_p}{(\sigma^э)_p}. \tag{40}$$

If, according to the results of statistical processing of values $\sigma^э$ and $\sigma_в$ the parameters of their probability distributions are obtained—(the coefficients of variation $V\sigma^э$ and $V\sigma_в$ and their average values $(\sigma^э_m)_m$ and $(\sigma_в)_m$, then the calculated values $(\sigma_в)_p$ and $(\sigma^э)_p$ for a given probability $P$ are obtained from the expressions.

$$\left\{(\sigma^э)p, (\sigma_в)_p\right\} = \{(\sigma^э)_m, (\sigma_в)_m\}(1 - Z_p\{V\sigma^э, V\sigma_в\}), \tag{41}$$

where $Z_p$ is distribution quantile depending on $P$.

For coefficients of variation in the range of $V\sigma^э$ и $V\sigma_в$ in the range of 0.03–0.1 the calculated probabilities $P_p$ are obtained when the margin factors $n_{\sigma в}$>1, 8 are in the range of $10^{-15}$ to $10^{-5}$.

With the currently existing banks of data on operational load $\sigma^э$ and mechanical properties of pipe steels $(\sigma_В, \sigma_Т)$, it is more reasonable to consider not determining the values of $P_p$ in the area of their low values, but determining the strength margins from (40) using the specified probability parameters $P$ in the range of $10^{-4}$ to $10^{-5}$ and above.

On the basis of (5) and (41), it is possible to analyze changes in the regulatory strength margin $n_в$ taking into account the probabilistic characteristics of the operational loading $\sigma^э$ and the limits of strength $\sigma_В$:

$$n_p = n_в \frac{1 - Z_p \cdot V\sigma_в}{1 + Z_p \cdot V\sigma^э} = n_в \cdot \overline{n}_p, \tag{42}$$

where $n_p$ is margin of strength for a given probability $P$ and $\overline{n}_p$ is margin reduction ration$_в$. The relationship between $\overline{n}_p$ and $P$ in (42) with $V\sigma_В = 0,05$ and $V\sigma^э = 0,08$ is shown in **Figure 6**. From the data in **Figures 5** and **6**, it can be seen that the greatest influence on the allowable change in the strength margins $n_в$ is observed when $P$ decreases from 0.5 to $10^{-3}$. Refinement of probabilistic calculations of strength at lower $P$ does not make much practical sense.

The probabilistic approach acquires its practical relevance in the critical sections of trunk pipelines:

- At their intersections with other transport systems (rail, high-voltage, pipeline), with non-compliance with the allowable distances from other facilities and infrastructures
- On water transitions
- In zones of geological faults, landslides, and seismic effects

This approach becomes significant and necessary for those cases when the assigned service lives and estimated durability are developed, and the in-line inspections show increased defectiveness.

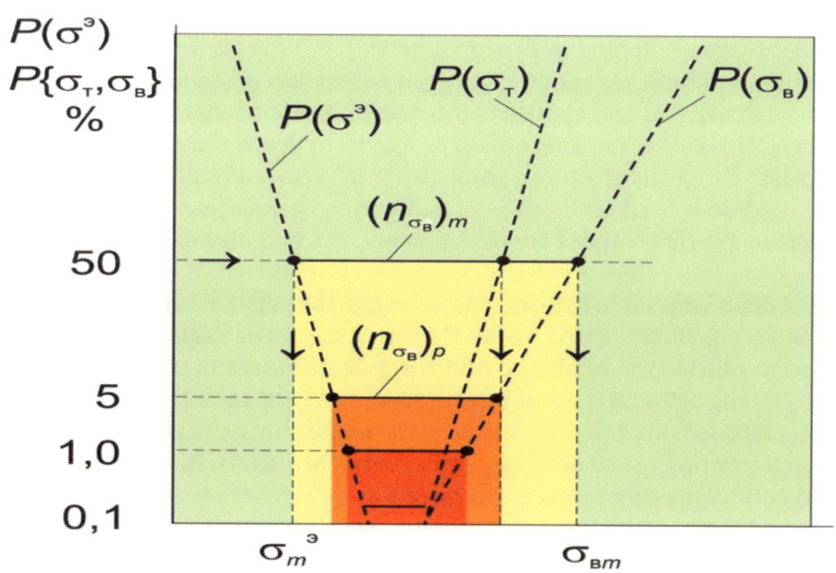


**Figure 5.**
*Scheme for assessing the impact of probability P on the strength margin.*

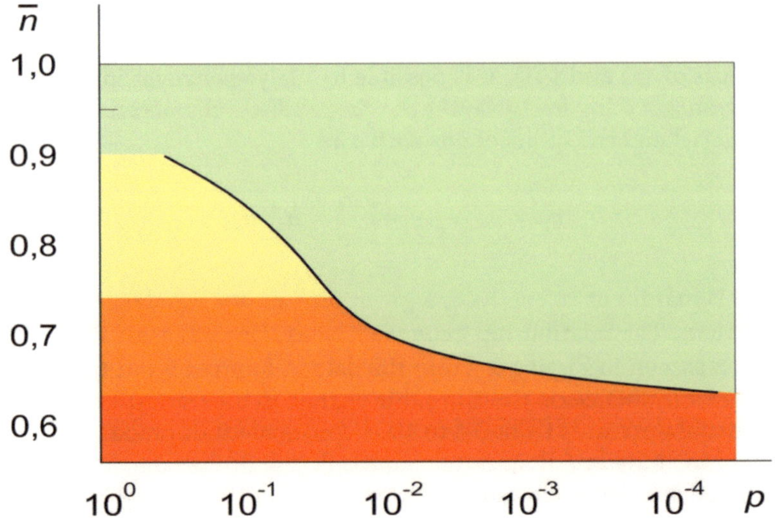


**Figure 6.**
*Relative decrease in strength margins $\bar{n}$ with changing probabilistic characteristics of loading and mechanical properties.*

## 3. Implementation of combined methods to substantiate strength

### 3.1 Formation of the structure of the combined methods

The generalized structure of the standard basic and calibration determination of the strength parameters of main oil pipeline systems discussed above (Section 2) reflects the theory and practice of computational and experimental substantiation of strength developed in our country and abroad for six to seven decades. The focus is on the trunk pipelines for the transportation of oil and oil products. The calculation of strength analysis is based on two methods—the method of calculation for permissible stresses (adopted in foreign practice) and the method of calculation for limiting states and limit resistances (adopted in Russian practice).

The most developed and applied is the deterministic strength calculation at the design stage. This solves the direct main problems of determining the wall thickness of the pipeline for given pressures, throughput of pipes and selected pipe steels. The same method is used at the stage of calibration calculations of the strength of the pipelines under construction and the majority of the pipelines being operated.

In those cases when it is necessary to calculate the substantiation of the strength of functioning pipelines with deviations from the design decisions and when defects in pipes occur outside the established norms, it is necessary to carry out calibration calculations using actual statistical information on all the calculated parameters. One of the tasks solved at the same time is the appointment of all the main design parameters according to the obtained statistical information. In these cases the preservation of regulatory reserves is typical.

For the most critical sections of pipelines, statistical strength analysis may be insufficient and unacceptable. Then probabilistic estimates of strength are required using the functions of the distribution of operational loading and the mechanical properties of pipe steels by the parameter of operation time. For these situations, it becomes possible to change the safety margins for the required probabilities of the occurrence of dangerous states.

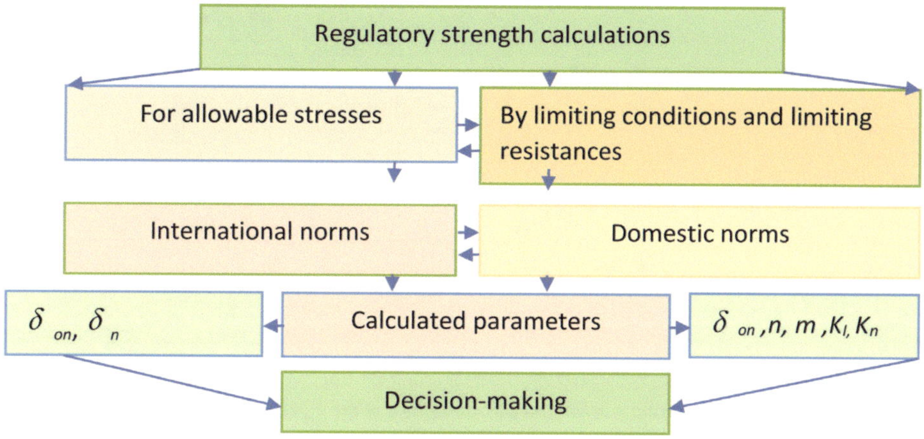


**Figure 7.**
*Block diagram of regulatory foreign and domestic calculations.*

The scientific basis of these calculations is the entire system of calculation expressions (1)–(42) (**Figure 7**).

This system has been and remains basic in all international practice [1–4] to the present time with the development of methods for the design, construction, and operation of trunk pipelines to ensure their strength and deformability expressed in a gradual decrease in margins $n$ ($1.8 \geq n_T \geq 1.2$; $2.5 \geq n_в \geq 1.7$) и and an increase in the strength characteristics $\sigma_{\text{т}}$, $\sigma_{\text{в}}$ ($200 \leq \sigma_T \leq 800$; $420 \leq \sigma_в \leq 920$ MPa).

All uncertainty factors included in the calculations and reflecting the operating conditions, design, and construction technologies were taken into account by the coefficients ($n_\sigma$, $n_T$, $n_в$) and the standard purpose of guaranteed mechanical properties ($\sigma_{\text{в}}$, $\sigma_{\text{T}}$).

A generalized analysis of trends and parameters of the development of pipeline transport of oil and oil pipelines and methods for calculating the strength is made in [6, 7].

Expressions (1) and (2) are initial in assessing the strength of pipelines at all the main stages of the life cycle—design, construction, operation, and decommissioning. Currently two tasks are being solved:

- The direct task of a deterministic basic calculation of the wall thickness $\delta$ of the pipeline at the design stage with a preliminary feasibility study of the diameter $D_в$ and pressure $p$ as well as with the selected structural material $\sigma_{on}$, $(\sigma_в, \sigma_T)$ and assigned margin $n_\sigma$, $(n_T, n_в)$:

$$\delta \geq \frac{pD_в}{2[\sigma]} = \frac{pD_в \cdot n_\sigma}{2\sigma_{on}}. \tag{43}$$

Under these conditions, the wall thickness $\delta$ cannot be less than the value calculated by expression (6) (**Figure 8**).

At the stages of construction, operation, and decommissioning on the basis of (43), deterministic calibration calculations are performed with the following objectives:

- Check of permissible operating pressure $[p]$ at specified

$$[p] \leq 2\frac{\sigma_{on}}{n_\sigma}. \tag{44}$$

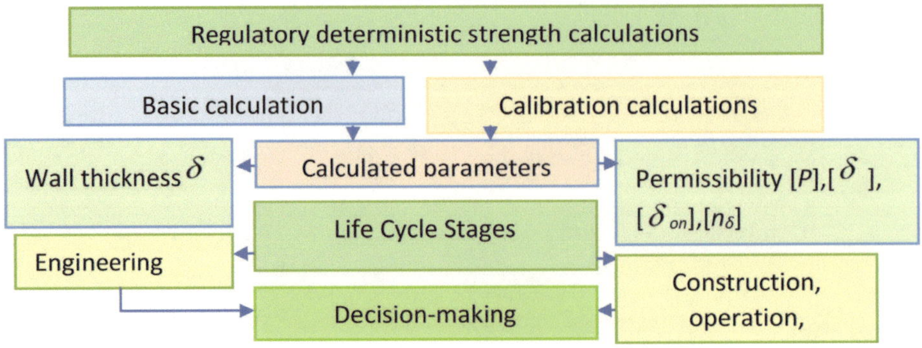


**Figure 8.**
*Regulatory basic and verification calculations for different stages of the pipeline life cycle.*

- Validation of selected and assigned mechanical properties $\sigma_{on}$, $(\sigma_T, \sigma_в)$ with known $p, D_в, \delta, n_\sigma$

$$\sigma_{on} \geq \frac{pD_в}{2\delta} \cdot n_\sigma. \tag{45}$$

- Check by (6) the permissible wall thickness $[\delta]$ at

$$[\delta] \geq \frac{pD_в n_\sigma}{2\sigma_{on}}. \tag{46}$$

- Checking the allowable strength margin$[n_\sigma]$with known

$$n_\sigma \geq \frac{\sigma_{on} \cdot 2\delta}{pD_в}. \tag{47}$$

In deterministic calculations according to (1) and (2), a systematic analysis of uncertainty factors affecting the quantities $n_\sigma$, $n$, $m$, $K_I$, and $K_н$ is carried out. These factors [6–10] included such factors as:

- The effect of the absolute dimensions of the sections $(D_в, \delta)$
- Type of stress–strain state (components of the main stress $\sigma_1, \sigma_2, \sigma_3$)
- Temperature–time change in mechanical properties $\sigma_T, \sigma_в$, which determines the processes of aging and degradation
- Availability of welded joints with altered properties

### 3.2 Databases for calculations

On the basis of statistical measurements and estimates of all specified design parameters (pressures $p$, mechanical properties $\sigma_T, \sigma_в$, geometrical dimensions $\delta$ and $D_в$ with variations within {min, max}), first of all, the determination of their average (median) values becomes important

$$R_{пm} = \frac{1}{n_п} \sum R_{пi}, \tag{48}$$

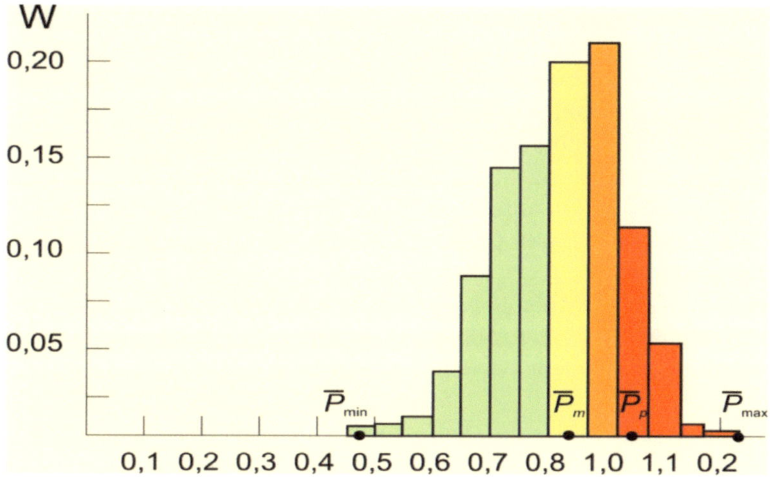


**Figure 9.**
*Histograms of pressures and strengths.*

where $n_{\text{п}}$ is the number of measurements of calculated parameters $R_{\text{п}i}$.

According to the obtained statistical information on the parameters $R_{ni}$, the corresponding histograms are constructed by the intervals of their values. For example, **Figure 9** shows the change in the main design parameters—pressure $p$ and ultimate strength $\sigma_{в}$ [10].

Inequality (47) with the parameters included in it, as well as the data from **Figure 9**, are the basis for calculating the determination of $R_p$ taking into account the statistical and probabilistic dispersion characteristics. At the same time, the assigned parameters $R_p$ should correspond to the inequality systems

$$P_{\min} \le P_m \le P_p \le P_{\max} \tag{49}$$

$$\{\sigma_T, \sigma_{в}\}_{\min} \le \{\sigma_T, \sigma_{в}\}_p \le \{\sigma_T, \sigma_{в}\}_m \le \{\sigma_T, \sigma_{в}\} \tag{50}$$

$$D_{\min} \le D_m \le D_p \le D_{\max} \tag{51}$$

$$\delta_{\min} \le \delta_p \le \delta_m \le \delta_{\max} \tag{52}$$

For the design stage, statistical analysis of the design parameters using expressions (8)–(12) is done using factory test data for sheet blanks for pipes ($\delta$), pipes ($\delta, D$), and laboratory samples for static tension ($\sigma_{\text{T}}, \sigma_{\text{в}}$).The values obtained $\delta_p, D_p$, $(\sigma_{\text{T}}, \sigma_{\text{в}})_p$ are entered in the technical conditions or standards. They are the basis of deterministic calculations.

If these measurements are carried out at the stage of construction or operation, then the data obtained from (49)–(52) are included in deterministic calibration calculations and expressions (44)–(47) (**Figure 10**).

Technical diagnostics of trunk pipelines (mainly using in-line diagnostics [11]) shows that the most significant from the point of view of strength is the decrease in time $\tau$ wall thickness parameters $\delta$ due to such processes as uniform and uneven corrosion, formation and development of cracks of corrosion, and cyclical nature. These processes, as a rule, increase the variation of values $\delta$ in (52).

The statistical variation of diameters in (11) at the stage of manufacturing, construction, and operation of the linear part of trunk pipelines is small ($0.99 \le D_i/D_m \le 1.01$) and can be neglected in deterministic and statistical strength calculations according to (1). However, if during operation there are significant reductions in wall thickness $\delta$, then a significant local increase in diameter $D_в$ (by 5–10 due to plastic deformations with the formation of shape defects) is possible. Similar processes of loss of shape and increase in $D_в$ are possible with nonstandard bending of pipelines with loss of stability and formation of corrugations.

The change in the average values and variation of the design characteristics of strength ($\sigma_B, \sigma_T$) according to (50) is associated with the instability of technological processes for the production of pipe steels, rolling and heat treatment of sheets, pipe manufacturing, construction of pipeline systems, as well as temporary processes of aging and degradation.

**Figures 11** and **12** show histograms and distribution functions of the mechanical properties of a long-term (up to 50 years) operated 17G1S tubular steel. This information is used in the implementation of calculations for paragraphs 2.1–2.3.

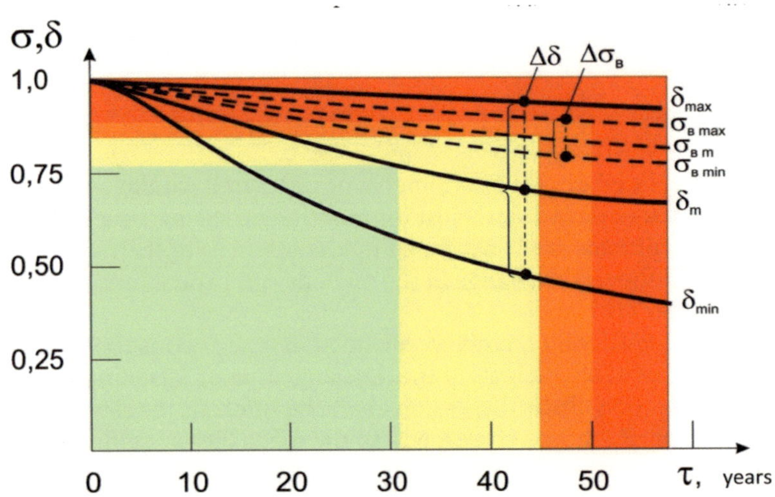


**Figure 10.**
*Statistics on the relative decrease or increase in operating time.*

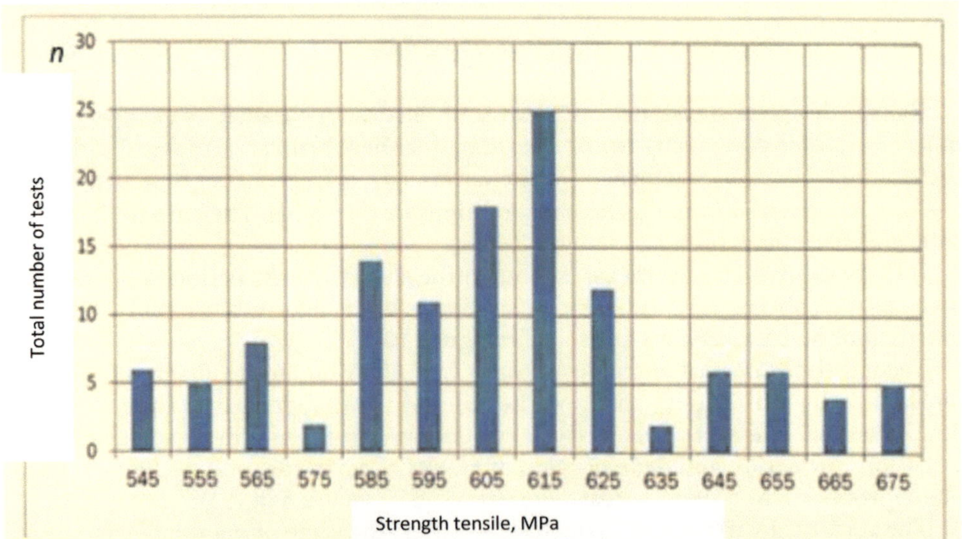


**Figure 11.**
*Strength tensile histograms $\sigma_B$ (total number of tests n = 160).*

In the process of development (in time $\tau$) of pipeline transportation of hydrocarbons in Russia and abroad, three trends remain dominant using deterministic (D), statistical (C), and probabilistic (P) methods (**Table 1** and **Figure 13**).

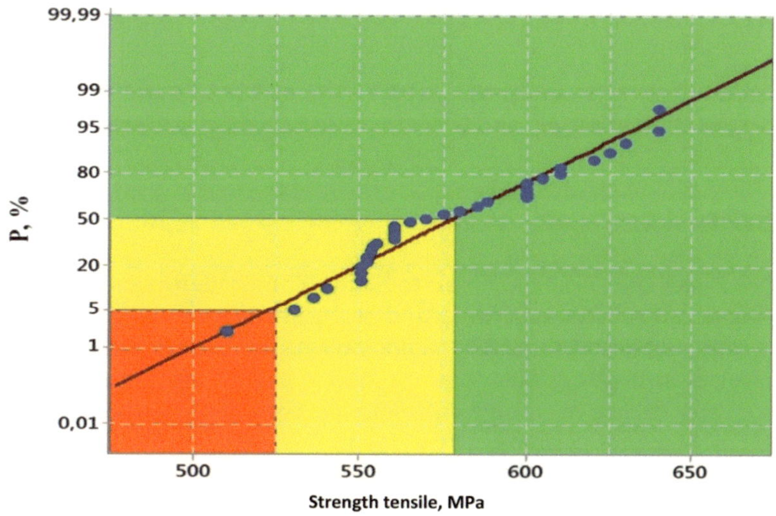


**Figure 12.**
*Distribution function of the ultimate strength of pipe steels 17G1S.*

| No. | Description | Symbols | Value |
|---|---|---|---|
| 1. | Coefficient of working conditions | $m$ | 0.6–0,9 |
| 2. | Load reliability factor | $K_1$ | 1.1–1,5 |
| 3. | Material reliability factor | $K_2$ | 1.34–1,55 |
| 4. | Reliability factor to destination | $K_н$ | 1.0–1,05 |

**Table 1.**
*Calculated standard values of coefficients.*

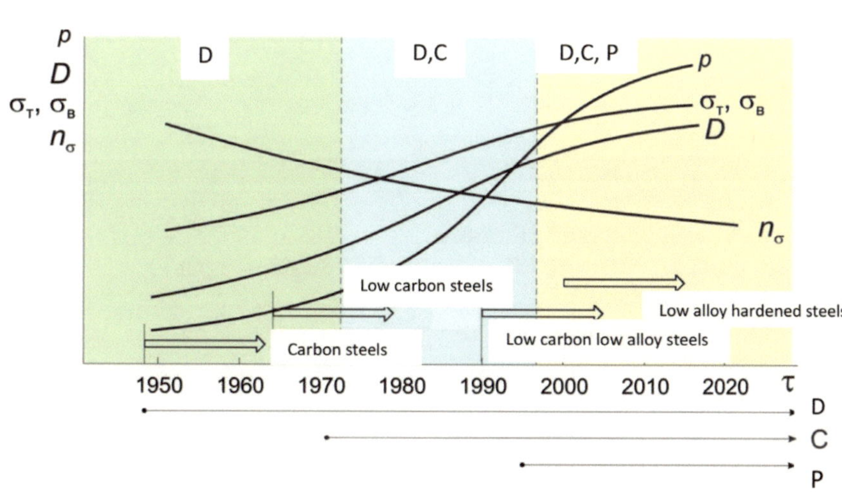


**Figure 13.**
*The main regularities of changes in the estimated parameters of pipelines.*

- Increase in diameters of pipelines D (from 250 to 300 mm to 1200 to 1400 mm) and pressures p (from 2.0 to 2.5 MPa to 14.0 to 16.0 MPa)
- Increase of mechanical properties of pipe hoists (yield strengths $\sigma_T$) from 200 to 250 МПа to 600 to 800 MPa and strength $\sigma_u$ from 400 to 450 MPa to 700 to 900 MPa
- Reduction of strength margins $n_T$ (от 1.8–3,2 до 1.2–1,5) and $n_u$ (от 2.4–3,5 до 1.6–1,8) in expression (3) and the estimated coefficients in expression (5)

## 4. Conclusion

The above data (pp. 1–3) allow you to build a comprehensive strength analysis system using deterministic, statistical, and probabilistic methods for various components, taking into account design, technological, and operational factors.

Deterministic strength calculations are used as part of regulatory national and international approaches for design calculations. They apply to the majority of functioning oil pipelines systems.

Statistical calculations become relevant in cases where during operational process the diagnostic studies of the condition of pipelines are carried out or during construction routine tests and during operation defects of technological and operational origin are detected. Probabilistic calculations are necessary on the most dangerous sections of the pipeline (in case of crossing water barriers, transport infrastructures, and laying offshore pipelines).

In all cases, safety margins are linked to the normative standard documentation (deterministic approach), the results of diagnostics and defect identification and measurement (statistical approach), and taking into account the most dangerous operating conditions (probabilistic approach).

## Author details

Dmitry Neganov* and Nikolay Makhutov
The Pipeline Transport Institute (Transneft R&D, LLC), Russia

*Address all correspondence to: neganovda@niitnn.transneft.ru

## References

[1] SNIP 2.05.06-85. Trunk Pipelines. USSR; 1986

[2] API 579/ASMEFFS-1 Fitness for Service. USA; 2007

[3] DIN. Stainless Welded Pipe DIN 17457. Germany; 1985

[4] Mazur II, Ivantsov OM. Safety of Pipeline Systems. Moscow: ITS-ELIMA; 2004. 1104p

[5] Radionova SG, Zhulina SA, Makhutov NA, Gadenin MM, Lisin YV, Neganov DA, et al. Prospects for research in the field of risk analysis to improve government regulation and improve the safety of petrochemical complex. Safety in Industry. 2017;**9**:5-13

[6] Makhutov NA. Strength and Safety: Basic and Applied Research. Novosibirsk: Nauka; 2008. 528p

[7] Lisin YV, Makhutov NA, Nadein VA, Neganov DA. In: Kostogryzov A, editor. Probabilistic Modeling in System Engineering. Probabilistic Analysis of Transportation Systems for Oil and Natural Gas. Rijeka, Croatia: InTechOpen; 2018. pp. 81-103

[8] Lisin YV, Makhutov NA, Neganov DA, Skorodumov SV, Studenov EP. Comprehensive mechanical tests to calculate the strength of the main pipeline transport and petroleum products. Factory Laboratory. Diagnostics of Materials. 2018;**84**(4): 47-59

[9] Security of Russia. Security of Energy Storage and Transportation Facilities. MGOF/Znanie; 2019. 928p

[10] Makhutov NA, Permyakov VV. Resource Safe Operation of Pipelines. Novosibirsk: Nauka; 2005. 516p

Section 2

# From a Formalization of Uncertainities to Probabilistic Modeling, Reasonable Control and Artificial Intelligence

Chapter 7

# From Asymptotic Normality to Heavy-Tailedness via Limit Theorems for Random Sums and Statistics with Random Sample Sizes

*Victor Korolev and Alexander Zeifman*

## Abstract

This chapter contains a possible explanation of the emergence of heavy-tailed distributions observed in practice instead of the expected normal laws. The bases for this explanation are limit theorems for random sums and statistics constructed from samples with random sizes. As examples of the application of general theorems, conditions are presented for the convergence of the distributions of random sums of independent random vectors *with finite covariance matrices* to multivariate elliptically contoured stable and Linnik distributions. Also, conditions are presented for the convergence of the distributions of asymptotically normal (in the traditional sense) statistics to multivariate Student distributions. The joint asymptotic behavior of sample quantiles is also considered.

**Keywords:** random sum, random sample size, multivariate normal mixtures, heavy-tailed distributions, multivariate stable distribution, multivariate Linnik distribution, Mittag-Leffler distribution, multivariate Student distribution, sample quantiles
**AMS 2000 Subject Classification**: 60F05, 60G50, 60G55, 62E20, 62G30

## 1. Introduction

In many situations related to experimental data analysis, one often comes across the following phenomenon: although conventional reasoning based on the central limit theorem of probability theory concludes that the expected distribution of observations should be normal, instead, the statistical procedures expose the noticeable non-normality of real distributions. Moreover, as a rule, the observed non-normal distributions are more leptokurtic than the normal law, having sharper vertices and heavier tails. These situations are typical in the financial data analysis (see, e.g., Chapter 4 in [1] or Chapter 8 in [2] and references therein), in experimental physics (see, e.g., [3]), and other fields dealing with statistical analysis of experimental data. Many attempts were undertaken to explain this heavy-tailedness. Most significant theoretical breakthrough is usually associated with the results of B. Mandelbrot and others who proposed, instead of the standard central limit theorem, to use reasoning based on limit theorems for sums of random summands with infinite variances (see, e.g., [4]) resulting in non-normal stable laws as heavy-tailed models of the distributions of experimental data. However, first, in most cases the key assumption within this approach, the infiniteness of the variances of elementary summands can hardly be believed to hold in practice and, second, although more heavy-tailed than the normal law, the real distributions often turn out to be more light-tailed than the stable laws.

In this work, in order to give a more realistic explanation of the observed non-normality of the distributions of real data, an alternative approach based on limit theorems for statistics constructed from samples with random sizes is developed. Within this approach, it becomes possible to obtain arbitrarily heavy tails of the data distributions without assuming the non-existence of the moments of the observed characteristics.

This work was inspired by the publication of the paper [5] in which, based on the results of [6], a particular case of random sums was considered. One more reason for writing this work was the recent publication [7], the authors of which reproduced some results of [8, 9] without citing these earlier papers.

Here we give a more general description of the transformation of the limit distribution of a sum of independent random variables or another statistic (i.e., of a measurable function of a sample) under the replacement of the non-random number of summands or the sample size by a random variable. General limit theorems are proved (Section 3). Section 4 contains some comments on heavy-tailedness of scale mixtures of normal distributions. As examples of the application of general theorems, conditions are presented for the convergence of the distributions of random sums of independent random vectors *with finite covariance matrices* to multivariate elliptically contoured stable and Linnik distributions (Section 5). Also, conditions are presented for the convergence of the distributions of asymptotically normal (in the traditional sense) statistics to multivariate Student distributions (Section 6).

In Section 7, the joint asymptotic behavior of sample quantiles is considered. In applied researches related to risk analysis, such characteristic as VaR (Value-at-Risk) is very popular. Formally, VaR is a certain quantile of the observed risky value. Therefore, the joint asymptotic behavior of sample quantiles in samples with random sizes is considered in detail in Section 7 as one more example of the application of the general theorem proved in Section 3. In this section, we show how the proposed technique can be applied to the continuous-time case assuming that the sample size increases in time following a Cox process. One more interpretation of this setting is related with an important case where the sample size has the mixed Poisson distribution.

In classical problems of mathematical statistics, the size of the available sample, that is, the number of available observations, is traditionally assumed to be deterministic. In the asymptotic settings, it plays the role of infinitely increasing *known* parameter. At the same time, in practice very often the data to be analyzed are collected or registered during a certain period of time and the flow of informative events each of which brings a next observation forms a random point process. Therefore, the number of available observations is unknown till the end of the process of their registration and also must be treated as a (random) observation. For example, this is so in insurance statistics where, during different accounting periods, different numbers of insurance events (insurance claims and/or insurance contracts) occur and in high-frequency financial statistics where the number of events in a limit order book during a time unit essentially depends on the intensity of order flows. Moreover, contemporary statistical procedures of insurance and financial mathematics do take this circumstance into consideration as one of possible ways of dealing with heavy tails. However, in other fields such as medical statistics or quality control, this approach has not become conventional; yet, the number of patients with a certain disease varies from month to month due to seasonal factors or from year to year due to some epidemic reasons and the number of failed items varies from lot to lot. In these cases, the number of available observations as well as the observations themselves is unknown beforehand and should be treated as random to avoid underestimation of risks or error probabilities.

Therefore, it is quite reasonable to study the asymptotic behavior of general statistics constructed from samples with random sizes for the purpose of construction of suitable and reasonable asymptotic approximations. As this is so, to obtain non-trivial asymptotic distributions in limit theorems of probability theory and mathematical statistics, an appropriate centering and normalization of random variables and vectors under consideration must be used. It should be especially noted that to obtain reasonable approximation to the distribution of the basic statistics, both centering and normalizing values should be non-random. Otherwise, the approximate distribution becomes random itself and, for example, the problem of evaluation of quantiles or significance levels becomes senseless.

In asymptotic settings, statistics constructed from samples with random sizes are special cases of random sequences with random indices. The randomness of indices usually leads to the limit distributions for the corresponding random sequences being heavy-tailed even in the situations where the distributions of non-randomly indexed random sequences are asymptotically normal (see, e.g., [2, 8, 10]).

Many authors noted that the asymptotic properties of statistics constructed from samples with random samples differ from those of the asymptotically normal statistics in the classical sense. To illustrate this, we will repeatedly cite [11] where the following example is given. Let $X_{(1)}, \dots, X_{(n)}$ be order statistics constructed from the sample $X_1, \dots, X_n$. It is well known (see, e.g., [12]) that in the standard situation the sample median is asymptotically normal. At the same time, in [11] it was demonstrated that if the sample size $N_n$ has the geometric distribution with expectation $n$, then the normalized sample median $\sqrt{n}\left(X_{([N_n/2]+1)} - \operatorname{med} X_1\right)$ has the limit distribution function

$$\Psi(x) = \frac{1}{2}\left(1 + \frac{x}{\sqrt{2 + x^2}}\right) \tag{1}$$

(the Student distribution with two degrees of freedom) which has such heavy tails that its moments of orders $\delta \geq 2$ do not exist. In general, as it was shown in [8], if a statistic that is asymptotically normal in the traditional sense is constructed on the basis of a sample with random size having negative binomial distribution, then instead of the expected normal law, the Student distribution with power-type decreasing heavy tails appears as an asymptotic law for this statistic.

## 2. Notation and definitions: auxiliary results

Let $r \in \mathbb{N}$. We will consider random elements taking values in the $r$-dimensional Euclidean space $\mathbb{R}^r$.

Assume that all the random variables and random vectors are defined on one and the same probability space $(\Omega, \mathcal{A}, \mathrm{P})$. By the measurability of a random field, we will mean its measurability as a function of two variates, an elementary outcome and a parameter, with respect to the Cartesian product of the $\sigma$-algebra $\mathcal{A}$ and the Borel $\sigma$-algebra $\mathcal{B}(\mathbb{R}^r)$ of subsets of $\mathbb{R}^r$.

The distribution of a random vector $\xi$ with respect to the measure P will be denoted $\mathcal{L}(\xi)$. The weak convergence, the coincidence of distributions, and the convergence in probability with respect to a specified probability measure will be denoted by the symbols $\Rightarrow$, $\stackrel{d}{=}$, and $\stackrel{P}{\longrightarrow}$, respectively.

Let $\Sigma$ be a positive definite matrix. The normal distribution in $\mathbb{R}^r$ with zero vector of expectations and covariance matrix $\Sigma$ will be denoted $\Phi_\Sigma$. This distribution is defined by its density

$$\phi(\mathbf{x}) = \frac{\exp\left\{-\frac{1}{2}\mathbf{x}^T\Sigma^{-1}\mathbf{x}\right\}}{(2\pi)^{r/2}|\Sigma|^{1/2}}, \qquad \mathbf{x}\in\mathbb{R}^r.$$

The characteristic function $\mathfrak{f}^Y(\mathbf{t})$ of a random variable $Y$ such that $\mathcal{L}(Y) = \Phi_\Sigma$ has the form

$$\mathfrak{f}^Y(\mathbf{t}) \equiv \mathrm{E}\exp\left\{i\mathbf{t}^\top Y\right\} = \exp\left\{-\frac{1}{2}\mathbf{t}^\top\Sigma\mathbf{t}\right\}, \quad \mathbf{t}\in\mathbb{R}^r. \tag{2}$$

Consider a sequence $\{S_n\}_{n\geq 1}$ of random elements taking values in $\mathbb{R}^r$. Let $\Xi(\mathbb{R}^r)$ be the set of all nonsingular linear operators acting from $\mathbb{R}^r$ to $\mathbb{R}^r$. The identity operator acting from $\mathbb{R}^r$ to $\mathbb{R}^r$ will be denoted $I_r$. Assume that there exist sequences $\{B_n\}_{n\geq 1}$ of operators from $\Xi(\mathbb{R}^r)$ and $\{a_n\}_{n\geq 1}$ of elements from $\mathbb{R}^r$ such that

$$Y_n \equiv B_n^{-1}(S_n - a_n) \Rightarrow Y \quad (n\to\infty) \tag{3}$$

where $Y$ is a random element whose distribution with respect to P will be denoted $H$, $H = \mathcal{L}(Y)$.

Along with $\{S_n\}_{n\geq 1}$, consider a sequence of integer-valued positive random variables $\{N_n\}_{n\geq 1}$ such that for each $n\geq 1$ the random variable $N_n$ is independent of the sequence $\{S_k\}_{k\geq 1}$. Let $c_n\in\mathbb{R}^r$, $D_n\in\Xi(\mathbb{R}^r)$, and $n\geq 1$. Now, we will formulate sufficient conditions for the weak convergence of the distributions of the random elements $Z_n = D_n^{-1}(S_{N_n} - c_n)$ as $n\to\infty$.

For $g\in\mathbb{R}^r$, denote $W_n(g) = D_n^{-1}(B_{N_n}g + a_{N_n} - c_n)$. In [13, 14], the following theorem was proved, which establishes sufficient conditions of the weak convergence of multivariate random sequences with independent random indices under operator normalization.

Theorem 1 [14]. *Let $\|D_n^{-1}\|\to\infty$ as $n\to\infty$ and let the sequence of random variables $\left\{\|D_n^{-1}B_{N_n}\|\right\}_{n\geq 1}$ be tight. Assume that there exist a random element $Y$ with distribution $H$ and an $r$-dimensional random field $W(g)$, $g\in\mathbb{R}^r$, such that* (3) *holds and*

$$W_n(g) \Rightarrow W(g) \qquad (n\to\infty)$$

*for $H$-almost all $g\in\mathbb{R}^r$. Then the random field $W(g)$ is measurable, linearly depends on $g$ and*

$$Z_n \Rightarrow W(Y) \qquad (n\to\infty),$$

*where the random field $W(\cdot)$ and the random element $Y$ are independent.*

Now, consider an auxiliary statement dealing with the identifiability of a special family of mixtures of multivariate normal distributions. Let $U$ be a nonnegative random variable. The symbol $\mathrm{E}\Phi_{U\Sigma}(\cdot)$ will denote the distribution which for each Borel set $A$ in $\mathbb{R}^r$ is defined as

$$\mathsf{E}\Phi_{U\Sigma}(A) = \int_0^\infty \Phi_{u\Sigma}(A)d\mathsf{P}(U < u).$$

Let $\mathcal{U}$ be the set of all nonnegative random variables.

It is easy to see that if $Y$ is a random vector such that $\mathcal{L}(Y) = \Phi_\Sigma$ independent of $U$, then $\mathsf{E}\Phi_{U\Sigma} = \mathcal{L}(\sqrt{U}Y)$.

Lemma 1. *Whatever nonsingular covariance matrix $\Sigma$ is, the family of distributions $\{\mathsf{E}\Phi_{U\Sigma}(\cdot) : U \in \mathcal{U}\}$ is identifiable in the sense that if $U_1 \in \mathcal{U}$, $U_2 \in \mathcal{U}$, and*

$$\mathsf{E}\Phi_{U_1\Sigma}(A) = \mathsf{E}\Phi_{U_2\Sigma}(A) \tag{4}$$

for any set $A \in \mathcal{B}(\mathbb{R}^r)$, then $U_1 \stackrel{d}{=} U_2$.

The proof of this lemma is very simple. If $U \in \mathcal{U}$, then the characteristic function $\mathfrak{v}^{(U)}(\mathbf{t})$ corresponding to the distribution $\mathsf{E}\Phi_{U\Sigma}(\cdot)$ has the form

$$\begin{aligned}\mathfrak{v}^{(U)}(\mathbf{t}) &= \int_0^\infty \exp\left\{-\frac{1}{2}\mathbf{t}^T(u\Sigma)\mathbf{t}\right\}d\mathsf{P}(U < u) = \int_0^\infty \exp\left\{-\frac{1}{2}u\mathbf{t}^T\Sigma\mathbf{t}\right\}d\mathsf{P}(U < u)\\ &= \int_0^\infty \exp\{-us\}d\mathsf{P}(U < u), \quad s = \frac{1}{2}\mathbf{t}^T\Sigma\mathbf{t}, \quad \mathbf{t} \in \mathbb{R}^r,\end{aligned} \tag{5}$$

But on the right-hand side of (5), there is the Laplace-Stieltjes transform of the random variable $U$. From (4), it follows that $\mathfrak{v}^{(U_1)}(\mathbf{t}) \equiv \mathfrak{v}^{(U_2)}(\mathbf{t})$ whence by virtue of (5) the Laplace-Stieltjes transforms of the random variables $U_1$ and $U_2$ coincide, whence, in turn, it follows that $U_1 \stackrel{d}{=} U_2$. The lemma is proved.

Remark 1. When proving Lemma 1, we established a simple but useful by-product result: if $\psi(s)$ is the Laplace-Stieltjes transform of the random variable $U$, then the characteristic function $\mathfrak{v}^{(U)}(\mathbf{t})$ corresponding to the distribution $\mathsf{E}\Phi_{U\Sigma}$ has the form

$$\mathfrak{v}^{(U)}(\mathbf{t}) = \psi\left(\frac{1}{2}\mathbf{t}^\top\Sigma\mathbf{t}\right), \quad \mathbf{t} \in \mathbb{R}^r. \tag{6}$$

## 3. General theorems

First, consider the case where the random vectors $\{S_n\}_{n\geq 1}$ are formed as growing sums of independent random variables. Namely, let $X_1, X_2, \ldots$ be independent $r$-valued random vectors, and for $n \in \mathbb{N}$ let

$$S_n = X_1 + \ldots + X_n.$$

Consider a sequence of integer-valued positive random variables $\{N_n\}_{n\geq 1}$ such that for each $n \geq 1$ the random variable $N_n$ is independent of the sequence $\{S_k\}_{k\geq 1}$. Let $\{b_n\}_{n\geq 1}$ be an infinitely increasing sequence of positive numbers such that

$$\mathcal{L}\left(\frac{S_n}{\sqrt{b_n}}\right) \Rightarrow \Phi_\Sigma \tag{7}$$

as $n \to \infty$, where $\Sigma$ is some positive definite matrix.

Let $\{d_n\}_{n\geq 1}$ be an infinitely increasing sequence of positive numbers. As $Z_n$ take the scalar normalized random vector

$$Z_n = \frac{S_{N_n}}{\sqrt{d_n}}.$$

Theorem 2. *Let $N_n \to \infty$ in probability as $n \to \infty$. Assume that the random variables $X_1, X_2, \ldots$ satisfy condition (6) with an asymptotic covariance matrix $\Sigma$. Then a distribution F such that*

$$L(Z_n) \Rightarrow F \quad (n \to \infty), \tag{8}$$

exists if and only if there exists a distribution function $V(x)$ satisfying the conditions

i. $V(x) = 0$ *for* $x < 0$;

ii. *for any* $A \in \mathcal{B}(\mathbb{R}^r)$,

$$F(A) = \mathrm{E}\Phi_{U\Sigma}(A) = \int_0^\infty \Phi_{u\Sigma}(A) dV(u), \quad x \in \mathbb{R}^1;$$

iii. $\mathrm{P}(b_{N_n} < d_n x) \Rightarrow V(x), n \to \infty.$

Proof. *The "if" part*. We will essentially exploit Theorem 1. For each $n \geq 1$, set $a_n = c_n = 0, B_n = D_n = \sqrt{d_n} I_r$. For the convenience of notation, introduce a random variable $U$ with the distribution function $V(x)$. Note that the conditions of the theorem guarantee the tightness of the sequence of random variables

$$\|D_n^{-1} B_{N_n}\| = \sqrt{\frac{b_{N_n}}{d_n}}, \quad n = 1, 2, \ldots$$

implied by its weak convergence to the random variable $\sqrt{U}$. Further, in the case under consideration, we have $W_n(g) = \sqrt{b_{N_n}/d_n} \cdot g, g \in \mathbb{R}^r$. Therefore, the condition $N_n/d_n \Rightarrow U$ implies $W_n(g) \Rightarrow \sqrt{U} g$ for all $g \in \mathbb{R}^r$. Condition (7) means that in the case under consideration, $H = \Phi_\Sigma$. Hence, by Theorem 1, $Z_n \Rightarrow \sqrt{U} Y$ where $Y$ is a random element with the distribution $\Phi_\Sigma$ independent of the random variable $U$. It is easy to see that the distribution of the random element $\sqrt{U} Y$ coincides with $\mathrm{E}\Phi_{U\Sigma}(\cdot)$ where the matrix $\Sigma$ satisfies (7).

*The "only if" part*. Let condition (8) hold. Make sure that the sequence $\{\|D_n^{-1} B_{N_n}\|\}_{n \geq 1}$ is tight. Let $Y$ be a random element with the distribution $\Phi_\Sigma$. There exist $\delta > 0$ and $R > 0$ such that

$$\mathrm{P}(\|Y\| > R) > \delta. \tag{9}$$

For $R$ specified above and an arbitrary $x > 0$, we have

$$\mathrm{P}(\|Z_n\| > x) \geq \mathrm{P}\left(\left\|\frac{S_{N_n}}{\sqrt{d_n}}\right\| > x; \left\|\frac{S_{N_n}}{\sqrt{b_{N_n}}}\right\| > R\right) =$$

$$= \mathrm{P}\left(\sqrt{\frac{b_{N_n}}{d_n}} > x \cdot \left\|\frac{S_{N_n}}{\sqrt{b_{N_n}}}\right\|^{-1}; \left\|\frac{S_{N_n}}{\sqrt{b_{N_n}}}\right\| > R\right) \geq \mathrm{P}\left(\sqrt{\frac{b_{N_n}}{d_n}} > \frac{x}{R}; \left\|\frac{S_{N_n}}{\sqrt{b_{N_n}}}\right\| > R\right) =$$

$$= \sum_{k=1}^\infty \mathrm{P}(N_n = k) \mathrm{P}\left(\sqrt{\frac{b_k}{d_n}} > \frac{x}{R}; \left\|\frac{S_k}{\sqrt{b_k}}\right\| > R\right) = \sum_{k=1}^\infty \mathrm{P}(N_n = k) \mathrm{P}\left(\sqrt{\frac{b_k}{d_n}} > \frac{x}{R}\right) P\left(\left\|\frac{S_k}{\sqrt{b_k}}\right\| > R\right) \tag{10}$$

(the last equality holds since any constant is independent of any random variable). Since by (7) the convergence $S_k/\sqrt{b_k} \Rightarrow Y$ takes place as $k \to \infty$, from (9) it follows that there exists a number $k_0 = k_0(R, \delta)$ such that

$$\mathsf{P}\left(\left\|\frac{S_k}{\sqrt{b_k}}\right\| > R\right) > \delta/2$$

for all $k > k_0$. Therefore, continuing (10) we obtain

$$\mathsf{P}(\|Z_n\| > x) \geq \frac{\delta}{2}\sum_{k=k_0+1}^{\infty} \mathsf{P}(N_n = k)\mathsf{P}\left(\sqrt{\frac{b_k}{d_n}} > \frac{x}{R}\right) =$$

$$= \frac{\delta}{2}\left[\mathsf{P}\left(\sqrt{\frac{b_{N_n}}{d_n}} > \frac{x}{R}\right) - \sum_{k=1}^{k_0} \mathsf{P}(N_n = k)\mathsf{P}\left(\sqrt{\frac{b_k}{d_n}} > \frac{x}{R}\right)\right] \geq \frac{\delta}{2}\left[\mathsf{P}\left(\sqrt{\frac{b_{N_n}}{d_n}} > \frac{x}{R}\right) - \mathsf{P}(N_n \leq k_0)\right].$$

Hence,

$$\mathsf{P}\left(\sqrt{\frac{b_{N_n}}{d_n}} > \frac{x}{R}\right) \leq \frac{2}{\delta}\mathsf{P}(\|Z_n\| > x) + \mathsf{P}(N_n \leq k_0). \tag{11}$$

From the condition $N_n \xrightarrow{P} \infty$ as $n \to \infty$, it follows that for any $\epsilon > 0$ there exists an $n_0 = n_0(\epsilon)$ such that $\mathsf{P}(N_n \leq n_0) < \epsilon$ for all $n \geq n_0$. Therefore, with the account of the tightness of the sequence $\{Z_n\}_{n\geq 1}$ that follows from its weak convergence to the random element $Z$ with $\mathcal{L}(Z) = F$ implied by (8), relation (11) implies

$$\lim_{x\to\infty} sup_{n\geq n_0(\epsilon)}\mathsf{P}\left(\sqrt{\frac{b_{N_n}}{d_n}} > \frac{x}{R}\right) \leq \epsilon, \tag{12}$$

whatever $\epsilon > 0$ is. Now assume that the sequence

$$\|D_n^{-1}B_{N_n}\| = \sqrt{\frac{b_{N_n}}{d_n}}, \quad n = 1, 2, \ldots$$

is not tight. In that case, there exists an $\alpha > 0$ and sequences $\mathcal{N}$ of natural and $\{x_n\}_{n\in\mathcal{N}}$ of real numbers satisfying the conditions $x_n \uparrow \infty$ $(n \to \infty, n \in \mathcal{N})$ and

$$\mathsf{P}\left(\sqrt{\frac{b_{N_n}}{d_n}} > x_n\right) > \alpha, \quad n \in \mathcal{N}. \tag{13}$$

But, according to (12), for any $\epsilon > 0$ there exist $M = M(\epsilon)$ and $n_0 = n_0(\epsilon)$ such that

$$sup_{n\geq n_0(\epsilon)}\mathsf{P}\left(\sqrt{\frac{b_{N_n}}{d_n}} > M(\epsilon)\right) \leq 2\epsilon. \tag{14}$$

Choose $\epsilon < \alpha/2$ where $\alpha$ is the number from (13). Then for all $n \in \mathcal{N}$ large enough, in accordance with (13), the inequality opposite to (14) must hold. The obtained contradiction by the Prokhorov theorem proves the tightness of the sequence $\left\{\|D_n^{-1}B_{N_n}\|\right\}_{n\geq 1}$ or, which in this case is the same as that, of the sequence $\{b_{N_n}/d_n\}_{n\geq 1}$.

Introduce the set $W(Z)$ containing all nonnegative random variables $U$ such that $\mathrm{P}(Z \in A) = \mathrm{E}\Phi_{U\Sigma}(A)$ for any $A \in \mathcal{B}(\mathbb{R}^r)$. Let $L(\cdot,\cdot)$ be any probability metric that metrizes weak convergence in the space of random variables, or, which is the same in this context, n the space of distribution functions, say, the Lévy metric or the smoothed Kolmogorov distance. If $X_1$ and $X_2$ are random variables with the distribution functions $F_1$ and $F_2$ respectively, then we identify $L(X_1, X_2)$ and $L(F_1, F_2)$. Show that there exists a sequence of random variables $\{U_n\}_{n\geq 1}$, $U_n \in \mathcal{W}(Z)$, such that

$$L\left(\frac{b_{N_n}}{d_n}, U_n\right) \to 0 \quad (n \to \infty). \tag{15}$$

Denote

$$\beta_n = \inf\left\{L\left(\frac{b_{N_n}}{d_n}, U\right) : \quad U \in \mathcal{W}(Z)\right\}.$$

Prove that $\beta_n \to 0$ as $n \to \infty$. Assume the contrary. In that case, $\beta_n \geq \delta$ for some $\delta > 0$ and all $n$ from some subsequence $\mathcal{N}$ of natural numbers. Choose a subsequence $\mathcal{N}_1 \subseteq \mathcal{N}$ so that the sequence $\{b_{N_n}/d_n\}_{n\in\mathcal{N}_1}$ weakly converges to a random variable $U$ (this is possible due to the tightness of the family $\{b_{N_n}/d_n\}_{n\geq 1}$ established above). But then $W_n(g) \Rightarrow \sqrt{U}g$ $(n \to \infty, n \in \mathcal{N}_1)$ for any $g \in \mathbb{R}^r$. Applying Theorem 1 to $n \in \mathcal{N}_1$ with condition (7) playing the role of condition (3), we make sure that $U \in \mathcal{W}(Z)$, since condition (8) provides the coincidence of the limits of all weakly convergent subsequences. So, we arrive at the contradiction to the assumption that $\beta_n \geq \delta$ for all $n \in \mathcal{N}_1$. Hence, $\beta_n \to 0$ as $n \to \infty$.

For any $n = 1, 2, \ldots$, choose a random variable $U_n$ from $\mathcal{W}(Z)$ satisfying the condition

$$L\left(\frac{b_{N_n}}{d_n}, U_n\right) \leq \beta_n + \frac{1}{n}.$$

This sequence obviously satisfies condition (15). Now consider the structure of the set $\mathcal{W}(Z)$. This set contains all the random variable's defining the family of special mixtures of multivariate normal laws considered in Lemma 1, according to which this family is identifiable. So, whatever a random element $Z$ is, the set $\mathcal{W}(Z)$ contains at most one element. Therefore, actually condition (15) is equivalent to

$$\frac{b_{N_n}}{d_n} \Rightarrow U \qquad (n \to \infty),$$

that is, to condition (iii) of the theorem. The theorem is proved.

Corollary 1. *Under the conditions of Theorem 2, non-randomly normalized random sums $S_{N_n}/d_n$ are asymptotically normal with some covariance matrix $\Sigma'$ if and only if there exists a number $c > 0$ such that*

$$\frac{b_{N_n}}{d_n} \Rightarrow c \qquad (n \to \infty).$$

Moreover, in this case, $\Sigma' = c\Sigma$.

This statement immediately follows from Theorem 2 with the account of Lemma 1.

Now consider a formally more general setting.

Let $N_1, N_2, \ldots$ and $W_1, W_2, \ldots$ be random variables and random vectors, respectively, such that for each $n \geq 1$ the random variable $N_n$ takes only natural values and is independent of the sequence $W_1, W_2, \ldots$. Let

$$T_n = T_n(W_1, \ldots, W_n) = (T_{n,1}(W_1, \ldots, W_n), \ldots, T_{n,r}(W_1, \ldots, W_n))$$

be a statistic taking values in $\mathbb{R}^r$, $r \geq 1$. For each $n \geq 1$ define a random vector (random element) $T_{N_n}$ by setting

$$T_{N_n}(\omega) = T_{N_n(\omega)}\big(W_1(\omega), \ldots, W_{N_n(\omega)}(\omega)\big)$$

for every elementary outcome $\omega \in \Omega$.

We shall say that a statistic $T_n$ is *asymptotically normal* with the asymptotic covariance matrix $\Sigma$ if there exists a non-random $r$-dimensional vector $t$ such that

$$\mathcal{L}\big(\sqrt{n}(T_n - t)\big) \Rightarrow \Phi_\Sigma \quad (n \to \infty). \tag{16}$$

Examples of asymptotically normal statistics are well known. Under certain conditions, the property of asymptotic normality is inherent in maximum likelihood estimators, sample moments, sample quantiles, etc.

Our nearest aim is to describe the asymptotic behavior of the random elements $T_{N_n}$, that is, of statistics constructed from samples with random sizes $N_n$.

Again let $\{d_n\}_{n \geq 1}$ be an infinitely increasing sequence of positive numbers. Now set

$$Z_n = \sqrt{d_n}(T_{N_n} - t).$$

Theorem 3. *Let $N_n \to \infty$ in probability as $n \to \infty$. Assume that a statistic $T_n$ is asymptotically normal in the sense of* (16) *with an asymptotic covariance matrix $\Sigma$. Then a distribution $F$ such that*

$$\mathcal{L}(Z_n) \Rightarrow F \quad (n \to \infty),$$

*exists if and only if there exists a distribution function $V(x)$ satisfying the conditions.*

(i) $V(x) = 0$ *for* $x < 0$;
(ii) *for any* $A \in \mathcal{B}(\mathbb{R}^r)$

$$F(A) = \int_0^\infty \Phi_{u^{-1}\Sigma}(A) dV(u), \quad x \in \mathbb{R}^1;$$

(iii) $\mathrm{P}(N_n < d_n x) \Rightarrow V(x)$, $n \to \infty$.

The proof of Theorem 3 relies on Theorem 1 with (16) playing the role of (3) and Lemma 1 and differs from the proof of Theorem 2 only by that $b_{N_n}/d_n$ is replaced by $d_n/N_n$.

Corollary 2. *Under the conditions of Theorem 3 the statistic $T_{N_n}$ is asymptotically normal with some covariance matrix $\Sigma'$ if and only if there exists a number $c > 0$ such that*

$$\frac{N_n}{d_n} \Rightarrow c \quad (n \to \infty).$$

Moreover, in this case, $\Sigma' = c^{-1}\Sigma$.

This statement immediately follows from Theorem 2 with the account of Lemma 1.

## 4. Some remarks on the heavy-tailedness of scale mixtures of normals

The one-dimensional marginals of the multivariate limit law in Theorems 2 and 3 are scale mixtures of normals with zero means of the form $\mathrm{E}\Phi(x/U)$, $x \in \mathbb{R}$, where $\Phi(x)$ is the standard normal distribution function and $U$ is a nonnegative random variable. It turns out, although absolutely not so evident, that these distributions are *always* leptokurtic having sharper vertex and heavier tails than the normal law itself.

It is easy to see that

$$\mathrm{E}\Phi(x/U) = \mathrm{P}(X \cdot U < x), \quad x \in \mathbb{R},$$

where $X$ is a standard normal variable independent of $U$. First, as a measure of leptokurtosity, consider the excess coefficient which is traditionally used in (descriptive) statistics. Recall that for a random variable $Y$ with $\mathrm{E}Y^4 < \infty$, the excess coefficient (kurtosis) $\kappa(Y)$ is defined as

$$\kappa(Y) = \mathrm{E}\left(\frac{Y - \mathrm{E}Y}{\sqrt{\mathrm{D}Y}}\right)^4.$$

If $\mathrm{P}(X < x) = \Phi(x)$, then $\kappa(X) = 3$. Densities with sharper vertices (and, respectively, with heavier tails) than the normal density, have $\kappa > 3$, and $\kappa < 3$ for densities with more flat vertices.

Lemma 2. *Let $X$ and $U$ be independent random variables with finite fourth moments; moreover, let $\mathrm{E}X = 0$ and $\mathrm{P}(U \geq 0) = 1$. Then*

$$\kappa(XU) \;\geq\; \kappa(X).$$

*Furthermore, $\kappa(XU) = \kappa(X)$ if and only if $\mathrm{P}(U = \mathrm{const}) = 1$.*

For the proof see [10].

So, if $X$ is a standard normal random variable and $U$ is a nonnegative random variable with $\mathrm{E}U^4 < \infty$ independent of $X$, then $\kappa(X \cdot U) \geq 3$ and $\kappa(X \cdot U) = 3$ if and only if $U$ is non-random.

Using the Jensen inequality, we can easily obtain one more inequality directly connecting the tails of the normal mixtures with the tails of the normal distribution.

Lemma 3. *Assume that the random variable $U$ satisfies the normalization condition $\mathrm{E}U^{-1} = 1$. Then*

$$1 - \mathrm{E}\Phi(x/U) \geq 1 - \Phi(x), \qquad x > 0.$$

From Lemma 3, it follows that if $X$ is the standard normal random variable and $U$ is a nonnegative random variable independent of $X$ with $\mathrm{E}U^{-1} = 1$, then for any $x \geq 0$

$$\mathrm{P}(|X \cdot U| \geq x) \;\geq\; \mathrm{P}(|X| \geq x) \;\; (= 2[1 - \Phi(x)]),$$

that is, scale mixtures of normal laws are always more leptokurtic and have heavier tails than normal laws themselves.

The class of scale mixtures of normal laws is very rich and involves distributions with various character of decrease of tails. For example, this class contains Student distributions with arbitrary (not necessarily integer) number of degrees of freedom (and the Cauchy distribution included), symmetric stable distributions (see the "multiplication theorem" 3.3.1 in [15]), symmetric fractional stable distributions (see [16]), symmetrized gamma distributions with arbitrary shape and scale parameters (see [10]), and symmetrized Weibull distributions with shape parameters belonging to the interval $(0, 1]$ (see [17, 18]). As an example, in the next section, we will discuss the conditions for the convergence of the distributions of the statistics constructed from samples with random sizes to the multivariate Student distribution.

## 5. Convergence of the distributions of random sums of random vectors with finite covariance matrices to multivariate elliptically contoured stable and Linnik distributions

### 5.1 Convergence of the distributions of random sums of random vectors to multivariate stable laws

Let $\Sigma$ be a positive definite $(r \times r)$-matrix, $\alpha \in (0, 2]$. A random vector $Z_{\alpha,\Sigma}$ is said to have the (centered) elliptically contoured stable distribution $G_{\alpha,\Sigma}$ with characteristic exponent $\alpha$, if its characteristic function $\mathfrak{g}_{\alpha,\Sigma}(\mathbf{t})$ has the form

$$\mathfrak{g}_{\alpha,\Sigma}(\mathbf{t}) \equiv \mathrm{E}\exp\left\{i\mathbf{t}^{\top}X\right\} = \exp\left\{-\left(\mathbf{t}^{\top}\Sigma\mathbf{t}\right)^{\alpha/2}\right\}, \quad \mathbf{t}\in\mathbb{R}^r.$$

Univariate stable distributions are popular examples of heavy-tailed distributions. Their moments of orders $\delta \geq \alpha$ do not exist (the only exception is the normal law corresponding to $\alpha = 2$). Stable laws and only they can be limit distributions for sums of a non-random number of independent identically distributed random variables with infinite variance under linear normalization. Here it will be shown that they also can be limiting for *random* sums of random vectors with *finite covariance matrices*. The result of this subsection generalizes the main theorem of [19] to a multivariate case.

By $\zeta_\alpha$, we will denote a positive random variable with the one-sided stable distribution corresponding to the characteristic function

$$\mathfrak{g}_\alpha(t) = \exp\left\{-|t|^\alpha \exp\left\{-\frac{1}{2}i\pi\alpha \mathrm{sign} t\right\}\right\}, \qquad t\in\mathbb{R},$$

with $0 < \alpha \leq 1$ (for more details see [15] or [4]).

Let $\alpha \in (0, 2]$. It is known that, if $Y$ is a random vector such that $\mathcal{L}(Y) = \Phi_\Sigma$ independent of the random variable $\zeta_{\alpha/2}$, then

$$Z_{\alpha,\Sigma} \stackrel{d}{=} \sqrt{\zeta_{\alpha/2}} \cdot Y \qquad (17)$$

(see Proposition 2.5.2 in [4]). In other words,

$$G_{\alpha,\Sigma} = \mathrm{E}\Phi_{\zeta_{\alpha/2}\Sigma}. \qquad (18)$$

As in Section 3, let $X_1, X_2, \ldots$ be independent $r$-valued random vectors. For $n \in \mathbb{N}$, denote $S_n = X_1 + \ldots + X_n$. Consider a sequence of integer-valued positive random variables $\{N_n\}_{n\geq1}$ such that for each $n\geq1$ the random variable $N_n$ is independent of the sequence $\{S_k\}_{k\geq1}$. Let $\{b_n\}_{n\geq1}$ be an infinitely increasing sequence of positive numbers providing convergence (6) with some positive definite matrix $\Sigma$.

Theorem 4. *Let $N_n \longrightarrow \infty$ in probability as $n \longrightarrow \infty$. Assume that the random variables $X_1, X_2, \ldots$ satisfy condition* (7) *with an asymptotic covariance matrix $\Sigma$. Then*

$$L\left(\frac{S_{N_n}}{\sqrt{d_n}}\right) \Rightarrow G_{\alpha,\Sigma} \qquad (n \longrightarrow \infty)$$

with some infinitely increasing sequence of positive numbers $\{d_n\}_{n\geq1}$ and some $\alpha\in(0,2]$, if and only if

$$\frac{N_n}{d_n} \Rightarrow \zeta_{\alpha/2,1}$$

as $n \longrightarrow \infty$.

Proof. This theorem is a direct consequence of Theorem 2 with the account of relations (17) and (18).

### 5.2 Convergence of the distributions of random sums of random vectors with finite covariance matrices to multivariate elliptically contoured Linnik distributions

In 1953, Yu. V. Linnik [20] introduced the class of univariate symmetric probability distributions defined by the characteristic functions

$$\mathfrak{f}_\alpha^L(t) = \frac{1}{1+|t|^\alpha}, \qquad t\in\mathbb{R},$$

where $\alpha\in(0,2]$. Later, the distributions of this class were called *Linnik distributions* [21] or *α-Laplace distributions* [22]. Here the first term will be used since it has become conventional. With $\alpha = 2$, the Linnik distribution turns into the Laplace distribution corresponding to the density

$$f^\Lambda(x) = \frac{1}{2}e^{-|x|}, \qquad x\in\mathbb{R}.$$

A random variable with the Linnik distribution with parameter $\alpha$ will be denoted $L_{1,\alpha}$.

The Linnik distributions possess many interesting analytic properties (see, e.g., [17, 18] and the references therein) but, perhaps, most often Linnik distributions are recalled as examples of geometric stable distributions often used as heavy-tailed models of some statistical regularities in financial data [23, 24].

The multivariate Linnik distribution was introduced by D. N. Anderson in [25] where it was proved that the function

$$\mathfrak{f}_{\alpha,\Sigma}^{(L)}(\mathbf{t}) = \frac{1}{1+(\mathbf{t}^\top\Sigma\mathbf{t})^{\alpha/2}}, \qquad \mathbf{t}\in\mathbb{R}^r, \quad \alpha\in(0,2), \tag{19}$$

is the characteristic function of an $r$-variate probability distribution, where $\Sigma$ is a positive definite $(r\times r)$-matrix. In [25], the distribution corresponding to the characteristic function (19) was called *the r-variate Linnik distribution*. For the properties of the multivariate Linnik distributions, see [25, 26].

The $r$-variate Linnik distribution can also be defined in another way. For this purpose, recall that the distribution of a nonnegative random variable $M_\delta$ whose Laplace transform is

$$\psi_\delta(s) \equiv \mathsf{E}e^{-sM_\delta} = \frac{1}{1+s^\delta}, \qquad s \geq 0, \tag{20}$$

where $0 < \delta \leq 1$, is called *the Mittag-Leffler distribution*. It is another example of heavy-tailed geometrically stable distributions; for more details see for example, [17, 18] and the references therein. The Mittag-Leffler distributions are of serious theoretical interest in the problems related to thinned (or rarefied) homogeneous flows of events such as renewal processes or anomalous diffusion or relaxation phenomena, see [27, 28] and the references therein. In [18], it was demonstrated that

$$L_{1,\alpha} \stackrel{d}{=} Y_1 \cdot \sqrt{2M_{\alpha/2}}, \tag{21}$$

where $Y_1$ is a random variable with the standard univariate normal distribution independent of the random variable $M_{\alpha/2}$ with the Mittag-Leffler distribution with parameter $\alpha/2$.

Now let $Y$ be a random vector such that $\mathcal{L}(Y) = \Phi_\Sigma$, where $\Sigma$ is a positive definite $(r \times r)$-matrix, independent of the random variable $M_{\alpha/2}$. By analogy with (21), introduce the random vector $L_{r,\alpha,\Sigma}$ as

$$L_{r,\alpha,\Sigma} = \sqrt{2M_{\alpha/2}} \cdot Y.$$

Then, in accordance with what has been said in Section 2,

$$\mathcal{L}(L_{r,\alpha,\Sigma}) = E\Phi_{2M_{\alpha/2}\Sigma}. \tag{22}$$

The distribution (14) will be called *the ( centered) elliptically contoured multivariate Linnik distribution*.

Using Remark 1, we can easily make sure that the two definitions of the multivariate Linnik distribution coincide. Indeed, with the account of (20), according to Remark 1, the characteristic function of the random vector $L_{r,\alpha,\Sigma}$ defined by (22) has the form

$$\mathsf{E}\exp\left\{i\mathbf{t}^\top L_{r,\alpha,\Sigma}\right\} = \psi_{\alpha/2}(\mathbf{t}^\top\Sigma\mathbf{t}) = \frac{1}{1+(\mathbf{t}^\top\Sigma\mathbf{t})^{\alpha/2}} = \mathfrak{f}^{(L)}_{\alpha,\Sigma}(\mathbf{t}), \quad \mathbf{t} \in \mathbb{R}^r,$$

that coincides with Anderson's definition (19).

Our definition (22) together with Theorem 2 opens the way to formulate a theorem stating that the multivariate Linnik distribution can not only be limiting for geometric random sums of independent identically distributed random vectors with infinite second moments [29], but it can also be limiting for random sums of independent random vectors with finite covariance matrices.

Theorem 5. *Let $N_n \to \infty$ in probability as $n \to \infty$. Assume that the random variables $X_1, X_2, \ldots$ satisfy condition (7) with an asymptotic covariance matrix $\Sigma$. Then*

$$L\left(\frac{S_{N_n}}{\sqrt{d_n}}\right) \Rightarrow \mathcal{L}(L_{r,\alpha,\Sigma}) \qquad (n \to \infty)$$

with some infinitely increasing sequence of positive numbers $\{d_n\}_{n\geq 1}$ and some $\alpha \in (0, 2]$, if and only if

$$\frac{N_n}{d_n} \Rightarrow 2M_{\alpha/2}$$

as $n \to \infty$.

Proof. This theorem is a direct consequence of Theorem 2 with the account of relation (22).

## 6. Convergence of the distributions of asymptotically normal statistics to the multivariate Student distribution

The multivariate Student distribution is described, for example, in [30] (also see [31]). Consider an $r$-dimensional normal random vector $Y$ with zero vector of expectations and covariance matrix $\Sigma$. Assume that a random variable $W_\gamma$ has the chi-square distribution with parameter (the "number of degrees of freedom") $\gamma > 0$ (not necessarily integer) and is independent of $Y$. The distribution $P_{\gamma,\Sigma}$ of the random vector

$$Q_{\gamma,\Sigma} = \sqrt{\gamma/W_\gamma} \cdot Y \tag{23}$$

is called the *multivariate Student distribution* (with parameters $\gamma$ and $\Sigma$). For any $\mathbf{x} \in \mathbb{R}^r$ the distribution density of $Z$ has the form

$$p_{\gamma,\Sigma}(\mathbf{x}) = \frac{\Gamma(r+\gamma)/2)}{|\Sigma|^{1/2}\Gamma(\gamma/2)(\pi\gamma)^{r/2}} \cdot \frac{1}{\left(1+\frac{1}{\gamma}\mathbf{x}^T\Sigma^{-1}\mathbf{x}\right)^{(r+\gamma)/2}}.$$

According to Theorem 3, the multivariate Student distribution is the resulting transformation of the limit distribution of an asymptotically normal (in the sense of (16)) statistic under the replacement of the sample size by a random variable whose asymptotic distribution is chi-square. Consider this case in more detail.

Let $G_{m,m}(x)$ be the gamma-distribution function with the shape parameter coinciding with the scale parameter and equal to $m$:

$$G_{m,m}(x) = \begin{cases} 0 & \text{if } x \leq 0, \\ \dfrac{m^m}{\Gamma(m)} \displaystyle\int_0^x e^{-my} y^{m-1} dy & \text{if } x > 0. \end{cases}$$

Theorem 6. *Let $\gamma > 0$ be arbitrary, $\Sigma$ be a positive definite matrix and let $\{d_n\}_{n\geq 1}$ be an infinitely increasing sequence of positive numbers. Assume that $N_n \to \infty$ in probability as $n \to \infty$. Let a statistic $T_n$ be asymptotically normal in the sense of* (16). *Then the convegence*

$$\mathcal{L}\left(\sqrt{d_n}(T_{N_n} - t)\right) \Rightarrow P_{\gamma,\Sigma} \qquad (n \to \infty),$$

takes place if and only if

$$P(N_n < d_n x) \Rightarrow G_{\gamma/2,\gamma/2}(x), \quad n \to \infty,$$

where $G_{\gamma/2,\gamma/2}(x)$ is the gamma-distribution function with coinciding shape and scale parameters equal to $\gamma/2$.

Proof. This statement is a direct consequence of Theorem 3, representation (23) and Lemma 1.

Let $N_{p,m}$ be a random variable with the negative binomial distribution

$$P(N_{p,m} = k) = C_{m+k-2}^{k-1} p^m (1-p)^{k-1}, \quad k = 1, 2, \ldots \tag{24}$$

Here $m > 0$ and $p \in (0, 1)$ are parameters; for non-integer $m$, the quantity $C_{m+k-2}^{k-1}$ is defined as

$$C_{m+k-2}^{k-1} = \frac{\Gamma(m+k-1)}{(k-1)! \cdot \Gamma(m)}.$$

In particular, for $m = 1$, relation (24) determines the geometric distribution. It is well known that

$$EN_{p,m} = \frac{m(1-p)+p}{p},$$

so that $EN_{p,m} \to \infty$ as $p \to 0$.

As is known, the negative binomial distribution with natural $m$ admits an illustrative interpretation in terms of Bernoulli trials. Namely, the random variable with distribution (24) is the number of the Bernoulli trials held up to the $m$th failure, if the probability of the success in a trial is $1-p$.

Lemma 4. *For any fixed* $m > 0$

$$\lim_{p\to 0} sup_{x\in\mathbb{R}} |P\left(\frac{N_{p,m}}{EN_{p,m}} < x\right) - G_{m,m}(x)| = 0,$$

where $G_{m,m}(x)$ is the gamma-distribution function with the shape parameter coinciding with the scale parameter and equal to $m$.

The proof is a simple exercise on characteristic functions; for more details, see [8].

Corollary 3. *Let* $m > 0$ *be arbitrary. Assume that for each* $n \geq 1$ *the random variable* $N_n$ *has the negative binomial distribution with parameters* $p = \frac{1}{n}$ *and* $m$. *Let a statistic* $T_n$ *be asymptotically normal in the sense of* (16). *Then*

$$\mathcal{L}(\sqrt{mn}(T_{N_n} - t)) \Rightarrow P_{2m,\Sigma} \qquad (n \to \infty)$$

where $P_{2m,\Sigma}$ is the $r$-variate Student distribution with parameters $\gamma = 2m$ and $\Sigma$.

Proof. By Lemma 4 we have

$$\frac{N_n}{nm} = \frac{N_n}{EN_n} \cdot \frac{EN_n}{nm} = \frac{N_n}{EN_n} \cdot \frac{m(n-1)+1}{mr} = \frac{N_n}{EN_n}\left[1 + O\left(\frac{1}{n}\right)\right] \Rightarrow U_m$$

as $n \to \infty$ where $U_m$ is the random variable having the gamma-distribution function with coinciding shape and scale parameters equal to $m$. Now the desired assertion directly follows from Theorem 6.

Remark 2. The $r$-variate Cauchy distribution ($\gamma = 1$) appears in the situation described in Corollary 2 when the sample size $N_n$ has the negative binomial distribution with the parameters $p = \frac{1}{n}$, $m = \frac{1}{2}$, and $n$ is large.

Remark 3. In the case where the sample size $N_n$ has the negative binomial distribution with the parameters $p = \frac{1}{n}$, $m = 1$ (that is, the geometric distribution with the parameter $p = \frac{1}{n}$), then, as $n \to \infty$, we obtain the limit $r$-variate Student distribution with parameters $\gamma = 2$ and $\Sigma$. Moreover, if $\Sigma = I_r$ (that is, the $r$-variate Student distribution is spherically symmetric), then its one-dimensional marginals have the form (1). As we have already noted, distribution (1) was apparently for the first time introduced as a limit distribution for the sample median in a sample with geometrically distributed random size in [11]. It is worth noticing that in the cited paper [11], distribution (1) was not identified as the Student distribution with two degrees of freedom.

Thus, the main conclusion of this section can be formulated as follows. If the number of random factors that determine the observed value of a random variable is random itself with the distribution that can be approximated by the gamma distribution with coinciding shape and scale parameters (e.g., is negative binomial with probability of success close to one, see Lemma 4), then those functions of the random factors that are regarded as asymptotically normal in the classical situation are actually asymptotically Student with considerably heavier tails. Hence, since gamma-models and/or negative binomial models are widely applicable (to confirm this it may be noted that the negative binomial distribution is mixed Poisson with mixing gamma distribution, this fact is widely used in insurance), the Student distribution can be used in descriptive statistics as a rather reasonable heavy-tailed asymptotic approximation.

## 7. The asymptotic distribution of sample quantiles in samples with sizes generated by a Cox process

Sometimes, when the performance of a technical or financial system is analyzed, a forecast of main characteristics is made on the basis of data accumulated during a certain period of the functioning of the system. As a rule, data are accumulated as a result of some "informative events" that occur during this period. For example, inference concerning the distribution of insurance claims, which is very important for the estimation of, say, the ruin probability of an insurance company, is usually performed on the basis of the statistic $W_1, W_2, \ldots, W_{N(T)}$ of the values of insurance claims arrived within a certain time interval $[0, T]$ (here $N(T)$ denotes the number of claims arrived during the time interval $[0, T]$). Moreover, this inference is typically used for the prediction of the value of the ruin probability for the next period $[T, 2T]$. But it is obvious (at least in the example above) that the observed number of informative events occurred during the time interval $[0, T]$ is actually a realization of a random variable, because both the number of insurance claims arrived within this interval follow a stochastic counting process. If the random character of the number of available observations is not taken into consideration, then all what can be done is the *conditional* forecast. To obtain a complete prediction with the account of the randomness of the number of "informative events," we should use the results similar to Theorems 2 and 3. One of rather realistic and general assumptions concerning $N(t)$, the number of observations accumulated by the time $t$, is that $N(t)$ is a Cox process. In this section, as an example, we will consider the asymptotic behavior of sample quantiles constructed from a sample whose size is determined by a Cox process. As we have already noted in the introduction, this problem is very important for the proper application of such risk measures as VaR (Value-at-Risk) in, say, financial engineering.

Let $W_1, \dots, W_n, n \geq 1$, be independent identically distributed random variables with common distribution density $p(x)$ and $W_{(1)}, \dots, W_{(n)}$ be the corresponding order statistics, $W_{(1)} \leq W_{(2)} \leq \dots \leq W_{(n)}$. Let $r \in \mathbb{N}$ $\lambda_1, \dots, \lambda_r$ be some numbers such that $0 < \lambda_1 < \lambda_2 < \dots < \lambda_r < 1$. The quantiles of orders $\lambda_1, \dots, \lambda_r$ of the random variable $W_1$ will be denoted $\xi_{\lambda_i}, i = 1, \dots, r$. The sample quantiles of orders $\lambda_1, \dots, \lambda_r$ are the random variables $W_{([\lambda_i n]+1)}$, $i = 1, \dots, r$, with $[a]$ denoting the integer part of a number $a$. The following result due to Mosteller [32] (also see [33], Section 9.2) is classical. Denote

$$Y^*_{n,j} = \sqrt{n}\Big(W_{([\lambda_j n]+1)} - \xi_{\lambda_j}\Big), \qquad j = 1, \dots, r.$$

Theorem 7 [32]. *If $p(x)$ is differentiable in some neighborhoods of the quantiles $\xi_{\lambda_i}$ and $p(\xi_{\lambda_i}) \neq 0$, $i = 1, \dots, r$, then, as $n \to \infty$, the joint distribution of the normalized sample quantiles $Y^*_{n,1}, \dots, Y^*_{n,r}$ weakly converges to the r-variate normal distribution with zero vector of expectations and covariance matrix $\Sigma = (\sigma_{ij})$,*

$$\sigma_{ij} = \frac{\lambda_i(1-\lambda_j)}{p(\xi_{\lambda_i})p\big(\xi_{\lambda_j}\big)}, \quad i \leq j.$$

To take into account the randomness of the sample size, consider the sequence $W_1, W_2 \dots$ of independent identically distributed random variables with common distribution density $p(x)$.

Let $N(t), t \geq 0$, be a Cox process controlled by a process $\Lambda(t)$. Recall the definition of a Cox process. Let $N_1(t), t \geq 0$, be a standard Poisson process (i.e., a homogeneous Poisson process with unit intensity). Let $\Lambda(t), \ t \geq 0$, be a random process with non-decreasing right-continuous trajectories, $\Lambda(0) = 0$, $\mathrm{P}(\Lambda(t) < \infty) = 1$ for all $t > 0$. Assume that the processes $\Lambda(t)$ and $N_1(t)$ are independent. Set

$$N(t) = N_1(\Lambda(t)), \qquad t \geq 0.$$

The process $N(t)$ is called a doubly stochastic Poisson process (or a Cox process) controlled by the process $\Lambda(t)$. The one-dimensional distributions of a Cox process are mixed Poisson. For example, if $\Lambda(t)$ has the gamma distribution, then $N(t)$ has the negative binomial distribution.

Cox processes are widely used as models of inhomogeneous chaotic flows of events, see, for example, [2].

Assume that all the involved random variables and processes are independent. In this section, under the assumption that $\Lambda(t) \to \infty$ in probability, the asymptotics of the joint distribution of the random variables $W_{([\lambda_i N(t)]+1)}$, $i = 1, \dots, r$ is considered as $t \to \infty$.

As we have already noted, it was B. V. Gnedenko who drew attention to the essential distinction between the asymptotic properties of sample quantiles constructed from samples with random sizes and the analogous properties of sample quantiles in the standard situation. Briefly recall the history of the problem under consideration. B. V. Gnedenko, S. Stomatovič, and A. Shukri [34] obtained sufficient conditions for the convergence of distribution of the sample median constructed from sample of random size. In the candidate (PhD) thesis of A. K. Shukri, these conditions were extended to quantiles of arbitrary orders. In [35], necessary and sufficient conditions for the weak convergence of the one-dimensional distributions of sample quantiles in samples with random sizes were obtained.

Our aim here is to give necessary and sufficient conditions for the weak convergence of the *joint* distributions of sample quantiles constructed from samples with random sizes driven by a Cox process and to describe the $r$-variate limit distributions emerging here, thus extending Mosteller's Theorem 4 to samples with random sizes. The results of this section extend those of [36] to the continuous-time case.

Lemma 5. *Let $N(t)$ be a Cox process controlled by the process $\Lambda(t)$. Then $N(t) \xrightarrow{P} \infty \ (t \to \infty)$ if and only if $\Lambda(t) \xrightarrow{P} \infty \ (t \to \infty)$.*

Lemma 6. *Let $N(t)$ be a Cox process controlled by the process $\Lambda(t)$. Let $d(t) > 0$ be a function such that $d(t) \to \infty \ (t \to \infty)$. Then the following conditions are equivalent:*

1. One-dimensional distributions of the normalized Cox process weakly converge to the distribution of some random variable $Z$ as $t \to \infty$:

$$\frac{N(t)}{d(t)} \Rightarrow Z \qquad (t \to \infty).$$

2. One-dimensional distributions of the controlling process $\Lambda(t)$, appropriately normalized, converge to the same distribution:

$$\frac{\Lambda(t)}{d(t)} \Rightarrow Z \qquad (t \to \infty).$$

For the proof of Lemmas 5 and 6 see [37].

Now we proceed to the main results of this section. In addition to the notation introduced above, for positive integer $n$ set $Q_j(n) = W_{([\lambda_j n]+1)}$, $j = 1, \dots, r$, $Q(n) = (Q_1(n), \dots, Q_r(n))$, $\xi = (\xi_{\lambda_1}, \dots, \xi_{\lambda_r})$. Let $d(t)$ be an infinitely increasing positive function. Set

$$Z(t) = \sqrt{d(t)}(Q(N(t)) - \xi).$$

Theorem 8. *Let $\Lambda(t) \xrightarrow{P} \infty$ as $t \to \infty$. If $p(x)$ is differentiable in neighborhoods of the quantiles $\xi_{\lambda_i}$ and $p(\xi_{\lambda_i}) \neq 0$, $i = 1, \dots, r$, then the convergence*

$$Z(t) \Rightarrow Z \quad (t \to \infty),$$

to some random vector $Z$ takes place, if and only if there exists a nonnegative random variable $U$ such that

$$\mathrm{P}(Z \in A) = \mathrm{E}\Phi_{U^{-1}\Sigma}(A), \ \ A \in \mathcal{B}(\mathbb{R}^r),$$

where $\Sigma = (\sigma_{ij})$,

$$\sigma_{ij} = \frac{\lambda_i(1-\lambda_j)}{p(\xi_{\lambda_i})p(\xi_{\lambda_j})}, \ \ i \leq j,$$

and

$$\frac{\Lambda(t)}{d(t)} \Rightarrow U \quad (t \to \infty).$$

The proof is a simple combination of Lemmas 1, 5, and 6 and Theorem 3.

Corollary 4. *Under the conditions of Theorem 8, the joint distribution of the normalized sample quantiles* $\sqrt{d(t)}\Big(W_{([\lambda_j N(t)]+1)} - \xi_{\lambda_j}\Big)$, $j = 1, \dots, r$, *weakly converges to the r-variate normal law with zero expectation and covariance matrix* $\Sigma$, *if and only if*

$$\frac{\Lambda(t)}{d(t)} \Rightarrow 1 \quad (t \to \infty).$$

This statement immediately follows from Theorem 8 with the account of Lemma 1.

Corollary 5. *Under the conditions of Theorem 8, the joint distribution of the normalized sample quantiles* $\sqrt{d(t)}\Big(W_{([\lambda_j N(t)]+1)} - \xi_{\lambda_j}\Big)$, $j = 1, \dots, r$, *weakly converges to the r-variate Student distribution with parameters* $\gamma > 0$ *and* $\Sigma$ *defined in Theorem 4, if and only if*

$$\mathrm{P}(\Lambda(t) < x d(t)) \Rightarrow G_{\gamma/2,\gamma/2}(x), \quad t \to \infty,$$

where $G_{\gamma/2,\gamma/2}(x)$ is the gamma-distribution function with coinciding shape and scale parameters equal to $\gamma/2$.

Let $0 < \lambda < 1$ and let $\xi_\lambda$ be the $\lambda$-quantile of the random variable $W_1$. As above, the standard normal distribution function will be denoted $\Phi(x)$.

## 8. Conclusion

The purpose of the chapter was to give a possible explanation of the emergence of heavy-tailed distributions that are often observed in practice instead of the expected normal laws. As the base for this explanation, limit theorems for random sums and statistics constructed from samples with random sizes were considered. Within this approach, it becomes possible to obtain arbitrarily heavy tails of the data distributions without assuming the non-existence of the moments of the observed characteristics. Some comments were made on the heavy-tailedness of scale mixtures of normal distributions. Two general theorems presenting necessary and sufficient conditions for the convergence of the distributions of random sums of random vectors and multivariate statistics constructed from samples with random sizes were proved. As examples of the application of these general theorems, conditions were presented for the convergence of the distributions of random sums of independent random vectors with finite covariance matrices to multivariate elliptically contoured stable and Linnik distributions. An alternative definition of the latter was proposed. Also, conditions were presented for the convergence of the distributions of asymptotically normal (in the traditional sense) statistics to multivariate elliptically contoured Student distributions when the sample size is replaced by a random variable. The joint asymptotic behavior of sample quantiles in samples with random sizes was considered. Special attention was paid to the continuous-time case assuming that the sample size increases in time following a Cox process resulting in the sample size having the mixed Poisson distribution.

## Acknowledgements

Supported by Russian Science Foundation, project 18-11-00155.

## Author details

Victor Korolev[1,2,3]* and Alexander Zeifman[3,4]

1 Faculty of Computational Mathematics and Cybernetics, Moscow State University, Moscow, Russia

2 Hangzhou Dianzi University, Hangzhou, China

3 Institute of Informatics Problems, Federal Research Center "Computer Science and Control" of the Russian Academy of Sciences, Moscow, Russia

4 Vologda State University, Vologda, Russia

*Address all correspondence to: vkorolev@cs.msu.ru

## References

[1] Shiryaev AN. Foundations of Financial Mathematics. Vol. 1. Facts, Models. – Singapore: World Scientific; 1998

[2] Bening V, Korolev V. Generalized Poisson Models and Their Application in Insurance and Finance. Utrecht: VSP; 2002

[3] Meerschaert MM, Scheffler H-P. Limit theorems for continuous-time random walks with infinite mean waiting times. Journal of Applied Probability. 2004;**41**(3):623-638

[4] Samorodnitsky G, Taqqu MS. Stable Non-Gaussian Random Processes, Stochastic Models with Infinite Variance. New York: Chapman and Hall; 1994

[5] Chen J. From the central limit theorem to heavy-tailed distributions. Journal of Applied Probability. 2003; **40**(3):803-806

[6] Korolev VY. Convergence of random sequences with independent random indices. I. Theory of Probability and its Applications. 1994;**39**(2):313-333

[7] Schluter C, Trede M. Weak convergence to the Student and Laplace distributions. Journal of Applied Probability. 2016;**53**:121-129

[8] Bening V, Korolev V. On an application of the Student distribution in the theory of probability and mathematical statistics. Theory of Probability and its Applications. 2005; **49**(3):377-391

[9] Bening V, Korolev V. Some statistical problems related to the Laplace distribution. Informatics and its Apllications. 2008;**2**(2):19-34

[10] Gnedenko BV, Korolev V. Random Summation: Limit Theorems and Applications. Boca Raton: CRC Press; 1996

[11] Gnedenko BV. On estimation of the unknown parameters of distributions from a random number of independent observations. Probability Theory and Mathematical Statistics. Proccedings of Tbilisi Mathematical Institute named after A. M. Razmadze. 1989;**24**:146-150 (in Russian)

[12] Kolmogorov AN. The method of median in the theory of errors. Matematicheskii Sbornik. 1931;**38**(3/4): 47-50

[13] Korolev VY, Kossova EV. On limit distributions of randomly indexed multidimensional random sequences with an operator normalization. Journal of Mathematical Sciences. 1992;**72**(1): 2915-2929

[14] Korolev VY, Kossova EV. Convergence of multidimensional random sequences with independent random indices. Journal of Mathematical Sciences. 1995;**76**(2): 2259-2268

[15] Zolotarev VM. One-Dimensional Stable Distributions. Providence: American Mathematical Society; 1986

[16] Kolokoltsov V, Korolev V, Uchaikin V. Fractional stable distributions. Journal of Mathematical Sciences. 2001;**105**(6):2569-2576

[17] Korolev VY, Zeifman AI. A note on mixture representations for the Linnik and Mittag-Leffler distributions and their applications. Journal of Mathematical Sciences. 2017;**218**(3): 314-327

[18] Korolev VY, Zeifman AI. Convergence of statistics constructed from samples with random sizes to the Linnik and Mittag-Leffler distributions

and their generalizations. Journal of the Korean Statistical Society. 2017;**46**(2): 161-181 Available online 25 July 2016. Also available on arXiv:1602.02480v1 [math.PR]

[19] Korolev VY. On the convergence of distributions of random sums of independent random variables to stable laws. Theory of Probability and its Applications. 1998;**42**(4):695-696

[20] Linnik YV. Linear forms and statistical criteria, I, II. Selected Translations in Mathematical Statistics and Probability. 1963;**3**:41-90 (Original paper appeared in: Ukrainskii Matematicheskii Zhournal, 1953. Vol. 5. pp. 207-243, 247-290)

[21] Kotz S, Kozubowski TJ, Podgorski K. The Laplace Distribution and Generalizations: A Revisit with Applications to Communications, Economics, Engineering, and Finance. Boston: Birkhauser; 2001

[22] Pillai RN. Semi-α-Laplace distributions. Communications in Statistical Theory and Methods. 1985;**14**: 991-1000

[23] Kozubowski TJ, Rachev ST. The theory of geometric stable distributions and its use in modeling financial data. European Journal of Operational Research. 1994;**74**(2):310-324

[24] Kozubowski T, Panorska A. Multivariate geometric stable distributions in financial applications 1. Mathematical and Computer Modelling. 1999;**29**:83-92

[25] Anderson DN. A multivariate Linnik distribution. Statistics & Probability Letters. 1992;**14**:333-336

[26] Ostrovskii IV. Analytic and asymptotic properties of multivariate Linnik's distribution. Mathematical Physics, Analysis and Geometry. 1995; **2**(3):436-455

[27] Weron K, Kotulski M. On the Cole-Cole relaxation function and related Mittag-Leffler distributions. Physica A: Statistical Mechanics and its Applications. 1996;**232**:180-188

[28] Gorenflo R, Mainardi F. Continuous time random walk, Mittag-Leffler waiting time and fractional diffusion: Mathematical aspects, chap. 4. In: Klages R, Radons G, Sokolov IM, editors. Anomalous Transport: Foundations and Applications. Weinheim, Germany: Wiley-VCH; 2008. pp. 93-127. Available at: http://arxiv.org/abs/0705.0797

[29] Kozubowski TJ, Rachev ST. Multivariate geometric stable laws. Journal of Computational Analysis and Applications. 1999;**1**(4):349-385

[30] DeGroot MH. Optimal Statistical Decisions. New York, London: McGraw-Hill Company; 1970

[31] Gupta SS. Bibliography on the Multivariate Normal Integrals and Related Topics. Providence: American Mathematical Society; 1963

[32] Mosteller F. On some useful "inefficient" statistics. Annals of Mathematical Statistics. 1946;**17**: 377-408

[33] David HA. Order Statistics. New York: Wiley; 1970

[34] Gnedenko BV, Stomatovič S, Shukri A. On the distribution of the median. Bulletin of Moscow University, Series Mathematics, Mechanics. 1984; **2**:59-63

[35] Selivanova DO. Estimates of Convergence Rate in Limit Theorems for Random Sums. PhD Thesis: Moscow State University; 1995

[36] Korolev VY. Asymptotic properties of sample quantiles constructed from samples with random sizes. Theory of

Probability and its Applications. 2000; **44**(2):394-399

[37] Korolev VY. On convergence of distributions of compound Cox processes to stable laws. Theory of Probability and its Applications. 1999; **43**(4):644-650

Chapter 8

# Probability Modeling Taking into Account Nonlinear Processes of a Deformation and Fracture for the Equipment of Nuclear Power Plants

*Nikolay Andreevich Makhutov, Mikhail Matveevich Gadenin, Igor Alexandrovich Razumovskiy, Sergey Valerievich Maslov and Dmitriy Olegovich Reznikov*

## Abstract

At the solution of integrated tasks of strength, safe life and service safety maintenance for the nuclear power plants (NPP) equipment with slow reactors—water-moderated power reactors (WMPR) of VVER type and channel-type graphite-moderated power reactors (GMPR) of RBMK type arise necessity of physical and mathematical modeling of nonlinear processes of a deformation, fracture and damage at nonlinear probability statement. First of all, it concerns deriving determined, statistical and probabilistic characteristics of mechanical properties of reactor materials. Expectations and variation factors of mechanical properties' characteristics obtained from experimental researches are inducted into the equations for the verification calculations at determination of static and cyclic strength margins with the use of nominal and local stresses and strains. For the improved determined and probability analysis of these margins modeling experimental researches of stress-strain states of the analyzed equipment are conducted. Special attention at such tests is given to concentration factors and variation factors of loading conditions. The final stage of estimation of basic normative and verification calculation accuracy at laboratory, modeling and test bench researches are full-scale pre-operational tests (cold-hot running-in) of pilot nuclear reactors with the use of the experimental mechanics methods. The conditions of safety service of the NPP equipment are estimated taking into account factors of reaching limiting states by criteria of risk of initiation of emergency situations.

**Keywords:** probability modeling, strength, deformation, damages, fracture, nuclear reactors, safety service, risk, structural materials

## 1. Introduction

The era of nuclear energy in the world started in 1954 by putting into service the first nuclear power plant (NPP)—the Obninsk NPP with a channel-type reactor and



IntechOpen

power of 5 MW. Since then, leading countries of the world (the USSR-Russia, the USA, Great Britain, France, etc.) have come up with a whole spectrum of a new type of power supply—nuclear-powered.

By 2019, in the Russian Federation, 10 NPPs with 35 power-generating units with a total power of 29 GW are operational. In model of the NPP of Russia, there are 20 pressurized water reactors, including water-moderated power reactors (12 units of VVER-1000 type, 1 unit of VVER-1100 type, 2 units of VVER-1200 type, 5 units of VVER-400 type, and 1 unit of VVER-417 type). There are also 13 units of channel boiling water reactors of a high power of RBMK type (channel-type graphite-moderated power reactor—GMPR)—(10 units of RBMK-1000 type and 3 units of type EPG-6 type with power of 12 MW) and 2 units of fast-neutron reactor (FNPR) of BN type (BN-600 type and BN-800 type).

In 56 states of the world, more than 430 nuclear reactors with a total power 370 GW is now operated. The NPPs in the world produce about 11% of the consumable electric power. Leaders in this production are France (80%), South Korea (32%), and Ukraine (30%). In Russia, this share amounts to 16%. In the long term of 20–25 years, probably accretion of this share will be about 25%.

On changeover to reactors of power plants of first generations of 1960–1970 reactors of new third and fourth breeds come. And if the first reactors were considered as "nuclear boilers" and designed on norms of boiler fabrication for thermal power, up-to-date reactors develop on these details both on scientifically well-founded norms and on methods of national (Russia, the USA, Great Britain, France, and Germany) and international levels (IAEA).

From stands of classes of hazards detection for technosphere objects, nuclear reactors undoubtedly fall into critically (CRO) and strategically (SRO) relevant objects. These are facts that demand the profound combined analysis and a justification of all design and service solutions for all stages of their life cycle.

In the proximal (till 2020), midrange (till 2030), and kept away (till 2050) prospects, the evolution of nuclear energetics will be carried out on the basis of operating, built, and designed nuclear power plants. Basis of the fundamental and application analysis of strength, life time, reliability, and safety of operation of NPP elements with reactors of VVER, RBMK, and BN types (**Figure 1**) in regular both emergency situations are the equations and criteria linear and nonlinear mechanics of deformation and fracture [1–11]. They contain in home and foreign strength standards and are used as at design, so at manufacture and operation of working in

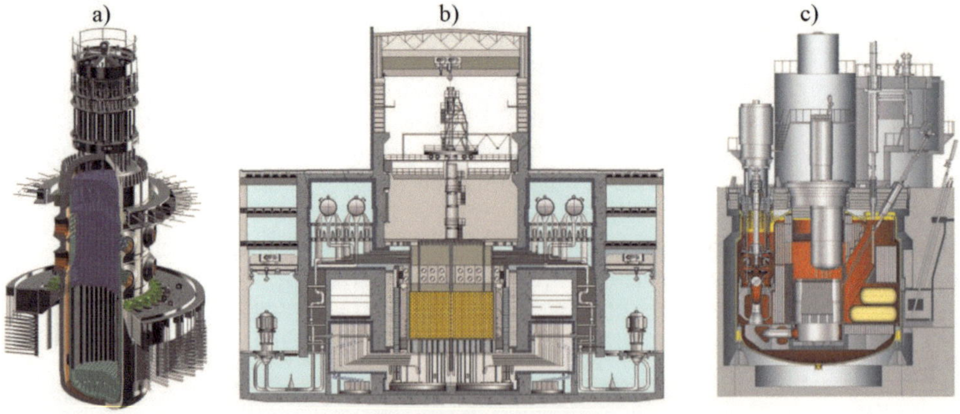


**Figure 1.**
*The Russian reactors of VVER (water-moderated power reactor) (a), RBMK (channel graphite-moderated power reactor) (b) and BN (fast-neutron reactor) (c) types.*

extremely conditions, a high-loaded power-generating plants with use physical and mathematical modeling [1, 12–17].

Results of traditional researches and a standardization of strength and life time of NPP in the determined statement in Russia and abroad are both initial scientific baseline of normative documents on design and actual baseline of making of perspective methods of a reliability estimate, survivability, initiation, and evolution of accidents and disasters by risk criteria, and also of makings of new principles, technologies, and engineering complexes ensuring safety service of NPP. These are conditions that are scientifically grounded to prevent initiation of the emergency and disastrous situations and also to minimize probable losses at their initiation at all stages of life cycle. Such situations within the limits of usual normative approaches and methods, as a rule, remained the least investigated from the scientific and application points of view owing to the complication, small predictability, and recurrence. At the same time, survivability of power-generating units in emergency situations and risk analysis of probable aftereffects should become weighable arguments in favor of building of nuclear units with a life expectancy from 60 to 100 years.

The analysis of sources, the reasons, and aftereffects of the heavy disasters occurring during installations of nuclear energetics display both their likeness and essential difference. Accidents known to the world on NPP with radioactivity ejection in a circumambient manner in the USA (the NPP "Three Mile Island (TMI)"—**Figure 2**), in the USSR (the Chernobyl NPP (CNPP)—**Figure 3**), and in Japan (the NPP "Fukushima-1—**Figure 4**) were the heaviest [3, 6, 8, 11, 18].

A common after effect of NPP accidents and disasters was that direct and indirect economical losses from them reached tens and hundreds of billions of USD. For their forestalling and preventing in the subsequent, the principal changes were made to designer, technological, and service solutions. Heavy emergency situations for NPP service arose earlier at the time of damage to their equipment, such as runners, steam plants, main coolant pumps, heat exchanger pipes, gate valves, and legs of reactor internals [11, 17].

The abovementioned NPP heavy accidents and disasters originated from unapproved impacts of human controllers, non-observance of technological discipline at emergency situation (TMI, CNPP), heavy-lift seismic loads, and a tsunami (Fukushima-1). Regular systems of the automatize guard of the NPP have been unreasonably disconnected (CNPP) or could not work in an emergency situation (TMI, Fukushima-1). Heavy emergency situations on turbine runners, steam plants,

**Figure 2.**
*The "Three Mile Island" NPP (TMI).*

**Figure 3.**
*The accident on the Chernobyl NPP (CNPP).*

**Figure 4.**
*The accident on the "Fukushima-1" NPP.*

gate valves, and legs arose due to the lack of suitable technical diagnostics of these situations [11, 19], when faults in the form of cracks because of technological or operational fault attained of the limiting, intolerable sizes ($10^2$ to $1.5 \times 10^3$ mm), affecting 50–70% of carrying cross-section and creating sharp magnification of runner chattering. Thus, the analysis of such situations was not envisioned by normative calculations.

## 2. Combined researches of strength and life time

For installations of a nuclear energy in our country and abroad in the second half of twentieth century, the whole complex of fundamental and application developments [1–7, 11–14, 20–23] on the creation of normative strength calculations of the equipment and pipelines for nuclear power plants has been executed. Thus in our country special meaning had the solution of policy-making bodies that the scientific adviser of research developments on a justification of norms had been defined the Academy of Sciences of the USSR (The A.A. Blagonravov Institute for Machine Sciences—the IMASH), and the head development engineer of norms—the Ministry of medium machine building of the USSR (The N.A. Dollezhal Research and Development Institute of Power Engineering—NIKIET).

The same organizations making all prototype models of reactors for the NPP established the total statement about the strength before starting a reactor in service. Such norms developed both in the USSR [1, 12] and in the USA [14] subsequently were developed according to international standards set by the International Atomic Energy Agency—IAEA [13]. Compared to home norms of an NPP design [1, 12], basic sections on calculations, monitoring, probability safety assessment, and a justification of life time extension have been included.

**Figure 5.**
*The structure of the main task solution at making and service of the NPP equipment.*

Long-term experience of home nuclear branch organizations and the academic institutes has allowed to form (**Figure 5**) the schematic diagram of the combined solution of tasks in view:

- The determined and statistical researches of deformation and fracture processes of laboratory specimens (with groups from 3–10 to 100–200 specimens of one steel)
- Model tests of the metallic specimens imitating most important parts (for example, studs of threaded connections with a diameter from 24 to 110 mm) and also nonmetallic specimens of studs with a diameter from 60 to 210 mm
- Tests of the modeling reactor vessels fabricated of nonmetallic materials in scale 1:10 and from metallic materials in scale 1:5
- Full-scale prestarting and starting tests of reactor prototype models of VVER, RBMK, and BN types

In considered norms, there are two cores sections: calculation of principal dimensions predominantly by criteria of a static strength and the verification calculations on a different combination of limiting states at low-cycle and high-cycle, long-term, vibration, seismic loads with initiation of static, cyclic, brittle, corrosion fracture, and also cyclic forming and radiation damage.

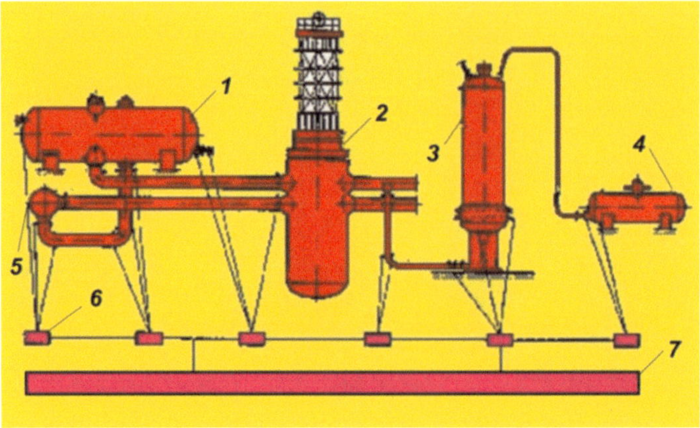


**Figure 6.**
*The flow chart of strain-gauging of power equipment: 1—steam generator, 2—reactor, 3—pressure compensator, 4—bubble tank, 5—a main coolant pump, 6—commutators, and 7—registering apparatuses.*

In the capacity of the most responsible and dangerous NPP components, nuclear reactor vessels, pipelines, pumps, steam generators, reactors, and machine halls have been accepted (**Figure 6**).

In an NPP with water-moderated power reactors (VVER) in the capacity of the major critical parts, it is possible to consider also the basic attachment fittings of reactor covers such as studs. Thus, the computational-experimental analysis of stress-strain states, strengths, and life times of a connection joint of reactor covers is conducted by improved methods in more detail (**Figure 7**).

For reactor installations of home production, such analysis was fulfilled [2–4, 11, 15, 16] jointly by the academic institutes, head branch research, and designer organizations on all prototype models of reactors in our country and abroad (Bulgaria, Finland, Hungary, Czech, and China) with application of the foremost methods: model researches of covers, studs, pressing rings on models from stress-optical and metallic materials, full-scale researches on reactors at preoperational tests on all regimes (including emergency), and also at an initial stage (till 1–3 years) of service.

In particular, the fifth unit of the Kozloduy NPP (Bulgaria) has been developed and implemented [15] after a most complicated program of full-scale researches by methods of a strain measurement, a thermometry, a vibrometry for all components of a primary loop with 1000 measuring points of local stresses, pressure pulsations, and temperatures (**Figure 8**).

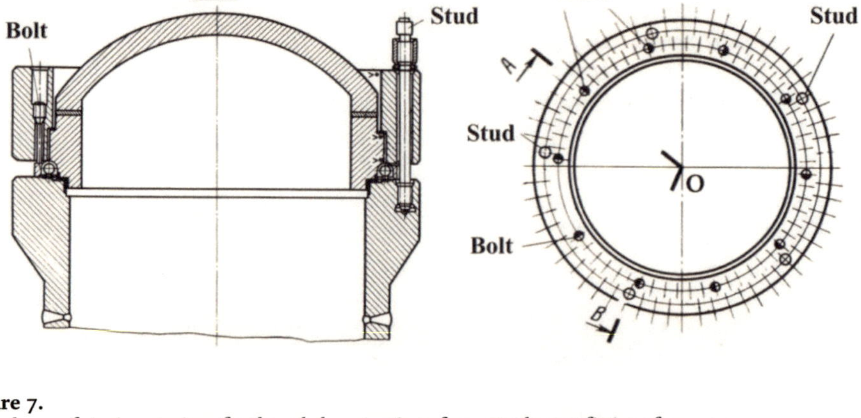


**Figure 7.**
*The scheme of strain-gauging of a threaded connection of an attachment fitting of a cover.*

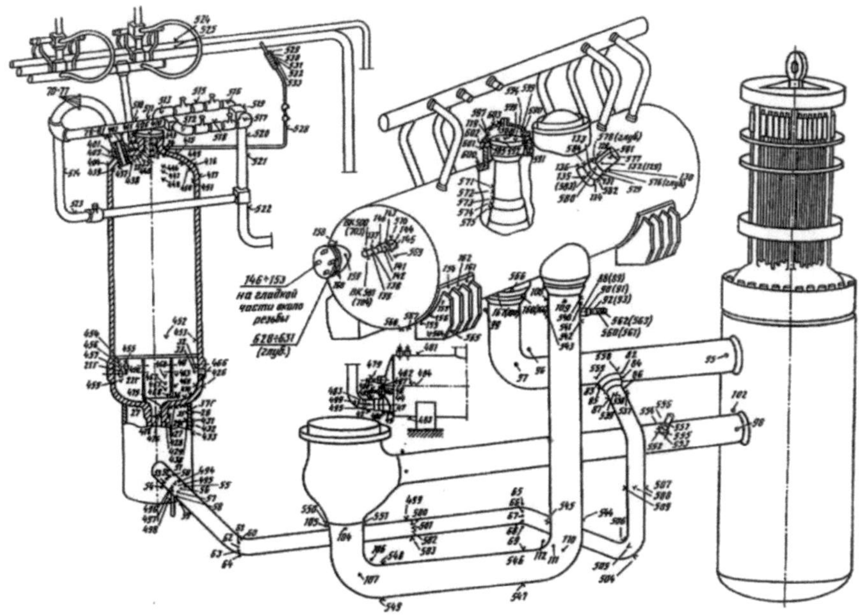

**Figure 8.**
*Zones and points of placing of measuring gauges on the NPP equipment.*

Modeling and full-scale researches have allowed to define detailed stress distributions on threads (**Figure 9**) and in a cover (**Figure 10**). These facts have given the chance to obtain real history of service impacts and nominal and local stresses on all parts of a reactor main joint.

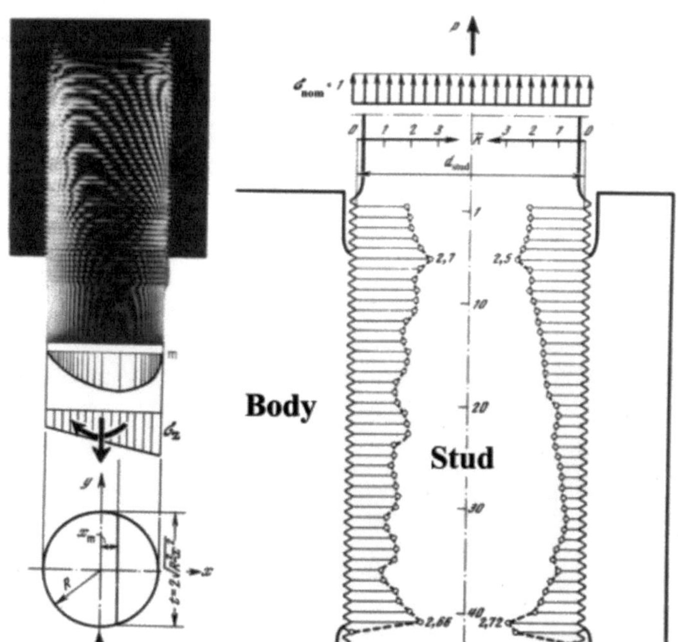


**Figure 9.**
*A stress loading of a stud attachment fitting of a reactor cover.*

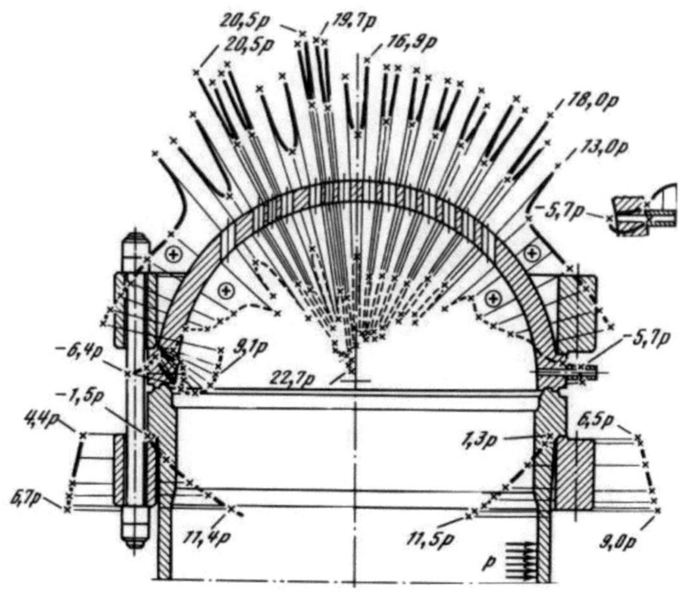


**Figure 10.**
*Stress distribution diagrams in a cover, flanges, and studs.*

Computational and special experimental test bench researches of a dynamic stress loading and cyclical damages from seismic loads had a particular actuality.

On metallic modeling studs with a diameter from M12 to M110, data about life time on the basis of $10^4$–$10^5$ cycles have been obtained. These data have allowed to justify improved margins on strength and life time of analyzed studs.

The principal great value in results these researches had that facts that the maximum accumulated damages (to 70%) arose in regimes multiple tightening and seal failure of caps (**Figure 11**). This fact has demanded work on special activities to decrease the indicated damages [15, 16].

Formation of development trends at the standardization instituting serviceability and safety of a nuclear (power-generating equipment went in a direction of specification and complicating of applied methods and criteria [1–3, 11, 20–23]. Thus, accidents and disasters (the TMI in the USA, the CNPP in the USSR, and

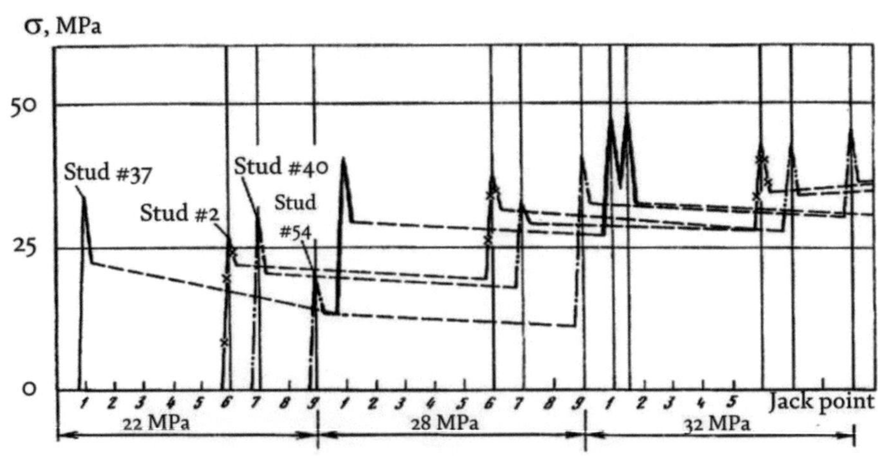


**Figure 11.**
*The diagram of stresses change in studs at sealing of the main joint of the VVER-1000 reactor.*

the Fukushima-1 in Japan) added additional information baseline for such development.

To the traditional solution of a problem of service safety [2, 6–10, 20–25], three groups of approaches had a direct ratio:

- From the position of strengths (in its multicriteria expression)
- From the position of life time (in time and cyclic statement)
- From the position of inadmissibility of large plastic strains

Traditional methods of strength justification were founded on a complex of determined characteristics of mechanical properties of materials and fracture criteria (yield point—$\sigma_y$, ultimate strength—$\sigma_u$, fatigue limit—$\sigma_{-1}$, and long-term strength—$\sigma_{lt}$). On the basis of these parameters of strength and fracture (present in standard and technical specifications for reactor structural materials), the status of safety and life time margins ($n_\sigma$, $n_N$, $n_\tau$) has been generated. These margins are included in the reference, educational, and standard literature [1, 2, 12, 20–26]. Today, a common system of criteria and strength margins guaranteeing a fracture of nonadmission for equipment components at observance of the given service conditions is developed.

Mathematical modeling at the determined normative requirements to strength and life time came down to two approaches:

- To modeling parts of rods, plates, and thin shell types on the basis of analytical solutions of the theory of a strength of materials and theory of elasticity
- To modeling real objects on the basis of numerical solutions by finite-element method, finite difference method, and integral equations method

Research of seismic impacts was the most complicated at computational and experimental modeling:

- By finite-element method (FEM) for all parts of the first circuit (**Figure 12**)
- By methods of physical modeling of a reactor with reactor internals (**Figure 13**)

It has thus appeared that most high stresses and damages from seismic loads occur at the zone of attaching of pipelines to a reactor vessel.

On the basis of such modeling, nominal $\sigma_n$ and maximum local $\sigma_{max}$ stresses in concentration zones were defined. However, in these traditional approaches, normative materials often did not contain the direct data quantitatively instituting strength and life time of considered objects taking into account a statistical property of parameters $\sigma_y$, $\sigma_u$, $\sigma_{-1}$, and $\sigma_{lt}$. Occurring actually dissipation of parameters for strength calculation and life time of a NPP environment is caused by instability of manufacturing procedures at production of structural materials and NPP bearing parts (reactor vessels, pipelines, pumps, and heat exchangers). In the last decades, this deficiency has been eliminated, and the sphere of the traditional analysis of serviceability of the NPP equipment includes the theory and criteria of life time and reliability [2, 20–27].

In addition to normative calculations of reactors on [1] at the complicated regimes (**Figure 14**) of an assembly, test and service loading (assembly, a

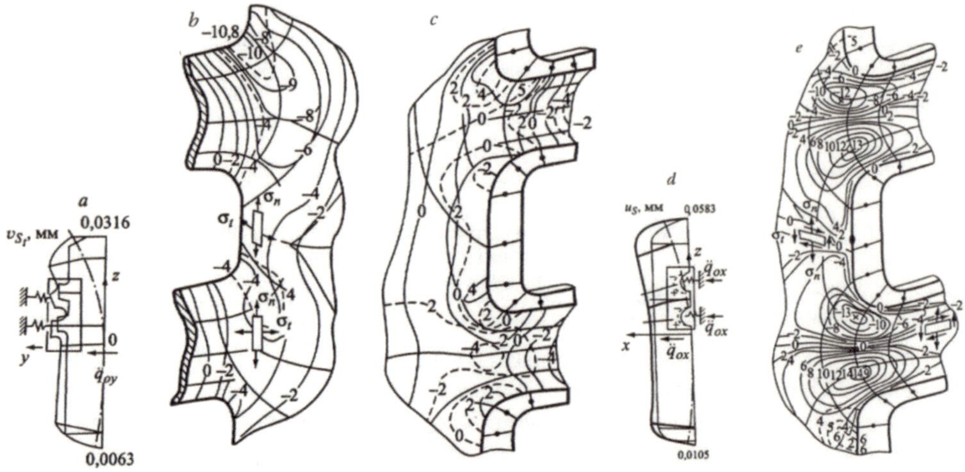


**Figure 12.**
*Loads and stresses in a connecting pipes zone of a reactor vessel at seismic impacts for YOZ plain—the computational scheme (a), response stresses (MPa) on outside (b) and interior (c) surfaces; for XOZ plain—the calculation scheme (d) and stresses (e) on an interior surface.*

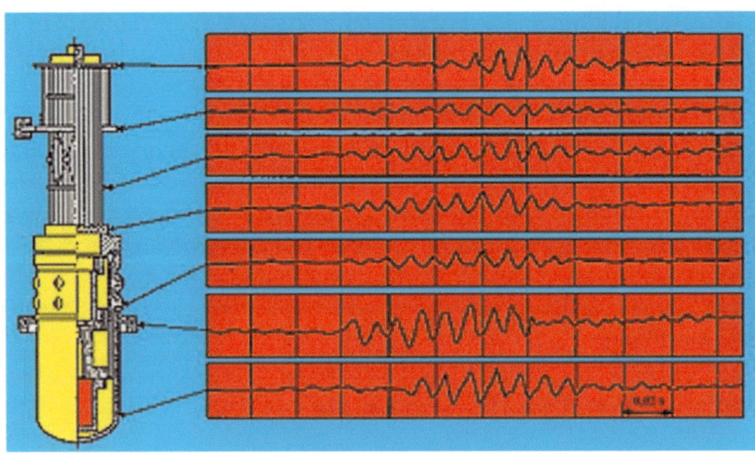

**Figure 13.**
*A research of a dynamic state of a reactor simulator at seismic excitation.*

tightening of studs, a hydroshaping testing, launch, capacity change, emergency operations, and shut-down) for events of occurrence of high levels of stresses improved strength, and life time calculations were carried out on the equations type

$$e_a = \frac{1}{(4N)^{m_e} + \frac{1+r_e}{1-r_e}} \cdot \ln \frac{1}{1-\psi_c} + 0,43 \frac{\sigma_b(1+\psi_c)}{E \cdot N^{m_\sigma}\left(1+\frac{1+r_\sigma}{1-r_\sigma}\right)}, \tag{1}$$

where $e_a$ is the amplitude of strain at a design regime; $N$ is the life time at a crack initiation stage, in cycles; $\sigma_b$ is the ultimate strength of a material ($400 \le \sigma_b \le 950$ MPa); $\psi_c$ is the reduction of area in a neck of a specimen at single-pass rupture ($0.3 \le \psi_c \le 0.7$); $r_e$, $r_\sigma$ are the cycle asymmetry parameters on strains and stresses, accordingly; and $m_e$, $m_\sigma$ are the characteristics of a real material ($0.5 \le m_e \le 0.6$), ($0.08 \le m_\sigma \le 0.12$). Values of parameters in Eq. (1) $e_a$, $\psi_c$, $r_e$, $r_\sigma$, $m_e$, and $m_\sigma$ are relative and dimensionless.

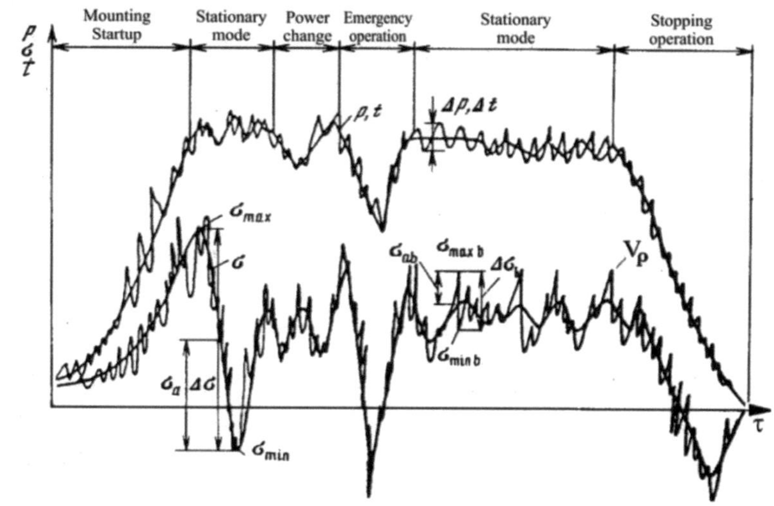


**Figure 14.**
*The diagram of change of service loading parameters.*

Calculation on Eq. (1) with the use of deformation criteria can be brought together to calculate by force criteria (on stresses) to accept $\sigma_a^* = e_a \cdot E$ ($E$—a modulus of elasticity).

Equation (1) is true for a wide band of life times ($10^0 \le N \le 10^{12}$). Permissible regimes of a stress loading are established in Eq. (1) with introduction of two margins $n_{\sigma_a}$ and $n_N$. Then, the computational curve of permissible values $[e_a]$ (or $[\sigma_a^*]$) and $[N]$ is accepted as lower enveloping curves on each of these margins.

For the complicated regimes of a two-frequency loading (low-frequency with frequency $f_l = f_0$ (hertz) and amplitude of stress $\sigma_{al}^* = \sigma_{a0}^*$ (MPa), and high-frequency with $f_h$ (hertz) and $\sigma_{ah}^*$ (MPa), accordingly) on the basis of generalization of experimental data, life time decrease from the number of cycles of basic loading $N_0$ (cycle) to two-frequency life time $N_2$ (cycle) is considered [1, 28] in equation

$$N_2 = N_0/\chi; \chi = \left(\frac{f_h}{f_0}\right)^{\eta \frac{\sigma_{a0}^*}{\sigma_{ah}^*}}, \tag{2}$$

where χ and η are dimensionless characteristics of a material and parameters of a two-frequency regime.

The same approach is used to calculate life time taking into account the presence of contact (wear resistance) and seismic impacts.

The presence of initial or service defects of cracks type with depth $l$ is reflected in calculations of survivability on the basis of the equations of linear and a nonlinear fracture mechanics by change of stresses $K_I$ (MPa·m$^{1/2}$) and strains $K_{Ie}$ intensity factors [2, 20, 29]. For one-time brittle or a ductile fracture,

$$K_I = \sigma\sqrt{\pi l} \cdot f_\kappa \le \frac{K_{Ic}}{n_{K_\sigma}}; K_{Ie} \le \frac{K_{Iec}}{n_{K_e}}, \tag{3}$$

where $K_{Ic}$ and $K_{Iec}$ are the critical (fracture) stresses and strains intensity factors, accordingly; $n_{K_\sigma}$ and $n_{K_e}$ are the dimensionless margins on stresses and strains intensity factors, accordingly ($n_{K_\sigma} \le n_{K_e}$).

Reliability of equipment $P_{QR}(\tau)$ along with the account of the probabilistic approach to estimations of mechanical properties of a structural material is defined also (**Figure 15**) on probabilistic characteristics of service stress loading $Q(\tau)$ and life time $R_{N\tau}(\tau)$ on the basis of distribution functions $f$ of service impacts $Q^s(\tau)$ and of ultimate loads $Q_c(\tau)$ for the given life times $N_c$, $\tau_c$. Thus, usually "trees of events" and "trees of failures" on experience of previous service of analogous technosphere objects are used. In such statement, the risk can be defined as

$$R(\tau) = 1 - P_{QR}(\tau). \qquad (4)$$

More oriented on the quantitative solution of a safety problem for complicated NPP installations, capable to cause severe accidents and disasters, are new methods and criteria of the following groups [2, 6–8, 11, 18–21, 24–26, 29–33]:

- Survivability (ability and steadiness of operation at occurrence of damages at different stages of accidents and disaster evolution)

- Safety (taking into account the risk criteria and characteristics of accidents and disasters)

- Risk (in probability-economic statement)

From the above-stated, the up-to-date justification of strength, life time, reliability, survivability, safety, and risks (**Figure 16**) should be based on results of corresponding calculations and tests with observance of the special and new requirements established by corresponding normative-legal documents.

For long-term operated high-risk installations of a nuclear energetic to which the NPPs with reactors of the VVER concern, the BN and the RBMK types' rate, initial parameters of strength, life time, risk, and safety were defined in an explicit and implicit kinds on stages of their design and commissioning on acting then norms and rules which place at the different displayed in **Figure 16** footsteps (on time and analysis level).

Thereupon, during estimations of their state, two scientific and application approaches are possible:

- To realize stage by stage an estimation of the initial, exhausted, and remaining life time

- To estimate current life time, as initial for the given level of the service damage that has been accumulated in the previous operating period

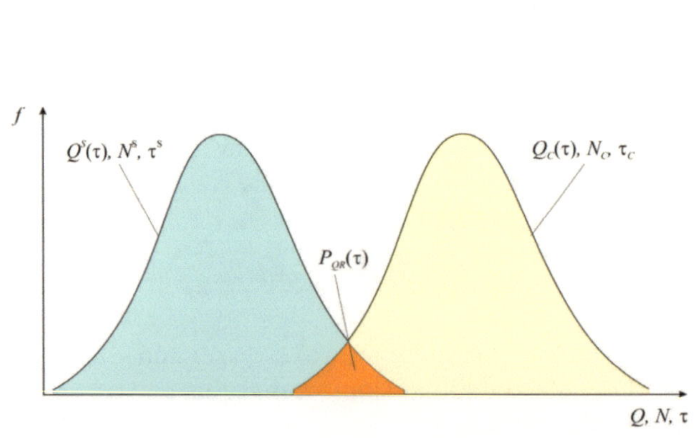


**Figure 15.**
*The scheme of determination of reliability, failures, accidents, and disaster probability $P_{QR}(\tau)$.*

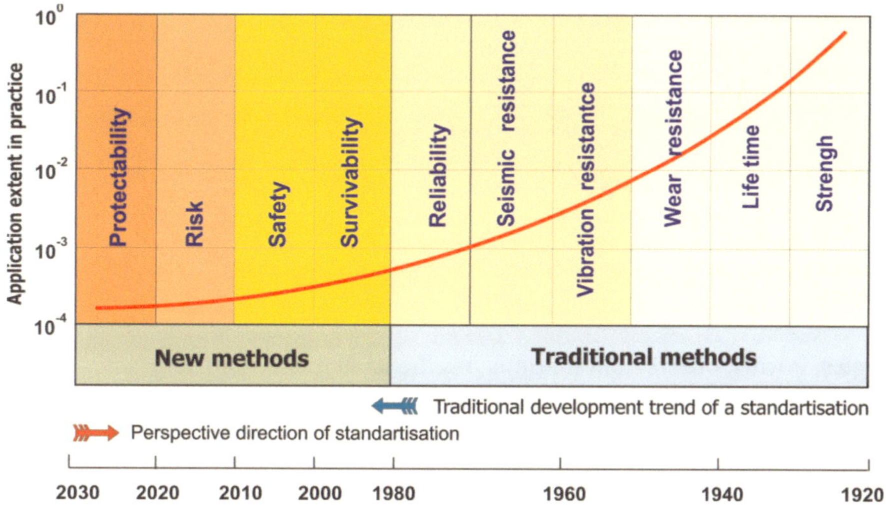


**Figure 16.**
*A structure and evolution of standardization methods on the determined justification of strength and life time with the use of physical modeling of materials behavior at a static, cyclic, and long-term loading.*

At the present time, the first approach was found to be the largest application. However, subsequently, the second approach appears to be deciding owing to its higher precision at estimations of the remaining strength, life time, and safety.

## 3. An estimation of risks and service safety

On the basis of the normative documents developed and accepted to present safety of power engineering as a whole, and NPPs in particular, the level of individual risks and risks of a possibility of accidents and disaster initiation should be estimated. In the process of perfecting NPPs and their nuclear reactors, these risks were reduced and will be reduced from $10^{-4}$ to $10^{-8}$ 1/year and less. For example, the reactor of natural safety with plumbeous heat-transfer agent will have a probability of fracture considerably below $10^{-8}$ 1/year [8, 11]. Individual risks of nonnuclear power engineering lay within the limits $10^{-4}$–$10^{-7}$ 1/year (**Table 1**).

The great importance for the analysis, support, and improvement of safety of the considered equipment within the limits of dominating and active concepts,

| Types of the electric power manufacture | Risk estimation on the person in a year |
|---|---|
| Hydropower plant | $10^{-5}$-$10^{-6}$ |
| Solar power plants | $10^{-4}$-$10^{-5}$ |
| Wind power plants | $10^{-4}$-$10^{-5}$ |
| Thermal power plants | $10^{-4}$-$10^{-5}$ |
| Nuclear power plant | $10^{-4}$-$10^{-8}$ |
| Reactors of the first generation | $10^{-4}$-$10^{-5}$ |
| Reactors of the second generation | $10^{-6}$ |
| Reactors of the third generation | $10^{-7}$ |
| Reactors of the fourth generation | $10^{-8}$ |
| Perspective reactors of the fifth generation | $<10^{-8}$ |

**Table 1.**
*Comparative data about a radiation-ecological risk for different directions of the electric power manufacture.*

| N | Analyzed situations | Protectability | Risks type |
|---|---|---|---|
| 1. | Service standard conditions | Heightened | Controlled |
| 2. | Shifts from standard conditions | Sufficient | Adjustabled |
| 3. | Design accidents | Partial | Analysed |
| 4. | Out of design accidents | Insufficient | Heightened |
| 5. | Hypothetical accidents | Low | High |

**Table 2.**
*Types extreme (emergency and catastrophic) situations and level of protectability from them of high-risk installations.*

strategies, norms, orders, and margins has the level of a scientific-practical justification of the predictable and acceptable risks characterizing generally regular and limiting states of these installations.

For all spectrum of technosphere installation types of emergency and catastrophic situations, the level of their protectability and types of accompanying risks at transition from standard conditions operation in regular states to emergency and catastrophic at service can be described (**Table 2**) as:

- Regular situations—occurring at installations operation in the breaking points established by norms and rules; risks for them controlled; and protectability from them increased
- Regime emergency situations—occurring at a shift from service standard conditions at regular operation of potentially dangerous installations; aftereffects from them predicted, risks for them controlled; and protectability from them sufficient
- Design emergency situations—arise at a runout of installation out of breaking points of regular regimes with predicted and acceptable aftereffects; risks for them analyzed; and protectability from them partial
- Out-of-design emergency situations—arise at nonreversible damages of important parts of installation with high losses and human sacrifices and with necessity of carrying out a recovery work; risks for them heightened; and the level of protectability from them insufficient
- Hypothetical emergency situations—can arise at the not forecast in advance scenarios of evolution with the greatest possible losses and sacrifices; are characterized by high risks; protectability from them low; and restoration of installations is impossible

The complex calculation-experimental analysis of the initial and remaining service life of an NPP is founded first of all on an estimation of service damages accumulation conditions at different service regimes taking into account corresponding state equations, and also on the study of conditions of transition in limiting states taking into account service kinetics of mechanical properties of materials, criteria of strength, crack resistance, and survivability.

Generally termed procedures are implemented with the use of a complex criteria equations, computational equations, and design parameters applied to the analysis and definition of regular and limiting states of engineering objects. The complex criteria include the following equations:

For an estimation of static and long-term strength,

$$F_Q\{\sigma, e, t, \tau\} = F_Q\left\{f_1\left(\frac{\sigma_y^\tau}{n_y}, \frac{\sigma_u^\tau}{n_u}, \frac{\sigma_{lt}^\tau}{n_\sigma}, \frac{e_c^\tau}{n_e}, \frac{\tau_c}{n_\tau}\right), f_2(m)\right\}, \tag{5}$$

where $F_Q$ is the functional characterizing dependence of stresses from actual force impacts $Q$; σ, $e$ are the operating in time $\tau$ at temperature $t$ stresses and strains; $f_1$ is the functional dependence, which includes $\sigma_y^\tau$, $\sigma_u^\tau$, and $\sigma_{lt}^\tau$ that are the yield, strength, and long-term stress points of a material for deformation time $\tau$, accordingly, $e_c^\tau$, $e_c^\tau$ is the critical values (at fracture) of strains at this time; $n_y$, $n_u$, and $n_\sigma$, $n_e$, $n_\tau$ are the margins accordingly on yield and strength stress points, on stresses, strains, and time; $f_2(m)$ is the functional dependence (in most cases, power) for a hardening parameter $m$ in elastoplastic field of a deformation [2, 20, 21].

Для оценки ресурса по параметрам числа $N$ циклов и времени $\tau$

$$F_L\{\sigma, e, N, \tau\} = \left\{f_1\left(\frac{\sigma_a}{n_\sigma}, \frac{e_a}{n_e}, \frac{N_f}{n_N}\right) f_2(\sigma_y, \psi_c, m_p, m_e)\right\}, \tag{6}$$

where $F_L$ is the functional characterizing dependence of life time from amplitudes of stresses $\sigma_a$, strains $e_a$, number of fracture cycles $N_f$, and margins corresponding to them, and from plasticity of material $\psi_c$ (the relative cross throat at fracture) and exponents for an equation of a fatigue curve for plastic $m_p$ and elastic $m_e$ components of cyclic strains $e_a$ [2, 7, 20, 21].

For a crack resistance estimation,

$$F_K\{\sigma, e, K_I, K_{Ie}, \tau, t\} = F_K\left\{\frac{\sigma}{n_\sigma}, \frac{e}{n_e}, \frac{K_{Ic}}{n_K}, \frac{K_{Iec}}{n_{Ke}}, \frac{\tau_c}{n_\tau}, \frac{t_c}{n_t}\right\}, \tag{7}$$

where $F_K$ is the functional characterizing dependence of stresses $K_I$ and strains $K_{Ie}$ intensity factors, from their critical values $K_{Ic}$ and $K_{Iec}$, from stresses $\sigma$ and strains $e$ levels, from critical time to fracture $\tau_c$ and critical temperature $t_c$ with corresponding margins [2, 7, 20, 21].

For a survivability estimation,

$$F_{L_{ld}}\{\sigma, e, l, N, \tau, K_I, K_{Ie}\} = F_{L_{ld}}\left\{(\Delta K_I, \Delta K_{Ie}), \left(\frac{dl}{dN}, \frac{dl}{d\tau}\right)\right\}, \tag{8}$$

where $F_{L_{ld}}$ is the functional characterizing dependence of survivability parameter from values of service stresses and strains, causing material damage $d$, from sizes of faults (cracks) $l$, from crack growth rates on number of cycles $dl/dN$, and time $dl/d\tau$ parameters, and also from values of ranges of stresses $K_I$ and strains $K_{Ie}$ intensity factors [2, 7, 20, 21].

For a risk and safety estimation,

$$F_R\{P(\tau), U(\tau)\} = R(\tau); \tag{9}$$

$$F_S\{R(\tau), n_R\} = S(\tau) \le \frac{1}{n_R} R_c(\tau) = [R(\tau)] = F_M\{R_c(\tau), n_R, M(\tau), m_M\}, \tag{10}$$

where $F_R$ is the functional, characterizing risk $R(\tau)$ as analytical dependence of probability $P(\tau)$ of occurrence on installation of an emergency situation of this or that type and probable loss $U(\tau)$ in case of its implementation; $F_S$ is the functional characterizing parameter of safety $S(\tau)$, which bundles parameters of really occurring risk with its critical $R_c(\tau)$ (limiting) and admitted $[R(\tau)]$ (acceptable) values through margin factor on risk $n_R$ defined in advance.

Thus, the level of installation safety functionally ($F_M$) depends on values of critical risk, from margin on a risk $n_R$, and also from costs $M(\tau)$ of carrying out steps to decrease danger (risk) of installation and from effectiveness factor of these costs $m_M$ [8, 18, 24].

The mentioned complex functional criteria in Eqs. (1)–(10) allow to implement the full sequence of installation calculation for the purpose of providing for its service safety, beginning from strength parameters and completing at protectability parameters with acceptable values of risk both on a design stage, and at concrete stages of service, including a decision made about life time extension.

At an estimation of the remaining life time on resistance to cyclic fracture, levels of cyclical stresses, cycle asymmetry parameters, a stress concentration, cyclical properties of a material, service temperatures, special conditions of loading, and residual stresses and strains are subject to analysis. Under these data calculation processes and parameters of impacts, fracture stresses and life time are defined. On the basis of such definition are the functionals that resulted above in Eqs. (4)–(10), which include calculation dependences (state equations, curve of deformations and fractures, and strain and force criteria). In improved calculation zones of welded joints, a plastic deformation in the most loaded zones, variety of operating conditions and impacts, and dispersion of characteristics of mechanical properties [2, 10, 20–29, 31, 34–36] are considered.

As appears from Eqs. (1)–(10) the computational-experimental justification of static, long-term, and cyclic strength, life time, and risks included in comprehensive analysis of conditions of safety service of the NPP equipment at regular and unnominal situations, sampling of types of limiting states, calculation schemes and calculation cases, methods of the analysis of stress-strain states, methods of preliminary diagnostics of technical state, assignment of margins on strength and on life times, study of probabilities of limiting states reaching, an estimation of risks of accidents and disasters [2, 9–11, 20–36].

The built-up calculation of curve (permissible amplitudes of stresses and life time at a cyclic loading, and also of the maximum stresses and time before fracture in the long term) is carried out for an estimation of initial and remaining life time on the basis of a schematization of history of loading, sampling of computational schemes, and computational cases. The calculation of initial and remaining life time is carried out in two alternatives: an approximate calculation and improved calculation.

The concept of an estimation, a diagnosis, and a prediction of service life of the NPP is correlated with the sampling of state variables of the equipment on the level of wearing and life time exhaustion. To define the factors and parameters influencing on life time, it is necessary to attribute maximum deviations of wall width and errors in measurement, a staging of prediction of life time, results of resource and strength researches, levels of diagnosing of installations, and influence of engineering preliminary diagnostics efficiency on the level of a fracture risk.

On the basis of summarizing of results of a life time design justification of reactors, it is possible to establish a dependence of life time on commissioning terms, for example, an NPP with VVER type reactor of all generations (**Figure 17**). To a twenty-first century kickoff in our country and abroad, the design life time (expected life) has increased to 40–60 years; by 2025, the design life time can increase to 100 years [1, 3, 7, 11, 24].

Thus, the key problems of design, manufacture, service, upgrading, and a leading-out from service of nuclear units of the following (the fourth and the fifth) generations with heightened characteristics of life time and safety are:

- Transition to new principles of reactor core build-up, sharply reducing severe accident possibility with its melting

- Use of joint guard from severe accidents by new organization of working master schedules both in regular and in the emergency situations promoting to decrease of negative and dangerous aftereffects of accident propagation

- Introduction in practice of making and service of reactors with an in-depth analysis of risks of occurrence and propagation of the emergency and catastrophic situations, considering both probabilities of these situations and their aftereffects

- Inclusion in the analysis of heightened life time, risks and safety of reactors of such base criteria as strength, life time, reliability, survivability, physical protectability, and economic justification

- Orientation to escalating requirements to safety of the NPP formed by national and international laws, norms, and rules

- Elimination of unreasonable conservatism in already accepted normative and technical documents and introduction in the safety analysis of new threats and risks (including risks of terrorism)

- Statement as the corner-stone fundamental and applied researches of safety of nuclear reactors of problems of forming of unified methodical baseline on integrated study of external and interior impacts of a wide spectrum, responses to these impacts of critical important bearing elements of the NPP in linear and nonlinear fields of a deformation, damages, and fractures

- Setting, justification, control, and monitoring of the major parameters of life time and safety of the NPP operation at regular and emergency situations for confinement of margins on strength, life time, and risks in safety breaking points

Problems of safety maintenance on the basis of the concept of risks generally should to be decided with the use of the determined, statistical, probability, and combined methods of fracture mechanics and mechanics of disasters. Probabilities $P_S$ of realization in an NPP of system threats can be presented with the use of functional $F_{PS}$ [2, 6, 8, 18, 24–26, 29, 32, 33]

$$P_S = F_{PS}\{P_N, P_T, P_O\}, \tag{11}$$

where $P_N$ is the probability of occurrence of the unfavorable event, stipulated by the human factor; $P_T$ is the probability of such event stipulated by a state of an NPP components; and $P_O$ is the probability of its occurrence stipulated by an environmental exposure.

The type of functional Eq. (11) remains the same and for probabilities of risks realization included in the analysis at design, making, and service of the NPP. The great importance thus has that facts that the role of the human factor in appraisal $P_S$ at change $P_N$ is defined not only human controllers and the personnel, their professional qualities and a physiological state, but the experts, making solutions on all level of the hierarchy by safety of the NPP.

Probabilities $P_T$ essentially depend on the level of protectability of the NPP from accidents and disasters. This protectability is defined by quality of their initial and current state, extent of degradation of installations at the given stage of service, and diagnosing and monitoring level. Such position indicates direct interacting of

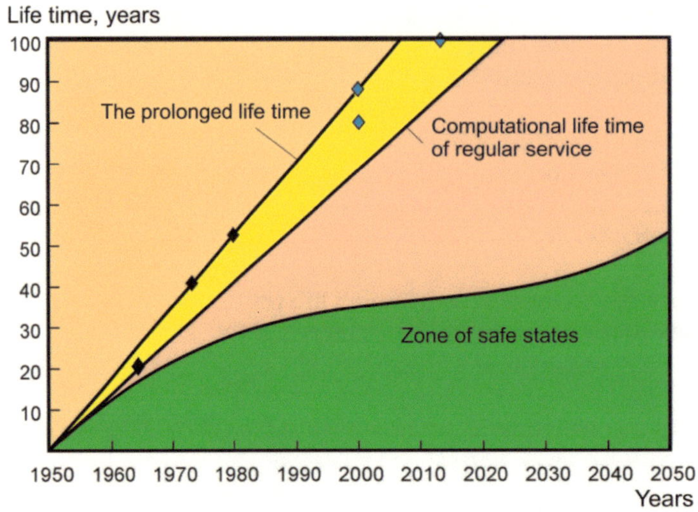


**Figure 17.**
*Characteristics of initial design (full line) and the prolonged expected life (lives times) of the NPPs with type reactors VVER of the first–the fifth generations.*

parameters $P_T$ and $P_N$ taking into account base parameters of reliability and quality of technosphere installations.

Probabilities $P_S$, as it is known, depend on occurrence of dangerous natural processes (earthquakes, floods, hurricanes, tsunami, landslides, etc.) and also from a state of the NPP installations and, hence, from $P_T$. Adoption unreasonable (from the point of view of risks) $R$ $(\tau)$ solutions on arrangement of technosphere installations and zones of population residing does parameter $P_S$ dependent and from $P_N$.

Losses $U_S$ from realization of system threats generally can be recorded through the functional $F_{US}$

$$U_S = F_{US}\{U_N, U_T, U_O\}, \qquad (12)$$

where $U_N$ is the losses caused to the population at interacting of primary and secondary knocking factors at realization of strategic system threats; $U_T$ is the losses caused to technosphere installations; and $U_O$ is the losses caused to an environment.

Values $U_N$, $U_T$, and $U_O$ can be measured both in natural units (for example, a death-roll of people, number of the blasted installations, and the square of injured territories) and in equivalents (for example, in economic, monetary parameters).

As a whole, in Russia, taking into account social and economic transformations, global processes to power supply and experience and prospects of nuclear energetics development based characteristics of risks $R$ of accidents and disasters of the natural-technogenic character, defined by their losses $U$ (or severity) and probability $P$ (or quantity), have rather complicated character of a time history $\tau$ with a common trend to increment (**Figure 18**).

Accepting that the relative risks $\overline{R}(\tau)$ increase eventually owing to natural aging processes, degradation, accumulation of damages, and level of safety $\overline{S}(\tau)$ depends on the relative protectability $\overline{Z}(\tau)$.

$$\overline{R}(\tau) = F_R\{\overline{U}(\tau), \overline{P}(\tau)\}; \overline{S}(\tau) = F_S\{\overline{R}(\tau), \overline{Z}(\tau)\}, \qquad (13)$$

where the fact of accident and disaster occurrence will correspond to the condition

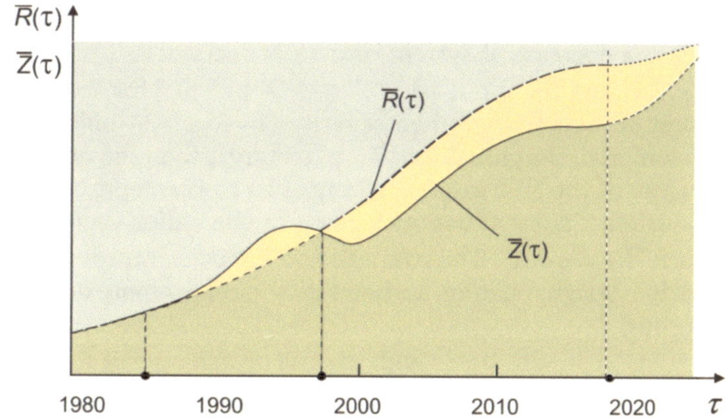


**Figure 18.**
*The time history of the relative risk levels and protectability.*

$$\overline{S}(\tau) = \overline{R}(\tau)/\overline{Z}(\tau) \leq 1. \tag{14}$$

Such conditions occurred at the moment of Chernobyl disaster (1986), last years the twentieth centuries at damages of collecting channels of steam generators PGV-1000 type, on boundary line of centuries at damages of welds to a weld zone of the principal circuital pipeline to the steam generator [4, 11].

In **Figure 18** the major role of improving of all service parameters of the NPP, and first of all life time and safety which promote decrease of probabilities of accidents and disasters occurrence $P(\tau)$ and accompanying them losses $U(\tau)$ is visible.

When for the equipment of the concrete NPP, the relative system risks $\overline{R}_S$ (for population $\overline{R}_N$, for technosphere installations $\overline{R}_T$, and for environment $\overline{R}_O$) are defined, the surface of limiting states on values of these system risks $\overline{R}_S$ varying on some random paths $V(R)$ can be plotted (**Figure 19**).

$$\overline{R}_S = \sqrt{\overline{R}_N^2 + \overline{R}_T^2 + \overline{R}_O^2}. \tag{15}$$

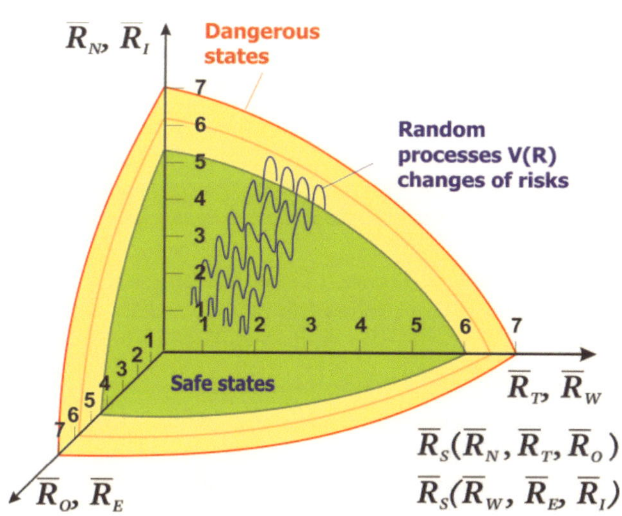


**Figure 19.**
*The surfaces dangerous and safe states on values of risks.*

To reach the acceptable protectability of the NPP equipment, implementation of complex steps on the decrease of system risks $\overline{R}_S$ is necessary.

If on axes $\overline{R}_T$, $\overline{R}_N$, and $\overline{R}_O$ to put aside classes from 1 to 7 for accidents and disasters on extent of increment of their severity (1—local, 2—object, 3—district, 4—regional, 5—national, 6—global, and 7—planetary), then the quantitative assessment of extent of the NPP safety and any of its components by criteria of risks is represented possible. Such estimation is given by the radius vector in three-dimensional space "$\overline{R}_T$-$\overline{R}_N$-$\overline{R}_O$". The strength and life time improvement on all stages of installation design, making, and service should promote decrease in danger of these installations.

For an NPP transfer in safe states with the use of risk criteria $\overline{R}_N$, $\overline{R}_T$, and $\overline{R}_O$ (**Figure 19**), it is necessary to reduce the possibility (risk $\overline{R}_S$) of uncontrollable emission of potentially dangerous substances $W$ and energies $E$ and also a loss of control (disruption of data flows $I$),

$$\overline{R}_S = \sqrt{\overline{R}_W^2 + \overline{R}_E^2 + \overline{R}_I^2}, \tag{16}$$

or to reduce the relative risks of accidents and disasters $\overline{R}_N$, $\overline{R}_T$, and $\overline{R}_O$ as in Eq. (15) and $\overline{R}_W$, $\overline{R}_E$, and $\overline{R}_I$ as in Eq. (16).

This result can be attained by the creation of monitoring systems for diagnostics and monitoring of risk parameters $\overline{R}_N$, $\overline{R}_T$, $\overline{R}_O$, $\overline{R}_W$, $\overline{R}_E$, and $\overline{R}_I$ and guard $Z(\tau)$, and also by the introduction in the analysis of safety $S(\tau)$ scenarios of occurrence and propagation of emergency and catastrophic situations.

The state, regional and object control, regulating and providing of safety $\overline{S}(\tau)$ by system risks criteria $\overline{R}_S(\tau)$ comes to the qualitative both quantitative statistical and determined analysis on the given interval of time $\Delta\tau$ of all service parameters and to implementation of complex activities on decrease of system risks from actual unacceptable $R_S$ to acceptable (admissible) levels $[R_S]$:

$$\overline{R}_S = F_R\{\overline{P}_S, \overline{U}_S\} \leq [\overline{R}_S] = (1/n_S) \cdot \overline{R}_{Sc} = F_R\{[\overline{P}_S], [\overline{U}_S]\} = F_M\{m_M{}^{-1}[\overline{M}]\}, \tag{17}$$

where $n_S$ is the safety factor on system risks; $\overline{R}_{Sc}$ is the unacceptable (critical) risk; $[\overline{P}_S]$ and $[\overline{U}_S]$ are the acceptable (permissible) probabilities and losses; $[\overline{M}]$ is the necessary acceptable expenditures for decrease of risks; and $m_M$ is the cost-effectiveness ratio ($1 \leq m_M \leq 10$).

Safety of the NPP by criteria of risks can be considered ensured if the inequality $n_S \geq 1$ is attained.

The interval of time $\Delta\tau$ for which risks $\overline{R}_S$ are defined usually is accepted to equal 1 year ($\Delta\tau$ = 1 year).

According to Eqs. (15) and (16), control and planning with the use of the criteria baseline grounded on risks come to following primal tasks:

To the development of scientifically well-founded methods of the analysis of risks $\overline{R}_S$ and their basic quantities $\overline{P}_S$ and $\overline{U}_S$

To decision making about the level of allowable values $[\overline{R}_S]$, $[\overline{P}_S]$, and $[\overline{U}_S]$ with an estimation of margin values $n_s$

To scientifically well-founded level of definition of necessary expenditures $[\overline{M}]$ on decreasing risks with sampling and improving of efficiency of these expenditures $m_M$

Thus, predicting, monitoring, and forestalling of accidents and disasters for an NPP (including by improving of all parameters of strength, life time and survivability) appear to be essentially more effective, than liquidating of aftereffects of catastrophic situations (type of the TMI, the CNPP, and the Fukushima-1). Values $\overline{M}$ at a suitable justification of activities on the decrease of risks can be considerable (in $m_M$ time) less losses $\overline{U}_S$ caused to economy by vulnerability of the equipment for all types of NPPs.

As it was already mentioned, safety of nuclear energy installations $S(\tau)$, as well as all other complicated engineering systems, on the given interval of time $\tau$ is defined in Eq. (13) by two basic quantities: probability $P(\tau)$ of unfavorable event occurrence (an unfavorable situation) and probable loss $U(\tau)$ from this event. Values $P(\tau)$ and $U(\tau)$ are generally statistically uncertain, demanding for their quantitative assessment of great volumes of the information on the nature, behaviors, sources, and scenarios of unfavorable events both for each of considered installations and for the given set of installations (group, batch, and series) at occurrence and propagation of unfavorable events and also the information on aftereffects for installations, persons, and an environment at occurrence, propagation, and liquidation of unfavorable events.

## 4. The analysis of limiting states

In nuclear energetics with reactors of all types and all generations (from the first to the fourth) prior to the beginning of the twenty-first century, at failure analysis, the basic attention was given to parameter $P(\tau)$ that defined reliability of safety operation of the NPP. Special meaning was added thus to the forestalling and prevention of the heaviest on the aftereffects of catastrophic situations with the peak damages—melting of the core and a radioactivity runout for breaking points of all guard barriers—casings of the fuel element, cartridge, reactor vessel, reactor hall, and containment. In this case, reactor vessel fracture is extremely dangerous. This event concerns the seventh group of limiting states.

Significant aftereffects arise also at fracture of the basic elements of the first circuit of a reactor vessel and collecting channels of steam generators, pumps, volume compensators, bubbler tanks, and also housings and runners of turbines in the second circuit. These fractures amount the sixth group of the limiting states creating threats to the population, the NPP, and the environment.

If while in service of the NPP because of occurrence of damages of parts of the first circuit has arisen a radioactivity outside breaking points of the NPP and there were thus threats of bombarding radiation for the population, then it is necessary to attribute these events to the fifth group of dangerous limiting states.

The leakages caused by partial damages (faults of crack type or depressurizations of connectors) and creating threats for human controllers and the personnel in the NPP concern the fourth group of limiting states.

The third group of limiting states should be bundled to the considerable damages of the above-termed parts of the first and the second circuit without a radioactivity runout for breaking points of an NPP, which are not demanding their mandatory substitution.

The second group of limiting states concern occurrence in bearing structures of the NPP of partial damages without a radioactivity runout for breaking points of the first circuit, not demanding their substitution, but demanding carrying out of repair-and-renewal operations.

The first group of limiting states is amounted by those of them which are bundled to damages and the faults that have fallen outside the limits admissible under inspection norms and calculation, but not demanding mandatory carrying

| Group | Types of limiting state | Danger extent | Objects for threats |
|---|---|---|---|
| LS-7 | Core damage. Fracture of a reactor vessel with a radioactivity runout in an environment | Extremely high | The population, NPP, environment |
| LS -6 | Fracture of the basic parts of the first and the second circuits with a radioactivity runout in an environment | Excessively high | The population, NPP, environment |
| LS -5 | Large leakages in the first circuit with a radioactivity runout for NPP breaking points | Very high | The population, NPP, environment |
| LS -4 | Damages and leakages in the first and the second circuits with probable runout of a radioactivity for NPP breaking points | High | The population, NPP, environment |
| LS -3 | Damage and leakages in the first and the second circuits with a runout of a radioactivity inside of the NPP | Heightened | Human controllers, NPP parts |
| LS -2 | Partial damages without a radioactivity runout for the breaking points of the first circuit demanding reconditioning | Not high | The damaged equipment |
| LS -1 | Partial damages without a runout of radioactivity which are not demanding reconditioning | Low | The damaged equipment |

**Table 3.**
*Groups of limiting states for the analysis of the NPP safety.*

out of repair-and-renewal operations and that can be admitted to prolongation of service before the next examination.

These facts allow to execute summary classification by groups of limiting states for the NPP equipment (**Table 3**) from the most dangerous admissible (the seventh group of limiting states LS-7) to the least dangerous admissible (the first group of limiting states LS-1).

For the groups of limiting states indicated in **Table 3** taking into account summarizing of great volume of normative and technical materials and results of the executed researches, it is possible to describe demanded (admissible) probabilities $[P(\tau)]$ occurrence of unfavorable events. To such probabilities there correspond their actual levels obtained from statistics of their occurrence while in service of NPPs of all generations. Each severe accident or disaster on an NPP, happening at the moment $\tau_c$, was accompanied by comprehensive analysis of their reasons and sources, and also realization of considerable on volumes and expenditures of activities for safety improving. Eventually, at $\tau_s > \tau_c$, after such accidents or disasters, decrease of probabilities from $P(\tau_c)$ to $P(\tau_s)$ was observed.

For values of probabilities $P(\tau_c)$ and $P(\tau_s)$ for all reactors operated in the world at $\tau \leq \tau_c$ and $\tau = \tau_s$, it is possible to estimate on ratios

$$P(\tau_c) = \frac{N_d}{N_{tc} \cdot \tau_c}; P(\tau_s) = \frac{N_d}{N_{ts} \cdot \tau_s}, \tag{18}$$

where $N_d$ is the quantity of the reactors that have obtained damages at the given $i$-th type of limiting state under **Table 3**; $N_{tc}$ is the total of reactors to the time $\tau_c$ of occurrence of the given $i$-th type of damage; $N_{ts}$ is the total of reactors to the time $\tau_s$; $\tau_c$ is the mean time (years) of service of one reactor to the time of reaching of the given $i$-th type of limiting state; and $\tau_s$ is the mean time of the service of one reactor.

As it was already mentioned, unfavorable events on an NPP (disasters, accidents, failures, and disruptions) are accompanied by corresponding losses $U(\tau)$ both at the moment of occurrence of these events $\tau_c$ and after them $(\tau \geq \tau_c)$. These losses are caused to the person (to human controllers, the personnel, and the population), to technosphere installations (to an NPP and other installations of its infrastructure), and also to the environment. Now while miss direct legal and normative documents by the quantitative definition of these losses. Some suggestions on this problem are stated below.

For a tentative estimation of loss $U(\tau)$, it is possible to use the simplified statistical and expert information on such losses. Generally, values of losses are defined by two basic parameters:

- Losses of human lives or health at occurrence and progressing of unfavorable situations

- Economical losses (for example, in Rubles or USD) from a loss of life, from maiming to people, and from fractures and damages of technosphere installations and the environment

Direct loss $U(\tau)$ for the LS-7 limiting state interlinked immediately to fracture of the NPP or full termination of its service. Then, the datum of loss $U(\tau)$ can be accepted to the equal cost of the NPP. In this, the loss can and should include charges $U(\tau_1)$ within 1–2 years on a primary elimination of the consequences of disaster or accident (realization of protective measures, evacuation of the population, and termination of infrastructure installation operation). These charges at $(\tau_1 \geq \tau)$ several times (2–4) can exceed the initial loss $U(\tau_c)$. Decrease of secondary consequences of heavy disasters on an NPP (making of shelters, recultivation, medical examination and the help, and compensating payments) demands complementary essential annual expenditures $U(\tau_2)$ for a long time $\tau_1 < < \tau_c \leq \tau_2$. In **Figure 20** is displayed schematization of the relative losses $\overline{U}(\tau) = U(\tau_i)/U(\tau_c)$ depending on time $\Delta\tau$ after the occurrence of heavy disaster ($\Delta\tau = \tau_s = \tau$) at reaching the most dangerous limiting state of the LS-7 type, summarized in **Table 3**.

With the reduction of the hazard level of accidents and disasters (at transition of limiting states from the LS-7 to the LS-1), value $U\ (\tau_c)$ and $\overline{U}(\tau)$ decrease because of decrease of losses $\overline{U}(\tau_1)$ and $\overline{U}(\tau_2)$.

From assemblage of tens methods for definition of risks parameters as the most simple is the statistical or determined-statistical method according to which it is possible to write

$$R(\tau_i) = P(\tau_i) \cdot U(\tau_i), \tag{19}$$

where $\tau_i$ is the time for which one the risk assessment is conducted and $P(\tau_i)$ and $U(\tau_i)$ are the probabilities and losses for time $\tau_i$.

If under $\tau_i$ is fathomed the time of unfavorable event occurrence of $\tau_c$, then according to Eq. (19), it is possible to obtain

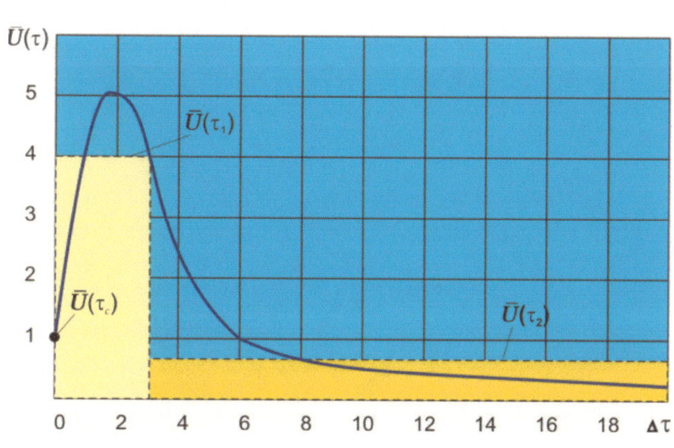


**Figure 20.**
*The time-history and schematization of losses* U(τ).

$$R(\tau_c) = P(\tau_c) \cdot U(\tau_c) = P^*(\tau_c) \cdot U(\tau_c). \tag{20}$$

Risk $R(\tau_c)$ is possible to consider as risks of the implemented unfavorable events at $\tau_i = \tau_c$ and to use them for prediction of events for times $\tau_i \geq \tau_c$. One such prospective risk appears as the risk for the current phase of service $\tau_i = \tau_s$. In this case, on the basis of Eq. (19), it is possible to write

$$R(\tau_s) = P^*(\tau_s) \cdot U(\tau_s), \tag{21}$$

where $\tau_s$ is the time after unfavorable event ($\tau_s \geq \tau_c$).

This time can be situated in the interval $\tau_c \leq \tau_1 \leq \tau_2$. Then, for one operated unit of the NPP, the common risk at reaching the given i-group of limiting state from the LS-7 to the LS-1 will constitute

$$R(\tau_c)_\Sigma = \sum_{i=1}^{7} R(\tau_c)_i \tag{22}$$

If at loss estimations to consider not only direct losses at occurrence of unfavorable event $U(\tau_\kappa)$ together with complementary losses $U(\tau_1)$ and $U(\tau_2)$, then it is possible to define common (integral) losses as

$$U(\tau)_\Sigma = U(\tau_c) + U(\tau_1) + U(\tau_2) \tag{23}$$

These integral losses respond to the appropriate risks

$$R(\tau_\Sigma)_\Sigma = \sum_{i=1}^{7} U(\tau_\Sigma)_i P^*(\tau_s)_i. \tag{24}$$

On the basis of results of an estimation considered above risk components, it is possible to build dependences between basic parameters of risk for the NPP—probabilities $P(\tau)$ occurrence of unfavorable situations and losses $U(\tau)$ from them (**Figure 21**).

The line had above and design points in the **Figure 21** belong to probabilities $P(\tau_c)$ and to losses $U(\tau_c)$ for the moment of accident or disaster occurrence on the NPPs. The lower line made like overhead characterizes a negligible zone of risk parameters $[P(\tau_c)]_{\min} - [U(\tau_c)]_{\min}$ and the midline characterizes a zone of acceptable risks $[P(\tau_c)] - [U(\tau_c)]$. If to allow common (near-term and long-time)

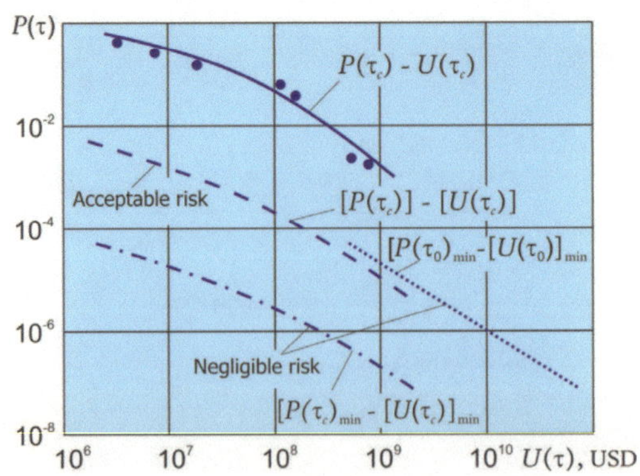


**Figure 21.**
*Parameters of risks for the NPP with reactors of VVER types.*

for negative consequences of accidents and disasters, it is possible to build a line of negligible risk parameters $[P(\tau_\Sigma)]_{\min} - [U(\tau_\Sigma)]_{\min}$.

## 5. Conclusion

From stated above follows that the major problems which have been not decided while to the full for a NPP there are problems of provision of their protectability and safety on the basis of new scientific fundamental and application researches on mechanics, hydrodynamics, economics, mathematical and physical modeling of dangerous processes resulting to heavy disasters, and also development of detailed methods of the analysis of risks for heavy disasters.

Results of the fulfilled scientific researches and developments in this direction, integrated [3–8, 15–17] in the serial of monographic publications on strength, life time, and safety of power nuclear reactors, are initial scientific baseline for the applicable normative, designer, technological solutions on provision of protectability of the NPP equipment from heavy disasters on the basis of criteria of acceptable risks.

The above-mentioned results of analytical and experimental researches can be considered in the capacity of a theoretical basis for the subsequent development of practical models of the computational analysis of risks for strategically relevant installations of a nuclear energetic on the basis of the complex Eqs. (1)–(24). Development of such models, and the most important—their filling up statistically reliable probability distribution of fractures on groups of limiting states (see **Table 3**) on the one hand, and economical computations of losses, with another, it is necessary to consider as the major task for a solution of a problem of safe development of power supply of human community.

At up-to-date and subsequent stages of evolution of power engineering in Russia in the capacity of a basic recommended position, it is necessary to use the position about provision of an acceptable risk level of occurrence of accidents and disasters. In this connection, it is not obviously possible to ensure from social-economic and technological stands the declared principle of absolute safety with null risks ($R(\tau) = 0$). Owing to it, the solution of the delivered problem is brought together to determination of scientifically well-founded admissibility of occurrence of the emergency situations with possible minimization of loss caused by them, with an estimation of the greatest possible, acceptable, and controlled risk both at probable occurrence of global and national accidents and disasters, and their realization at regional and local levels.

## Author details

Nikolay Andreevich Makhutov*, Mikhail Matveevich Gadenin,
Igor Alexandrovich Razumovskiy, Sergey Valerievich Maslov and
Dmitriy Olegovich Reznikov
The A.A. Blagonravov Institute for Machine Sciences of the RAS, Moscow, Russia

*Address all correspondence to: safety@imash.ru

## References

[1] Strength Calculation Norms for Equipment and Pipelines of Nuclear Power Plants. Moscow: Energoatomizdat; 1989. 525p

[2] Makhutov NA. Strength and Safety: Fundamental and Applied Researches. Novosibirsk: Nauka; 2008. 528p

[3] Makhutov NA, Frolov KV, Stekolnikov VV, et al. Strength and Life-Time of Water-Moderated Power Reactors. Researches of Stresses and Strength of Nuclear Reactors. Moscow: Nauka; 1988. 312p

[4] Makhutov NA, Dragunov Yu G, Frolov KV, et al. Carrying Capability of Steam Generators of Water-Moderated Power Reactors. Researches of Stresses and Strength of Nuclear Reactors. Moscow: Nauka; 2003. 440p

[5] Makhutov NA, Dragunov Yu G, Frolov KV, et al. Dynamics and Strength of Water-Moderated Power Reactors. Researches of Stresses and Strength of Nuclear Reactors. Moscow: Nauka; 2004. 440p

[6] Makhutov NA, Frolov KV, Dragunov Yu G, et al. Strength and Safety Problems of Water-Moderated Power Reactors. Researches of Stresses and Strength of Nuclear Reactors. Moscow: Nauka; 2008. 464p

[7] Makhutov NA, Frolov KV, Dragunov Yu G, et al. Life Time and Survivability Provision of Water-Moderated Power Reactors. Researches of Stresses and Strength of Nuclear Reactors. Moscow: Nauka; 2009. 343p

[8] Makhutov NA, Frolov KV, Dragunov Yu G, et al. Analysis of Risk and Safety Increase of Water-Moderated Power Reactors. Researches of Stresses and Strength of Nuclear Reactors. Moscow: Nauka; 2009. 499p

[9] Makhutov NA, Gadenin MM, Cherniavskii OF, Cherniavskii AO, Evropin SV. Low-cycle deformation of the structures in a nuclear power plant and methods for calculating them. Atomic Energy. 2009;**107**(3):173-179

[10] Dragunov YG, Evropin SV, Gadenin MM, Makhutov NA, Rebiakov YN, Cherniavskii AO, et al. Stress-strain kinetics in calculations of high-temperature strength and longevity of reactor structures. Atomic Energy. 2016; **119**(3):177-189. DOI: 10.1007/s10512-015-0046-y

[11] Dranchenko BN, Dragunov YG, Portnov BB, et al. Experimental Researches of Stress State and Strength of the VVER Equipment. Moscow: IKTs Akademkniga; 2004. 640p

[12] Norms of Strength Calculation of Parts of Reactors, Steam Generators, Pressure Vessels and Pipelines of Nuclear Power Plants, Experimental and Research Nuclear Reactors and Installations. Moscow: Metallurgia; 1973. 408p

[13] Norms of the IAEA on Safety. Safety of Nuclear Power Plants: Design. Concrete Safety Requirements. No. SSR-2/1. Vienna: IAEA; 2012. 80p

[14] ASME. Boiler and Pressure Vessel Code. 2010th ed. New York: ASME; 2011. [Retrieved: November 9, 2011]

[15] Makhutov NA, Frolov KV, Dragunov Yu G, et al. Modeling Researches and Full-Scale Tensometry of Power Reactors. Researches of Stresses and Strength of Nuclear Reactors. Moscow: Nauka; 2001. 293p

[16] Makhutov NA, Frolov KV, Stekolnikov VV, et al. Experimental Researches of Strains and Stresses in Water-Moderated Power Reactors.

Researches of Stresses and Strength of Nuclear Reactors. Moscow: Nauka; 1990. 296p

[17] Makhutov NA, Stekolnikov VV, Frolov KV, et al. Constructions and Computational Methods of Water-Moderated Power Reactors. Researches of Stresses and Strength of Nuclear Reactors. Moscow: Nauka; 1987. 232p

[18] Makhutov NA, Gadenin MM. Fundamental and applied researches of safety and risks for power engineering installations. In: The Federal Hand-Book: Information-Analytical Edition. Vol. 25. Moscow: Center of Strategic Partnership; 2011. pp. 439-446

[19] Makhutov NA, Gadenin MM. Technogenic safety: Diagnostics and monitoring of states of potentially dangerous equipment and risks of its service. In: The Federal Hand-Book: Information-Analytical Edition. Vol. 26. Moscow: Center of Strategic Partnership; 2012. pp. 307-314

[20] Makhutov NA. Structural Strength, Life-Time and Technogenic Safety. Part 1: Strength and Life-Time Criteria. Novosibirsk: Nauka; 2005. 494p

[21] Makhutov NA, Gadenin MM. Fatigue resistance at variation of temperature-time factors. International Journal of Fracture. 2008;**1-2**(150): 105-127

[22] Makhutov NA, Vorobiev AZ, Gadenin MM, et al. Strength of Structures at Low-Cycle Loading. Moscow: Nauka; 1983. 271p

[23] Makhutov NA, Gadenin MM, Gokhfeld DA, et al. The State Equations at a Low-Cycle Loading. Moscow: Nauka; 1981. 245p

[24] Makhutov NA. Safety and Risks: System Researches and Developments. Novosibirsk: Nauka; 2017. 724p

[25] Makhutov NA, Gadenin MM. Basic researches of safety service problems for the technosphere objects on the basis of strength, safe life and risk. Engineering and Automation Problems. 2018;**3**:7-24

[26] Makhutov NA, Gadenin MM. Engineering Diagnostics of Remaining Resource and Safety. Safety Diagnostics. Moscow: Publishing House "Spektr"; 2011. 187p

[27] Makhutov NA, Zatsarinnyi VV, Gadenin MM, et al. Statistical Regularities of Low-Cycle Fracture. Moscow: Nauka; 1989. 253p

[28] Gadenin MM. Study on damaging and fatigue life of constructions under single- and two-frequency loading modes based on deformational and energy approaches. Inorganic Materials. 2018;**54**(15):1543-1550. DOI: 10.1134/S0020168518150049

[29] Makhutov NA, Gadenin MM. Development of fundamental and applied researches in the field of machine sciences with use of strength, life-time, survivability and safety criteria. Industrial Laboratory (Diagnostics of Materials). 2018;**84**(10): 41-52. DOI: 10.26896/1028-6861-2018-84-10-41-52

[30] Lepikhin AM, Makhutov NA, Moskvichov VV, Cherniaev AP. Probability Risk-Analysis of Structures of Engineering Systems. Novosibirsk: Nauka; 2003. 174p

[31] Gadenin MM. Multiparameter analysis of safety service and protectability conditions of machines and structures by strength, life-time and survivability criteria. Problems of Safety and Emergency Situations. 2012;**6**:22-36

[32] Makhutov NA, scientific editor. Safety of Russia. Legal, social-economic and scientifically-engineering aspects. In: Analysis of Risk and Safety Problems. Part 1. Grounds of the

Analysis and Safety Control. Moscow: MGOF Znanie; 2006. 640p

[33] Makhutov NA, scientific editor. Safety of Russia. Legal, social-economic and scientifically-engineering aspects. In: Technogenic, Technological and a Technospheric Safety. Moscow: MGOF Znanie; 2018. 1016p

[34] Gadenin MM. Estimation of the effect of loading modes on the conditions of attainment of marginal states and resource assignment. Inorganic Materials. 2014;**50**(15): 1537-1542. DOI: 10.1134/S0020168514150035

[35] Gadenin MM. Characteristics of mechanical properties of materials in studies of conditions of attainment of marginal states. Inorganic Materials. 2013;**49**(15):1352-1356. DOI: 10.1134/S0020168513150053

[36] Makhutov NA, Gadenin MM. Structure of the main calculations for determination of the initial and remaining safe service. Problems of Safety and Emergency Situations. 2018; **2**:21-33

Chapter 9

# Combinatorial Enumeration of Graphs

*Carlos Rodríguez Lucatero*


## Abstract

In this chapter, I will talk about some of the enumerative combinatorics problems that have interested researchers during the last decades. For some of those enumeration problems, it is possible to obtain closed mathematical expressions, and for some other it is possible to obtain an estimation by the use of asymptotic methods. Some of the methods used in both cases will be covered in this chapter as well as some application of graph enumeration in different fields. An overview about the enumeration of trees will be given as an example of combinatorial problem solved in a closed mathematical form. Similarly, the problem of enumeration of regular graphs will be discussed as an example of combinatorial enumeration for which it is hard to obtain a closed mathematical form solution and apply the asymptotic estimation method used frequently in analytic combinatorics for this end. An example of application of the enumerative combinatorics for obtaining a result of applicability criteria of selection nodes in a virus spreading control problem will be given as well.


**Keywords:** combinatorial graph enumeration, generating functions, probability

## 1. Introduction

Enumerating a finite set of objects is one of the most basic tasks that can be done by any person. Anybody begins to do it at first while learning the basic arithmetic operations at school. In a very general setting, to enumerate a finite set of things is to put in bijective relation a finite subset of the natural numbers with the things to be counted. As simple as it can be at first sight, the enumeration of things can become a very interesting mathematical activity when we start to count, for instance, in how many ways can we dispose or can we select the $n$ elements of set in $k$ places without repeating them, sometimes allowing repetitions of all of them, or just some of them, etc. That is the starting point of the combinatorics subject. Combinatorics is the field of discrete mathematics that allows us, among other things, to calculate in how many ways some objects can be selected or arranged to comply with any given property.

Within the topic of combinatorial analysis, there is a subfield that is interested in accurately predicting large structured configurations under the analytical method approach and that uses the tool known as generating functions. The analytic combinatorics is devoted to the study of finite structures whose construction follows a certain finite number of given rules. On the other hand, the use of generating



IntechOpen

functions is a tool that allows to relate discrete analysis proper to discrete mathematics with continuous analysis.

One of the most beautiful and ingenious applications of combinatorial enumeration is the probabilistic method. The probabilistic method is intimately related to the important role that randomness plays in the field of theoretical computer science. The utility and beauty of the probabilistic method consists of being an indirect or nonconstructive proof method. This method has been used successfully during the last 60 years and constitutes one of the most important scientific contributions of the great Hungarian mathematician Erdös [1, 2]. Commonly, this method has been used to prove the existence of a certain mathematical object by showing that if we choose some object of a given class in a random way, the probability that this mathematical object complies with certain property is greater than zero.

The probabilistic method has been used with great success to obtain important results in fields as diverse as number theory, combinatorics, graph theory, linear algebra, computational sciences, or information theory.

One aim of combinatorial analysis is to count the different ways of arranging objects under given constraints. Sometimes the structures to be counted are finite and some other times they are infinite. To enumerate is very important in many scientific fields because it allows to evaluate and compare different solutions to a given problem. For example, in computer science, if I want to compare different algorithms that solve a given kind of information processing problem, it is necessary to enumerate the number of steps taken by each one of them in the worst case and to choose the one whose performance is the best. The task of enumerating things can evolve in complexity to some point that the elementary arithmetical operations are not enough to reach the goal. For that reason many enumeration problems have inspired the most talented mathematicians for developing very ingenious methods for solving them. Some of these techniques for solving combinatorics enumeration problems are going to be exposed in the next section. One very interesting subject of the discrete mathematics is the graph theory. It is in the graph theory where many of the most interesting enumerations of graphs arise that have some given structural property that require the utilization of mathematical tools that facilitate the discovery of closed mathematical forms for the calculation of the number of graphs that accomplish some topological property. In this context it can be important to enumerate how many labeled graphs can be constructed with $n$ vertex or how many connected graphs with $n$ vertex exist, etc. For some of these problems, clever methods have been devised for their calculation that lead to closed mathematical forms by the use of generating functions. For other problems, it is very hard to obtain a closed mathematical form solution, and in that case some asymptotic methods have been developed for estimating a bound when the number of vertex is very large by means of the Cauchy theorem. These mathematical tools are going to be covered in the following sections of the present chapter.

## 2. Body of the manuscript

The chapter will have the following structure. In Section 3, we will make a quick revision of some mathematical tools that are frequently used for solving combinatorial enumeration problems. The first tool described will be the ordinary generating functions as well as the exponential generating functions. These tools are used when a closed form solution for an enumeration problem can be obtained. The second tool that will be described in this section will be the analytic combinatorics method that is normally used for those enumeration problems whose closed mathematical form are hard to be calculated. The analytic combinatorics techniques allow to estimate an upper bound of that kind of enumeration problems. We take the enumeration of labeled trees as an example of combinatorial enumeration problem that can be solved in a closed mathematical form and make an overview of some different methods devised for that end in Section 4. In Section 5, we roughly describe how the generating functions can be used to solve graph enumeration problems. In Section 6, we take the problem of enumerating regular graphs as an example of a problem whose closed mathematical form is hard to obtain and apply the analytic combinatorics techniques for estimating an upper bound. In Section 7, I give an application of the combinatorial enumeration for proving the almost sure applicability of a node selecting criteria for controlling virus spreading in a complex network. Finally in Section 8, we make some comments about the possible future applications of the combinatorial enumeration methods.

## 3. Generating functions and analytic combinatorics overview

In this section of the chapter, we are going to make a revision of the basic mathematical tools that have been developed and that facilitate the solution of many combinatorial enumeration problems. We are going to start with some basic conceptual definitions that can be found in many textbooks about this topic. In order to be able to clearly expose the mathematical tools used, it will be necessary to make use of some basic concepts about them that can be consulted more widely in texts such as [3]. One of these basic notions that can be found in [3] is the concept of *combinatorial class* whose definition is as follows:

Definition 1.1 A combinatorial class is a finite or denumerable set in which a size function is defined, satisfying the following conditions:

i. The size of an element is a nonnegative integer.

ii. The number of elements of any given size is finite.

On the other hand, it is necessary to know if two *combinatorial classes* are related in any way which can be determined using the concept of *isomorphism* and that can be defined as follows:

Definition 1.2 The combinatorial classes $\mathcal{A}$ and $\mathcal{B}$ are said to be (combinatorially) isomorphic which is written $\mathcal{A} \cong \mathcal{B}$ if and only if their counting sequences are identical. This condition is equivalent to the existence of a bijection from $\mathcal{A}$ to $\mathcal{B}$ that preserves size, and one also says that $\mathcal{A}$ and $\mathcal{B}$ are bijectively equivalent.

The notion of ordinary generating functions (OGF) as $\sum_{i=0}^{\infty} a_i x^i$ where the coefficients $a_i$ are elements of the sequence $\mathcal{A} = \{a_0, a_1, \dots\}$ or combinatorial class $\mathcal{A}$. This function is also the generating function of the numbers $\mathcal{A}_n$ whose sizes $a_n = \text{card}(\mathcal{A}_n)$ such that the OGF of class $\mathcal{A}$ admits the combinatorial form

$$A(x) = \sum_{\alpha \in \mathcal{A}} x^{|\alpha|}. \tag{1}$$

This means that the variable $x$ marks size in the generating function. The OGF form (1) can be easily interpreted by observing that the term $x^n$ occurs as many times as there are objects in $\mathcal{A}$ of size $n$. A basic operation that can be defined is the one that allows to extract the coefficient of the term $x^n$ in the power series $A(x) = \sum a_n x^n$ as follows:

$$[x^n]\left(\sum_{n\geq 0} a_n x^n\right) = a_n. \tag{2}$$

The concept of generating function can then be defined as:
Definition 1.3 Let be $a_0, a_0, \ldots$ a succession of real numbers. The function

$$f(x) = a_0 + a_1 x + a_2 x^2 + \ldots = \sum_{i=0}^{\infty} a_i x^i \tag{3}$$

is the generating function of the given succession.

The idea comes from the development of Newton's binomial that con be defined as:

Definition 1.4

$$(1+x)^n = \binom{n}{0} + \binom{n}{1}x + \binom{n}{2}x^2 + \cdots + \binom{n}{n}x^n \tag{4}$$

which is the generating function of the succession

$$\binom{n}{0}, \binom{n}{1}, \binom{n}{2}, \binom{n}{3}, \ldots, \binom{n}{n}, 0, 0, 0, \ldots \tag{5}$$

We can say, for example, that

$$\sum_{i=0}^{n} x^i = 1 + x + x^2 + x^3 + \ldots + x^n = \frac{1 + x^{n+1}}{1-x} \tag{6}$$

is the generating function of the succession $1, 1, 1, \ldots, 1, 0, 0, 0, \ldots$ where the first $n+1$ terms are equal to 1. The upper limit of the generating function 6 can be extended to $\infty$ which is known as geometric series. This series is known to be convergent if $x < 1$ and that case can be defined as

$$\sum_{i=0}^{\infty} x^i = 1 + x + x^2 + x^3 + \ldots + = \frac{1}{1-x} \tag{7}$$

which is the geometric series of the succession $1, 1, 1, \ldots$. One of the nice properties of the generating functions is that they can be easily manipulated due to the fact that they are infinite polynomials.

For instance, if we take the first derivative of the generating function 7, we get

$$\frac{d}{dx}\frac{1}{1-x} = \frac{d}{dx}(1 + x + x^2 + x^3 + \ldots) = 1 + 2x + 3x^2 + 4x^3 + \ldots = \frac{1}{(1-x)^2} \tag{8}$$

then $\frac{1}{(1-x)^2}$ is the generating function of the succession $1, 2, 3, 4, \ldots$, while $\frac{x}{(1-x)^2}$ is the generating function of the succession $0, 1, 2, 3, 4, \ldots$. Similarly if we take the first derivative of the generating function $\frac{x}{(1-x)^2}$, we get

$$\frac{d}{dx}\frac{x}{(1-x)^2} = \frac{d}{dx}(0 + x + 2x^2 + 3x^3 + 4x^4 + \ldots) = \frac{(x+1)}{(1-x)^3} \tag{9}$$

then $\frac{(x+1)}{(1-x)^3}$ is the generating function of the succession $1^2, 2^2, 3^2, \ldots$, and $\frac{x(x+1)}{(1-x)^3}$ is the generating function of the succession $0^2, 1^2, 2^2, 3^2, \ldots$ Other simple manipulations that can be done with the generating functions allow us to cancel some element of an associated succession. For example, if $g(x) = \frac{1}{(1-x)}$ and $h(x) = g(x) - x^2 = \frac{1}{(1-x)} - x^2$, then $h(x)$ becomes the generating function of the succession $1, 1, 0, 1, 1, 1, \ldots$. By the other side, if we know that $\frac{x(x+1)}{(1-x)^3}$ is the generating function of the succession $0^2, 1^2, 2^2, 3^2, \ldots$ and that $\frac{x}{(1-x)^2}$ is the generating function of the succession $0, 1, 2, 3, 4, \ldots$, then if we add these two generating functions, we get

$$\frac{x(x+1)}{(1-x)^3} + \frac{x}{(1-x)^2} = \frac{2x}{(1-x)^3} \tag{10}$$

where $\frac{2x}{(1-x)^3}$ becomes the generating function of the succession $0, 2, 6, 12, 20, 30, 42, \ldots$ whose $n$th element can be expressed as $a_n = n^2 + n$. If we use the coefficient extracting operator defined by 2, we can say that

$$[x^n]\frac{2x}{(1-x)^3} = n^2 + n \tag{11}$$

As an example, let us take the sequence $0, 1, 1, 2, 3, 5, 8, 13, 21, 34, 55, \ldots$ know as Fibonacci numbers. This succession can be generated by applying the recurrence relation

$$F_{n+1} = F_n + F_{n-1} \text{ where } n \geq 1, \; F_0 = 0, \; F_1 = 1. \tag{12}$$

The goal is to obtain $n$th element of the succession of numbers generated by 12 that is the coefficient of $x^n$ in the expansion of the function

$$\frac{x}{1-x-x^2} \tag{13}$$

as a power series about the origin. The roots of $1 - x - x^2$ are $x_1 = \frac{1+\sqrt{5}}{2}$ and $x_2 = \frac{1-\sqrt{5}}{2}$. Given that the generating function of the Fibonacci succession is 13, we have that

$$f(x) = \frac{x}{1-x-x^2} = \frac{1}{\sqrt{5}}\left[\frac{1}{1-\frac{1+\sqrt{5}}{2}x} - \frac{1}{1-\frac{1-\sqrt{5}}{2}x}\right] \tag{14}$$

and the $n$th term of the Fibonacci succession is expressed in closed form as

$$F_n = \frac{1}{\sqrt{5}}\left[\left(1-\frac{1+\sqrt{5}}{2}\right)^n - \left(1-\frac{1-\sqrt{5}}{2}\right)^n\right] \tag{15}$$

The calculation of the n-th term of the Fibonacci sequence by using power series allows us to obtain a mathematical closed formula. Because of that we can calculate the n-th term of the Fibonacci sequence in a more efficient way given that the computer program to do it will consist in the direct application of the closed mathematical expression obtained, which is much more efficient than using a computer program based on the application of the Fibonacci recurrence.

In [4] the problem of the various results that can be obtained using generating functions is addressed. The authors of [4] say that by using the tool of generating functions for obtaining the $n$th element of a succession, sometimes an exact formula can be obtained easily, and if it is not the case, a good estimation of the $n$th element can be obtained. It can happen also that we get a recurrence formula from where a generating function can be obtained, or it can happen that from the generating function a new recurrence is obtained giving us a deeper understanding about the succession. The use of generating functions provides statistical information about the succession. The authors of [4] also point out that when it is very difficult to mathematically obtain the n-th term of a given sequence as a closed mathematical expression, a good option is to use asymptotic methods to obtain an estimate of that term. For example, the $n$th prime number is approximately $n \log n$ when $n$ is very big. The authors of [4] also also point out that using generating functions some properties of a succession such as unimodality or convexity can be proven. Another advantage of using generating functions is that some identities as, for example,

$$\sum_{j=0}^{n} \binom{n}{j}^2 = \binom{2n}{n} \quad (n = 0, 1, 2, \ldots), \tag{16}$$

are easy to obtain. Finally, the authors of [4] pointed out that by using generating function, the relationship between problems that have similar generating functions can be discovered.

As was mentioned in the last paragraph, sometimes it is hard to obtain mathematically closed formulas when using generating functions, and in that case, a good alternative is to the use asymptotic formulas. The mathematical tools used for this purpose are part of the field known as analytic combinatorics, and a good reference of this topic is [3]. The main objective of analytic combinatorics is to estimate with a high level of precision the properties of large structured combinatorial configurations by the use of mathematical analysis tools [3].Under this approach, we begin with an exact enumerative description of the combinatorial structure using the generating functions. This description is considered as an algebraic object. The next step is to take the generating function as an analytical object which is a mapping from the complex plane to itself. The singularities found in such a mapping allow to obtain the coefficients of the function in its asymptotic form, resulting in an excellent estimate on the count of the sequences. With this purpose, the authors of [3] classify the analytic combinatorics in the next three topics:

1. Symbolic methods that establish systematically relations of discrete mathematics constructions and operations on generating functions that encode counting sequences

2. Complex asymptotics that allow for extracting asymptotic counting information from the generating functions by the mapping to the complex plane mentioned above

3. Random structures concerning the probabilistic properties accomplished by large random structures

A large material concerning the subject of complex asymptotic analysis is addressed in [5]. A highly recommended text to consult because it covers the applications of the combinatorial tools enumerative to the analysis of algorithms is [6].

In this chapter, there will be a particular interest in the application of generating functions tool for the enumeration of graphs that have some given properties.

Let us start with the definition of a graph. A graph $G = \langle V, E \rangle$ is a structure with a set of vertices $V = \{v_1, v_2, \dots, v_p\}$ whose size $|V| = p$ it is called the order of $G$ and a set of unordered pairs of adjacent vertices, called edges, $E = \left\{\{v_{i_1}, v_{j_1}\}, \{v_{i_2}, v_{j_2}\}, \dots, \left\{v_{i_q}, v_{j_q}\right\}\right\}$ if $G$ is undirected or a set of ordered pairs of adjacent vertices $E = \left\{(v_{i_1}, v_{j_1}), (v_{i_2}, v_{j_2}), \dots, \left(v_{i_q}, v_{j_q}\right)\right\}$, if $G$ is a directed graph, whose edge set size is $|E| = q$, that has no loops and that has no loops multiple edges. A graph $G$ with $p$ vertices and $q$ edges is called a $(p, q)$ graph. In a labeled graph of order $p$, each vertex has a label that is an integer from 1 to $p$. Typical questions such as how many graphs can be constructed with $n$ vertices, how many trees with $n$ vertices can be obtained, and how many of these trees are binary can be answered using the generating functions, and a mathematically closed formula can be obtained. In the case that a closed mathematical formula is very hard to be obtained, an accurate asymptotic estimation formula can be a good option.

There are two commonly used generating functions: the first is the ordinary generating functions (OFG) and the other is the exponential generating functions (EGF).

An ordinary generating function is a mathematical function defined by the following expression:

$$a(x) = \sum_{k=0}^{\infty} a_k x^k, \tag{17}$$

where the coefficients are elements of the succession of numbers $a_0, a_1, a_2, \dots$. The ordinary generating functions are used for enumeration problems where the order of the objects is not important. An exponential generating function is a mathematical function defined by the following expression:

$$b(x) = \sum_{k=0}^{\infty} b_k \frac{x^k}{k!}, \tag{18}$$

where the coefficients are elements of the succession of numbers $b_0, b_1, b_2, \dots$. The exponential generating function are used for enumeration problems where the order of the objects matter. Graphs, words, trees, or integer partitions are some of the kinds of objects with which combinatorics deals. Combinatorics deals with discrete objects as, for example, graphs, words, trees, and integer partitions. Counting such objects is one of the most interesting tasks. Enumeration of graphs that have some structural property is the main purpose of the present chapter.

The great mathematician George Pólya made huge contributions to the field of graph enumeration graphs. He obtained closed mathematical expressions for the enumeration of graphs with a given number of vertices and edges for many graph counting problems using group theory [7]. Pólya's formulas greatly facilitated the enumeration of rooted graphs, connected graphs, etc. The enumeration of number of triangulations of certain plane polygons was one of the first problems of enumeration that attracted the attention of the great mathematician Leonhard Euler [8] in the eighteenth century. Some years later, Kirchhoff in [9] discovered a method for enumerating spanning trees in a connected graph. After that, Arthur Cayley obtained a closed mathematical formula that enumerates labeled trees, rooted trees, and ordinary trees in [10]. The Cayley discovery will be covered with more detail in

Section 4. The brilliant and unknown mathematician John Howard Redfield [11] discovered many enumeration formulas anticipating some of Pólya's contributions.

Some enumeration problems of objects that are not graphs (automata, Boolean functions, or chemical isomers) can be solved by cleverly transforming them to graphs. The generating functions are the tool used for enumerating graphs. There are two types of graphs associated with combinatorial graph enumeration problems:

1. Labeled graph problems

2. Unlabeled graph problems

Some enumeration problems of labeled graphs are normally addressed by applying the exponential generating functions tool. The enumeration problems of unlabeled graphs are normally addressed by applying the ordinary generating functions but require the application of Pólya's theorem.

One of the first problems of enumerating labeled graphs that may arise is that of how many graphs with $p$ vertices and $q$ edges can be obtained.

To solve this enumeration problem, let $G_p(x)$ be the polynomial or ordinary generating function whose coefficient of the term $x^k$ represents the number of labeled graphs with $p$ vertices and $k$ edges. If $V$ is the set of vertices of cardinality $p$, there are $q = \binom{p}{2}$ pairs of these vertices. In every vertex set $V$, each pair is adjacent or not adjacent. The number of labeled graphs with $k$ edges is therefore $\binom{q}{k} = \binom{\binom{p}{2}}{k}$. Thus, the ordinary generating function $G_p(x)$ for labeled graphs with $p$ vertices is given by

$$G_p(x) = \sum_{k=0}^{m} \binom{m}{k} x^k = (1+x)^m, \tag{19}$$

where $m = \binom{p}{2}$. Then, the number of labeled graphs with $p$ vertices is $G_p(1)$; so we have:

$$G_p = 2^m = 2^{\binom{p}{2}}. \tag{20}$$

For example, if we want to know how many labeled graphs with $p = 3$ vertices can be obtained, we apply Formula (20), and we get:

$$G_3 = 2^{\binom{3}{2}} = 2^{\frac{3!}{2!1!}} = 2^3 = 8. \tag{21}$$

If we want to know how many labeled graphs with $p = 4$ vertices and exactly $q = 5$ edges exist, before expression (19), we use the coefficient of the term $x^5$

$$[x^5]G_p(x) = [x^5]\sum_{k=0}^{6} \binom{6}{k} x^k = \binom{6}{5} = \frac{6!}{5!1!} = 6. \tag{22}$$

A question that arises naturally when working with labeled graphs is in how many ways can this be constructed from a certain number of vertices and edges. In order to give a possible answer, the number of symmetries or automorphisms must be taken into account. It is said that there is isomorphism between two graphs $G$ and $G_1$ if there is a one-to-one map $A : V(G) \mapsto V(G_1)$ between them that preserve adjacency. When $G_1 = G$, then the mapping $A$ is called automorphism of $G$. The set of all the automorphisms of $G$, represented by $\Gamma(G)$, is called the group of $G$. The elements of $\Gamma(G)$ are permutations over $V$. Let $s(G) = |\Gamma(G)|$ be the order of the group or number of symmetries of $G$. Therefore, the number of ways in which a graph $G$ of order $p$ can be labeled is

$$l(G) = \frac{p!}{s(G)}. \tag{23}$$

Another example of graph enumeration problem is to count the number of differently labeled connected graphs. The notion of path of length $n$ is defined as a sequence of vertices $\{v_0, v_1, \dots, v_n\}$ where the edges $\{v_i, v_{i+1}\}$ for $i = 0, \dots n$ that belong to the path are distinct. It is said that a graph is connected if for any pair of vertices there is a path between them. To obtain the formula that counts all the connected graphs $C_p$ of order $p$, it will be necessary to make use of the concept of subgraph. It is said that a $H$ is a subgraph of a graph $G$ if $V(H) \subset V(G)$ and $E(H) \subset E(G)$. A subgraph that is maximally connected is called component. A rooted graph is a graph that has a distinguished vertex called root. When there exists an injective mapping $f : V(H_1) \mapsto V(H_2)$ between two rooted graphs $H_1$ and $H_2$ that preserves the adjacency among vertices as well as the roots, it is said that the two rooted graphs are isomorphic. Let us make the assumption that $a_k$ for $k = 1, 2, 3, \dots$ represents the number of ways in which we can label all graphs of order that accomplish the property $P(a)$ and whose exponential generating function is

$$a(x) = \sum_{k=1}^{\infty} \frac{a_k x^k}{k!}. \tag{24}$$

Let us also assume that $b_k$ for $k = 1, 2, 3, \dots$ is the number of ways of labeling all graphs of order that accomplish the property $P(b)$ and whose exponential generating function is

$$b(x) = \sum_{k=1}^{\infty} \frac{b_k x^k}{k!}. \tag{25}$$

If we make the product of series (24) and (25), the coefficients of $\frac{x^k}{k!}$ in $a(x)b(x)$ is the number of ordered pairs $(G_1, G_2)$ of two disjoint graphs, where $G_1$ meets the property $P(a)$, $G_2$ fulfills the property $P(b)$, $k$ is the number of vertices in $G_1 \cup G_2$, and the labels from 1 to $k$ have been distributed over $G_1 \cup G_2$. If $C(x)$ is the exponential generating function for labeled connected graphs

$$C(x) = \sum_{k=1}^{\infty} \frac{C_k x^k}{k!}, \tag{26}$$

then $C(x)C(x)$ becomes the generating function that counts all the ordered pairs of connected graphs with labels. If you divide by 2, equation (26) that is, we divide it by 2, we get the generating function for labeled graphs that have exactly two components.

By applying this operation $n$ times, the coefficient of $\frac{x^k}{k!}$ that corresponds to the number of labeled graphs of order $k$ is obtained with exact $n$ components

$$G(x) = \sum_{n=1}^{\infty} \frac{C^n(x)}{n!}. \tag{27}$$

From (27), we obtain the following relation

$$1 + G(x) = e^{C(x)}. \tag{28}$$

Riordan in [12] obtained the relation $C_p = J_p(2)$, where $J_p(x)$ enumerates the trees by inversions, and then deduced

$$C_p = \sum_{k=1}^{p-1} \binom{p-2}{k-1} (2^k - 1) C_k C_{p-k}. \tag{29}$$

From (29), it should be noted that if the exponential generating function for the graph class is known in advance, then the exponential generating function for the class of graphs will be the logarithm of the first series, just as in (28). It should also be mentioned that another equivalent recurrence function can be obtained for the enumeration of the connected graphs that have order tags $p$ (p. 7 in [13]) and that can be expressed mathematically as

$$C_p = 2^{\binom{p}{2}} - \frac{1}{p} \sum_{k=1}^{p-1} k \binom{p}{k} 2^{\binom{p-k}{2}} C_k. \tag{30}$$

From (30), the following MATLAB code can be implemented for calculating the number of connected graphs $C_p$ of orders going from $p = 1$ to $p = 20$.

```
function y = CuentaGrafConnEtiq( p )
% recurrence satisfied by the number of connected graphs
% Harary Graph Enumeration pag. 7
% C_p= 2^{combinations(p,2)}-1/p* \sum_{k=1}^{p-1} k*
% combinations(p,k)*2^{combinations(p-k,2)}*C_{k}

C(1:p)=0;
C(1,1)=1;

for k=2:p
   acum=0;
   for j=1:k
      acum = acum + j * combinaciones(k,j) * CuentaGrafEtiq(k-j) * C(1,j);
   end
   C(1,k) = CuentaGrafEtiq(k)-(1/k)*acum;
end
y=C(1,p);
sprintf('end function z=combinaciones(n,k) z= factorial(n)/(factorial(k)
*factorial(n-k));
end

function z=combinaciones(n,k)
    z= factorial(n)/(factorial(k)*factorial(n-k));
```

```
end

%% calling the function from the matlab prompt for the calculation of the
%% evaluation from graphs of order 1 to 20

>> for i=1:20
     R(1,i)=CuentaGrafConnEtiq( i );
end
```

The calculations obtained by the execution of MATLAB code are shown in **Table 1**.

From the results of 1, it can be observed that the number of possible connected graphs $C_p$ grows very fast in terms of the number $p$ of vertices. It should be mentioned that (29) and (30) are recurrence relations instead of a closed formula. The recurrences (29) or (30) can be used for the calculation of $C_p$ with a computer program. The generating functions can be used to solve recurrences and obtain a closed mathematical expression for the $n$th term of the succession associated with the recurrence. It can happen that calculation of the solution of some recurrences becomes very hard to be solved and in the worst cannot be solved at all. An alternative method for obtaining an approximate value for big values of $p$ is to recur to the application of methods used in analytic combinatorics and calculate accurate approximations of the $p$th coefficient of the generating function. The generating functions algebraic structure allows to reflect the structure of combinatorial classes.

| $p$ | $C_p$ |
|---|---|
| 1 | 1 |
| 2 | 1 |
| 3 | 4 |
| 4 | 38 |
| 5 | 728 |
| 6 | 26,704 |
| 7 | 1,866,256 |
| 8 | 251,548,592 |
| 9 | 66,296,291,072 |
| 10 | 34,496,488,594,816 |
| 11 | 35,641,657,548,953,344 |
| 12 | $7.335460 \times 10^{19}$ |
| 13 | $3.012722 \times 10^{23}$ |
| 14 | $2.471649 \times 10^{27}$ |
| 15 | $4.052768 \times 10^{31}$ |
| 16 | $1.328579 \times 10^{36}$ |
| 17 | $8.708969 \times 10^{40}$ |
| 18 | $1.41641 \times 10^{46}$ |
| 19 | $2.992930 \times 10^{51}$ |
| 20 | $1.569216 \times 10^{57}$ |

**Table 1.**
*Order 1–20.*

The analytic combinatorics method consists in examining the generating functions from the point of view of the mathematical analysis by giving not only real value values to its variables but also values in the complex plane. When complex values are assigned to the variables of the generating functions, the function is converted in a geometric transformation of the complex plane. This kind of geometrical mapping is said to be regular (holomorphic) near the origin of the complex plane. When we move away from the origin of the complex plane, some singularities appear that are related with the absence of smoothness of the function and give a lot of information about the function coefficients and their asymptotic growth. It can happen that elementary real analysis is enough for estimating asymptotically enumerative successions. If this is not the case, the generating functions are still explicit, but its form does not allow the easy calculation of the coefficients of the series. The complex plane analysis however is a good option for asymptotic estimation of these coefficients. In order to give an example of the use of the notion of singularities, let us take the ordinary generating function of the Catalan numbers

$$f(x) = \frac{1}{2}\left(1 - \sqrt{1-4x}\right). \tag{31}$$

Eq. (31) expresses in a compact way the power series of the form

$$(1-y)^{1/2} = 1 - \frac{1}{2}y - \frac{1}{8}y^2 - \dots. \tag{32}$$

The generating function (31) coefficients can be explicitly expressed as

$$f_n = [x^n]f(x) = \frac{1}{n}\binom{2n-2}{n-1}. \tag{33}$$

Using the Stirling formula, we can get the asymptotic approximation of (33) that is expressed as

$$f_n \sim \lim_{n\to\infty} \frac{4^n}{\sqrt{\pi n^3}}. \tag{34}$$

If the generating function is used as an analytic object, the approximation (34) can be obtained.

In order to do it, we substitute in the power series expansion of the generating function $f(x)$ any real or complex value $\rho_f$ whose modulus is small enough, for example, $\rho_f = 4$. The graph that we get by the use of (31) is smooth and differentiable in the real plane and tends to the limit $\frac{1}{2}$ as $x \to \left(\frac{1}{4}\right)^-$, but, if we calculate its derivative, we obtain the following function

$$f(x) = \frac{1}{1-\sqrt{1-4x}}, \tag{35}$$

and it can be noticed that the derivative (35) becomes infinite in $\rho_f = \frac{1}{4}$. The singularities will correspond to those points where the graph is not smooth.

It should be pointed out that that the region where function (31) is still being continuous can be extended. Let us take for example the value $x = -1$

$$f(-1) = \frac{1}{2}\left(1 - \sqrt{5}\right). \tag{36}$$

We can evaluate in the same manner (31) giving to $x$ values in the complex plane whose modulus is less than the radius of convergence of the series defined by (31) and realize that the orthogonal and regular grid in it transforms the real plane in a grid on the complex plane that preserves the angles of the curves of the grid. This property corresponds to the complex differentiability property, which also is equivalent to the property of analyticity. Concerning the asymptotic behavior of the coefficients $f_n$ of the generating function, it should be observed that it has a general asymptotic pattern composed by an exponential growth factor $A^n$ and a subexponential factor $\theta(n)$.

For the case of the expression (34) $A = 4$ and $\theta(n) \sim \frac{1}{4}\left(\sqrt{\pi n^3}\right)^{-1}$, the exponential growth factor can be put in relation with the radius of convergence of the series by $A = \frac{1}{\rho_f}$ that is the singularity that can be observed along the positive real axis of the complex plane that normally corresponds to the pole of the generating function, and the subexponential part $\theta(n) = O\left(n^{-\frac{3}{2}}\right)$ arises from the singularity of the square root type. This asymptotic behavior can be compactly expressed as

$$[x^n]f(x) = A^n\theta(n). \tag{37}$$

The exponential growth part of (37) is known as first principle of coefficient asymptotics and the subexponential growth part as second principle of coefficient asymptotics. By recalling to the results that can be found in the field complex variable theory, more general generating functions can be obtained. One of those results is the Cauchy residue theorem that relates global properties of a meromorphic function (its integral along closed curves) to purely local characteristics at the residues poles. An important application of the Cauchy residue theorem concerns a coefficient of analytic functions. This is stated in the following theorem [3]:

Theorem 1.5 (Cauchy's coefficient formula). Let $f(z)$ be analytic in a region containing 0, and let $\lambda$ be a simple loop around 0 that is positively oriented. Then, the coefficient $[z^n]f(z)$ admits the integral representation:

$$f_n \equiv [z^n]f(z) = \frac{1}{2i\pi}\int_\lambda f(z)\frac{dz}{z^{n+1}}. \tag{38}$$

For more details about analytic combinatorics, we recommend to consult [3] as well as [5].

## 4. Enumeration of trees

A graph is a structure composed by a set of vertices $V = \{v_1, v_1, ...v_n\}$ and a set of pairs of connected vertices $E = \{e_1, e_2, ..., e_m\}$ called edges where $E \subset V \times V$ and each edge $e_k \in E$ is composed as pair of vertices $e_k = (v_i, v_j)$. A tree is a special type of graph that do not have cycles. A tree have no loops or edges that connect a vertex with himself. The subject of combinatorial graph enumeration has been the center of interest of many mathematicians a long time ago. The enumeration of total possible labeled trees with $n$ nodes being $n^{n-2}$ was one of first results obtained by Cayley in [10]. Cayley's formula for enumerating trees is one of the simple and elegant mathematical results in enumeration of graphs. He detected that from $n+1$ vertices, the number of possible trees that can be built is equal to $(n+1)^{n-1}$. Cayley [10] gives an example for the case of four vertices $4 = n + 1$ then $n = 3$, and the total number of possible trees calculated using his formula gives $(4)^2 = 16$. In this same publication, Cayley gives another example having $n+1=6$ vertices (he used the term knots in his publication) with labels *a, b, c, d, e,f* and related the concatenation of vertices given by the edges for obtaining sequences of labels representing a given tree. For instance, if the tree is a chain of vertices connected by edges starting with the vertex $a$ and ending with the vertex $f$ and having as connecting edges $(a,b)$, $(b,c)$, $(c,d)$, $(d,e)$, $(e,f)$, the corresponding label sequence is *abcdef* as shown in **Figure 1**. As another example of sequence of vertex labels, if the root of the tree is $a$ and it is connected directly with the other five vertices, then the connecting edges will be $(a,b)$, $(a,c)$, $(a,d)$, $(a,e)$, $(a,f)$, and the corresponding sequence of vertex labels will be in that case $a^5bcdef$ as shown in **Figure 2**. As can be noticed, the exponent of $a$ is 5 that represents the number of occurrences of this label in the set of connecting edges.

After that Cayley states the theorem for this particular case as follows:

Theorem 1.6 The total number of trees $T(n+1)$ that can be built with $n+1=6$ vertices can be calculated as follows

$$T(n+1) = (a+b+c+d+e+f)^4 abcdef = 6^4 = 1296 \qquad (39)$$

This calculation relates the sum of the products of the coefficients of the multinomial $(a+b+c+d+e+f)^4$ with the number of terms of its corresponding type. Each term obtained by multiplying *abcdef* with the vertex label inside $(a+b+c+d+e+f)^4$ corresponds to different trees.

At the end of [10], Cayley generalizes his theorem by recalling a result obtained by C.W. Borchardt in [14] that relates some particular kind of determinants that represent spanning trees and whose product represents the branches of those spanning trees. Given that the number of terms of these determinants is $(n+1)^{n-1}$, Cayley can conclude that the number of spanning trees is the same. Since these first results, many other methods have been proposed for obtaining the same result. One way to enumerate a collection of objects is to find a bijection between a set of objects whose enumeration is known and the set of objects that we want to enumerate. This was the method used by Prüfer in [15] for enumerating the set of possible spanning trees with $n$ vertices. The set whose number was known beforehand was a sequence of length $n-2$ of numbers from 1 to $n$. For this end the autor of article [15] encoded the trees as Prüfer sequences. In [16] Moon generalized the result derived by Clarke in [17] by induction on $d$ the degree, making induction on $n$, and the number of vertices and obtains a new enumeration method for $n$-labeled $k$ trees.

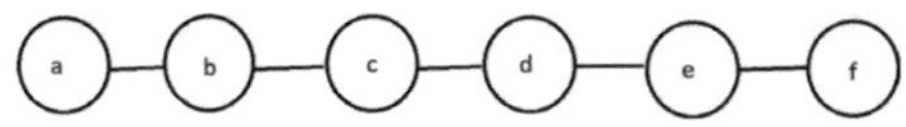


**Figure 1.**
*Example of tree of one branch.*

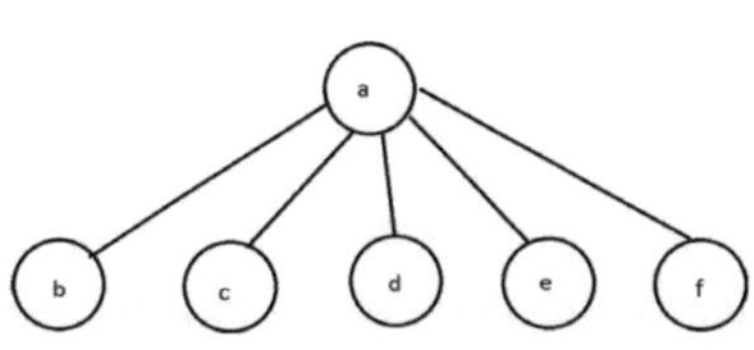


**Figure 2.**
*Example of tree with more branches.*

## 5. Enumeration and generating functions

The field of combinatorial enumeration has aroused enormous interest among mathematicians who have worked in the area of discrete mathematics during the last decades [18–21]. Combinatorial enumerative technique developed by these brilliant researchers have allowed us to count words, permutations, partitions, sequences, and graphs. As was mentioned in Section 3, the mathematical tool frequently used for this purpose is the generating functions or formal power series.

The generating functions allow to connect discrete mathematics and continuous analysis in a very special way with complex variable theory. The typical situation that someone faces when trying to solve an enumeration problem is that you want to know the mathematically closed form that has the $n$th term of a given sequence of numbers $a_0, a_1, a_2, \ldots$ For some sequences, we can do by inspection. For example, if the numerical sequence $1, 3, 5, 7, 9, \ldots$, it is easy to see that it is a sequence of odd numbers whose $n$th element is $a_n = 2n - 1$.

A more complicated sequence is the set of prime numbers $2, 3, 5, 7, 11, 13, 17, 19, \ldots$, whose $a_n$ is the $n$th prime number. A closed mathematical formula for $n$th prime number is not known, and it seems impossible to obtain in general.

In many cases it is very hard to get a simple formula just by inspection. However it can be very useful to use the generating functions whose coefficients are the elements of that sequence transforming it as follows:

$$\sum_{i=0}^{\infty} a_i x^i. \tag{40}$$

Eq. (40) defines an ordinary generating function. As mentioned in Section 3, since they are infinite polynomials, they can be algebraically manipulated easily.

In this chapter, the main interest will be the application of the generating functions tool as well as the analytic asymptotic methods for the enumeration of graphs accomplishing some given properties. Many questions about the number of graphs that have some specified property can be answered by the use of generating functions. Some typical questions about the number of graphs that fulfill a given property are, for example: How many different graphs can I build with $n$ vertices? How many different connected graphs with $n$ vertices exist? How many binary trees can be constructed with $n$ vertices? [18, 19], etc. For some of these questions, the application of generating functions allows us to easily obtain a simple formula. For some other questions, the answer is an asymptotic estimation formula. The most commonly used generating functions are the *ordinary generating functions* and the *exponential generating functions*. The generating functions are the tool used for enumerating graphs. From the point of view of the generating functions, there are two types of graph enumerating problems:

1. Labeled graph problems

2. Unlabeled graph problems

The labeled graph problems can be easily solved with the direct application of the exponential generating functions. The case of the unlabeled enumeration problems can be solved by using ordinary generating functions but require the use of more combinatorial theory and the application of Pólya's theorem.

## 6. Enumerating regular graphs

As we mentioned in Section 6, some enumerating problems can be solved easily using generating tools for obtaining a closed formula. Some other problems are more hard to deal with for obtaining a closed mathematical expression, but we can resort in such a case to the asymptotic approximation of the coefficients of the power series [22–25]. It was also mentioned in Section 6 that there are some graph enumerating problems where the nodes are labeled, and in such a case the use of the *exponential generating functions* is well adapted for these kinds of problems. The other case of graph enumerating problems is when we are dealing with graphs whose nodes do not have an assigned label. Then, we can resort in such case to Pólya's enumerating method [7, 13], and the best choice is to use *ordinary generating functions*. It should also be mentioned that the edges of the graphs to be enumerated can be directed or undirected.

One of the seminal articles of enumerating graphs is [26], where a fundamental theorem was proven in the theory of random graphs on $n$ unlabeled nodes and with a given number $q$ of edges.

In [26], the authors obtained a necessary and sufficient condition for relating asymptotically the number of unlabeled graphs with $n$ nodes and $q$ edges with the number of labeled graphs with $n$ nodes and $q$ edges. Let $T_{nq}$ be the number of different graphs with $n$ nodes and $q$ edges, $F_{nq}$ the corresponding to number of labeled graphs, $N = \frac{n(n-1)}{2}$ the possible edges, and $F_{nq} = \begin{pmatrix} N \\ q \end{pmatrix} = \frac{N!}{q!(N-q)!}$. The result obtained in [26] can be stated as the following theorem:

Theorem 1.7 The necessary and sufficient conditions that

$$T_{nq} \sim \frac{F_{nq}}{n!} \tag{41}$$

as $n \to \infty$ is that

$$\min(q, N-q)/n - (\log n/2) \to \infty. \tag{42}$$

The formal result expressed in Theorem 1.7 for unlabeled graphs is a starting point on the enumeration of regular graphs because it allows for enumerating those unlabeled graphs that have some number of edges. In fact, the author of [26] proved that if a graph with $|E| = E(n)$ edges, where $n$ is the number of vertices or order of such a graph, has no isolated vertices or two vertices of degree $n-1$, then the number of unlabeled graphs of order $n$ and number of edges $|E|$ divided by the number unlabeled graphs is asymptotic to $n!$.

Another interesting article on asymptotic enumeration of labeled graphs having a given degree sequence was [27]. The authors of [27] obtained their asymptotic result for $n \times n$ symmetric matrices subject to the following constraints:

i. Each row sum is specified and bounded.

ii. The entries are bounded.

iii. A specified *sparse* set of entries must be zero.

The authors of [27] mentioned that their results can be interpreted in terms of incidence matrices for labeled graphs. The results of [27] can be stated as follows. Let $\mathcal{M}(n,z)$ be the set of all $n \times n$ symmetric $(0,1)$ matrices with at most $z$ zeroes in each row, $\mathbf{r}$ a vector over $[d] = \{0, 1, ...d\}$, and $G(M, r, t)$ the number of $n \times n$ symmetric matrices $(g_{ij})$ over $[t] = \{0, 1, ...t\}$ such that

i. $g_{ij} = 0$ if $m_{ij} = 0$.

ii. $\sum_j g_{ij} = r_i$.

Theorem 1.8 For given $d, t$ and $z$,

$$G(M, \mathbf{r}, t) \sim \frac{T(f, \delta)e^{\varepsilon a - b}}{\prod r_i!}. \tag{43}$$

Uniformly, for $(M, \mathbf{r}) \in \cup_{n=1}^{\infty}(\mathcal{M}(n, z) \times [0, d]^n)$ as $f \to \infty$, where $f = \sum_i r_i$, $\varepsilon = -1$ if $t = 1$ and $+1$ if $t>1$, for

$$a = \left(\frac{\sum_i \binom{r_i}{2}}{f}\right)^2 + \left(\frac{\sum_{m_{ij}} \binom{r_i}{2}}{f}\right), b = \left(\sum_{m_{ij}=0,\, i<j} r_i r_j + \sum_i \frac{\binom{r_i}{2}}{f}\right), \delta = \sum_{m_{ij}=0} r_i$$

and $T(f, \delta)$ being the number of involutions on $[1, f]$ such that no element in some specified set of size $\delta$ is fixed.

Three years later, the article [28] appeared giving a different approach of [27] allowing for obtaining a more general asymptotic formula without reference to an exact formula. The asymptotic result obtained by Bela Bollobás in [28] for enumerating labeled regular graphs is proven by a probabilistic method. This result can be stated as follows. Let $\Delta$ and $n$ be natural numbers such that $\Delta n = 2m$ is even and $\Delta \leq (2\log n)^{\frac{1}{2}}$, where $n$ is the number of vertices and $m$ is the number of edges of the graph $G$. Then, as $n \to \infty$, the number of labeled $\Delta$-regular graphs on $n$ vertices is asymptotic to

$$e^{-\lambda - \lambda^2} \frac{(2m)!}{m! 2^m (\Delta!)^m}, \tag{44}$$

where $\lambda = \frac{(\Delta - 1)}{2}$.

The authors of [27] affirm that the asymptotic Formula (44) holds not only for $\Delta$ constant but also for $\Delta$ growing slowly as $n \to \infty$ and summarized this in the following theorem.

Theorem 1.9 Let $d_1 \geq d_2 \geq ...d_n$ be natural numbers with $\sum_{i=1}^n d_i = 2m$ even. Suppose $\Delta = d_1 \leq (2\log n)^{\frac{1}{2}} - 1$ and $m \geq \max\{\varepsilon \Delta n, (1+\varepsilon)n\}$ for some $\varepsilon>0$. Then, the number $L(\mathbf{d})$ of labeled graphs with degree sequence $\mathbf{d} = (d_i)_1^n$ satisfies

$$L(\mathbf{d}) \sim e^{-\lambda - \lambda^2} \frac{(2m)_m}{\{2^m \prod_{i=1}^n d_i!\}}, \tag{45}$$

where $\lambda = \frac{1}{2m} \sum_{i=1}^n \binom{d_i}{2}$.

In the next year, the author on [27] extended this result to the case of unlabeled graphs in [29]. The result of Theorem 1.9 extended for the case of unlabeled graphs can be summarized in the following theorem.

Theorem 1.10 If $\Delta \geq 3$ and $L_\Delta = e^{-\lambda - \lambda^2} \frac{(2m)!}{m! 2^m (\Delta!)^m}$, then

$$U_\Delta \sim \frac{L_\Delta}{n!} \sim e^{-\frac{(\Delta^2-1)}{4}} \frac{(2m)!}{2^m m!} \frac{(\Delta!)^{-n}}{n!}, \tag{46}$$

where $m = \frac{\Delta n}{2}$.

For the details of the proof of Theorems 1. 9 and 1.10, see [28, 29], respectively.

## 7. Example of proof of a result on control using combinatorial enumeration

In [30] I used the combinatorial enumeration methods and probability for showing the applicability of the selection node criteria in a virus spreading control problem in complex networks. The main purpose of this section is to illustrate how the enumerative combinatorics in combination with probability theory can be used for demonstrating mathematically a result in an application field. We can mention that the case of homogeneity in the behavior of the nodes and their interaction cannot be discarded given what has been observed in the reaction of the agents in the context of social networks is that they try to minimize the conflict. Many successful models can, for example [30–33], base their predicting effectiveness on the homogeneity of the behavior of the nodes and their interaction. By the other side, in [30] we have obtained a criteria for selecting the nodes to be controlled, but such criteria fail if we have homogeneity in the behavior of the nodes and, at the same time, the topology of the network is regular. Then, what we want to do here is to justify the applicability of node selection criteria, keeping the homogeneity of the nodes and trying to compare the number of regular graphs with $n$ vertices with the total of graphs that can be constructed with $n$ vertices. For this end, based on the results on combinatorial graph enumeration mentioned on Theorems 1.9 and 1.10, we can state our main result as follows. First of all, we suppose that our graph is labeled, $G = (V, E)$ is $r$-regular with $r \geq 3$ constant and $rn = 2m$, where $n = |V|$ corresponds to the number of vertices and $m = |E|$ corresponds to the number of edges. Let $L_r$ the number of labeled regular graphs of degree $r$ whose asymptotic value is [28]

$$L_r \sim e^{-\frac{r^2-1}{4}} \frac{(2m)!}{2^m m!} (r!)^n. \tag{47}$$

Let $G_n$ be the number of all possible graphs with $n$ vertices whose value is

$$G_n = 2^{\binom{n}{2}}. \tag{48}$$

Theorem 1.11 If $r \geq 3$ and $nr = 2m$, then

$$\lim_{n\to\infty} \frac{L_r}{G_n} = 0. \tag{49}$$

Proof of Theorem 1.11. If $nr = 2m$ and $r$ is constant of value $c_1$, then we can deduce that $m = \frac{r}{2}n = \frac{c_1}{2}n$, and this implies that $m = O(n)$ so let us say that $m = c_2 n$; then,

$$\lim_{n\to\infty} \frac{L_r}{G_n} = \lim_{n\to\infty} \frac{e^{-\frac{c_1^2-1}{4}} \frac{(2c_2 n)!(c_1!)^{-n}}{2^{c_2 n}(c_2 n)!}}{2^{\binom{n}{2}}} \tag{50}$$

$$= \lim_{n\to\infty} \frac{\frac{e^{-\frac{c_1^2-1}{4}}}{(c_1!)^n}\frac{(2c_2n)!}{2^{c_2n}(c_2n)!}}{2^{\frac{n(n-1)}{2}}}, \tag{51}$$

applying the approximation Stirling formula $n! \sim \sqrt{2\pi n}\left(\frac{n}{e}\right)^n$

$$(c_1!)^n = \left(\sqrt{2\pi c_1}\left(\frac{c_1}{e}\right)^{c_1}\right)^n \tag{52}$$

then

$$= \lim_{n\to\infty} \frac{\frac{e^{-\frac{c_1^2-1}{4}}}{\left(\sqrt{2\pi c_1}\left(\frac{c_1}{e}\right)^{c_1}\right)^n}\frac{(2c_2n)!}{2^{c_2n}(c_2n)!}}{2^{\frac{n(n-1)}{2}}} \tag{53}$$

simplifying

$$= \lim_{n\to\infty} \frac{\frac{1}{\left(\sqrt{2\pi c_1}\right)^n (c_1)^{c_1n} e^{\frac{c_1^2+4c_1n+1}{4}}}\frac{(2c_2n)!}{2^{c_2n}(c_2n)!}}{2^{\frac{n(n-1)}{2}}} \tag{54}$$

applying the approximation Stirling formula

$$(2c_2n)! = \sqrt{2\pi(2c_2n)}\left(\frac{(2c_2n)}{e}\right)^{(2c_2n)} \tag{55}$$

and

$$(c_2n)! = \sqrt{2\pi(c_2n)}\left(\frac{(c_2n)}{e}\right)^{(c_2n)} \tag{56}$$

we get

$$= \lim_{n\to\infty} \frac{\frac{1}{\left(\sqrt{2\pi c_1}\right)^n (c_1)^{c_1n} e^{\frac{c_1^2+4c_1n+1}{4}}}\frac{\sqrt{2\pi(2c_2n)}\left(\frac{(2c_2n)}{e}\right)^{(2c_2n)}}{2^{c_2n}\sqrt{2\pi(c_2n)}\left(\frac{(c_2n)}{e}\right)^{(c_2n)}}}{2^{\frac{n(n-1)}{2}}}, \tag{57}$$

simplifying

$$= \lim_{n\to\infty} \frac{\frac{\sqrt{2}}{\left(\sqrt{2\pi c_1}\right)^n c_1^{c_1n} e^{\left(c_1^2+4c_1n+1\right)/4}}\left(\frac{c_2n}{e}\right)^{c_2n}}{2^{\frac{n(n-1)}{2}}} \tag{58}$$

given that $r \geq 3$, for which we assumed that $r$ is a constant $c_1$, and that $nr = 2m$, then we have that $c_1 = 2c_2$, and replacing that, in (58), we can express it in terms of $c_1$, which is the regular degree $r$ assumed as fixed; then, we get

$$\lim_{n\to\infty} \frac{\frac{\sqrt{2}}{\left(\sqrt{2\pi c_1}\right)^n c_1^{c_1n} e^{\left(c_1^2+4c_1n+1\right)/4}}\left(\frac{c_1n/2}{e}\right)^{c_1n/2}}{2^{\frac{n(n-1)}{2}}}, \tag{59}$$

and, replacing $c_1$ by $r$ in (59), we get

$$\lim_{n\to\infty} \frac{\frac{\sqrt{2}}{\left(\sqrt{2\pi r}\right)^n r^{rn} e^{\left(r^2+4rn+1\right)/4}} \left(\frac{rn/2}{e}\right)^{rn/2}}{2^{\frac{n(n-1)}{2}}}. \tag{60}$$

Therefore, if the degree $r$ is constant, the $\lim_{n\to\infty} \frac{L_r}{G_n} = 0$.

Now, our main result can be stated as a consequence of Theorem 1.11.

Theorem 1.12 If we assume that all graphs are uniformly distributed and that the nodes have homogeneous behavior, then the criteria for selecting nodes to be controlled are almost always applicable.

**Proof of Theorem 1.11.** As a consequence of Theorem 11, we know that the probability that a regular graph appears tends to zero as $n \to \infty$. Then, the mentioned criteria are almost always applicable.

## 8. Conclusions

In the present chapter, some problems of combinatorial graph enumeration as well as some useful techniques for obtaining a closed mathematical expression were addressed. When it is not possible to obtain a closed expression, asymptotic estimations of the kind used in analytic combinatorics can be used. In Section 10 the use of these techniques for proving a result in the field of virus spreading control was illustrated [31–34].

This allowed to explore the application of combinatorial techniques to control problems in networks and thus verify the goodness of said methods for network analysis and the control of virus propagation in them. This still needs to be studied by applying the combinatorial methods discussed above, if there are any other types of topologies that prevent the application of the selection criteria of nodes to be controlled under the hypothesis of behavior of partially heterogeneous nodes, that is, if in the network we have subsets of nodes with the same behavioral parameters. The selecting node criteria described in [30] are based on the combination of the parameter values of the selected nodes as well as their degrees. In many recent publications [35–42], interesting and elaborated methods for detecting the influencer nodes in complex networks have been proposed that I will try to apply in combination with the mentioned criteria in the future in order to reduce the number of nodes to be controlled.

## Author details

Carlos Rodríguez Lucatero
Departamento de Tecnologías de la Información, Universidad Autónoma Metropolitana Unidad Cuajimalpa, Mexico City, Mexico

*Address all correspondence to: profesor.lucatero@gmail.com

## References


[1] Erdös P. Graph theory and probability. Canadian Journal of Mathematics. 1959;**11**:34-38

[2] Alon N, Spencer JH. The Probabilistic Method. 2nd ed. New York: Wiley-Interscience; 2000

[3] Sedgewick R, Flajolet P. Analytic Combinatorics. 0th ed. Cambridge, UK: Cambridge University Press; 2005. Fifth printing

[4] Wilf HS. Generating Functionology. 3rd ed. Wellesley, MA, USA: A K Peters Ltd.; 2006

[5] Louis C. Advanced Combinatorics: The Art of Finite and Infinite Expansions. Dordrecht-Holland, The Netherlands; Boston, MA, USA: D. Reidel Publishing Company; 1974

[6] Sedgewick R, Flajolet P. An Introduction to the Analysis of Algorithms; 2nd printing. Boston, MA, USA: Addison-Wesley; 2001

[7] Pólya G. Kombinatorische Anzahlbestimmungen für Gruppen, Graphen und chemische Verbindungen. Acta Mathematica. 1937;**68**:145-254

[8] Euler L. Journal Novi Comment Acad. Sci. Imperialis Petropolitanae, Holding Institution: American Museum of Natural History Library, 1758–1759, 7, pp. 13–14. Available online: https:// www.biodiversitylibrary.org/ bibliography/9527#/summary [Accessed: 21 March 2018]

[9] Kirchhoff G. Über die Auflösung der Gleichungen, auf welche man bei der Untersuchung der linearen Verteilung galvanischer Ströme gefuhrt wird. Annals of Physical Chemistry. 1847;**72**: 497-508

[10] Cayley A. The collected mathematical papers of Arthur Cayley: A theorem on trees. Cambridge University Press. 1898;**13**:26-28

[11] Redfield JH. The theory of group-reduced distributions. American Journal of Mathematics. 1927;**49**:433-455

[12] Mallows CL, Riordan J. The inversion enumerator for labeled trees. Bulletin of the American Mathematical Society. 1968;**74**:92-94

[13] Harary F, Palmer EM. Graph Enumeration. New York, NY, USA; London, UK: Academic Press; 1973

[14] Borchardt CW. Ueber eine der Interpolation entsprechende Darstellung der Eliminations-Resultante. Journal für die reine und angewandte Mathematik. 1860;(57): 111-121

[15] Prüfer H. Neuer Beweis cines Satzesuber Permutationen. Archiv für Mathematik und Physik. 1918;**27**: 142-144

[16] Moon JW. The number of labeled k-trees. Journal of Combinatorial Theory. 1969;**6**:196-199

[17] Clarke LE. On Cayley formula on counting trees. Journal of the London Mathematical Society. 1958;**33**:471-475

[18] Graham RL, Knuth DE, Patashnik O. Concrete Mathematics; 6th Printing with Corrections. Boston, MA, USA: Addison-Wesley; 1990

[19] Nijenhuis A, Wilf HS. The enumeration of connected graphs and linked diagrams. Journal of Combinatorial Theory. 1979;**27**:356-359

[20] Courcelle B, Makowsky JA, Rotics U. On the fixed parameter complexity of graph enumeration problems definable in monadic second

order logic. Discrete Applied Mathematics. 2001;**108**:23-52

[21] Read RC. Some unusual enumeration problems. Annals of the New York Academy of Sciences. 1970; **175**:314-326

[22] Mackay Brendan D. Asymptotics for symmetric 0–1 matrices with prescribed row sums. Ars Combinatoria. 1985;**19A**: 15-25

[23] Mackay Brendan D, Wormald Nicholas C. Uniform generation of random regular graphs of moderate degree. Journal of Algorithms. 1990;**11**: 52-67

[24] Mackay Brendan D, Wormald Nicholas C. Asymptotic enumeration by degree sequence of graphs of high degree. European Journal of Combinatorics. 1990;**11**:565-580

[25] Mackay Brendan D, Wormald Nicholas C. Asymptotic enumeration by degree sequence with degrees $O\left(n^{\frac{1}{2}}\right)$. Combinatorica. 1991;**11**:369-382

[26] Wright EM. Graphs on unlabelled nodes with a given number of edges. Acta Mathematica. 1971;**126**:1-9

[27] Bender EA, Canfield ER. The asymptotic number of labeled graphs with given degree sequences. Journal of Combinatorial Theory. 1978;**24**:296-307

[28] Bollobás B. A probabilistic proof of an asymptotic formula for the number of labelled regular graphs. European Journal of Combinatorics. 1980;**1**: 311-316

[29] Bollobás B. The asymptotic number of unlabelled regular graphs. Journal of the London Mathematical Society. 1981; **1**:201-206

[30] Rodrguez-Lucatero C, Alarcón L. Use of enumerative combinatorics for proving the applicability of an asymptotic stability result on discrete-time SIS epidemics in complex networks. MDPI Mathematics Open Access Journal. 2019;**7**(1). DOI: 10.3390/ math7010030

[31] Chakrabarti D, Wang Y, Wang C, Leskovec J, Faloutsos C. ACM Transactions on Information and System Security. 2008;**10**:1-26

[32] Galam S. Rational group decision making: A random field Ising model at T = 0, arXiv:cond-mat/9702163v1; 1997

[33] Axelrod R. The dissemination of culture: A model with local convergence and global polarization. Journal of Conflict Resolution. 1997;**41**(2):203-226

[34] Gonzalez-Avella JC, Eguiluz VM, Cosenza MG, Klemm K, Herrera JL, San Miguel M. Nonequilibrium transition induced by mass media in a model for social influence. Physical Review E. 2005;**72**(6) 065102(1-4)

[35] Pei S, Morone F, Makse HA. Theories for influencer identification in complex networks. arXiv 2018. arXiv: physics.soc-ph/1707.01594v2

[36] Cha M, Haddadi H, Benevenuto F, Gummandi PK. Measuring user influence in twitter: The million follower fallacy. In: Proceedings of the 4th International AAAI Conference on Weblogs and Social Media, Washington, DC, USA, 23–26 May 2010; Vol. 10. 2010. pp. 10-17

[37] Watts DJ, Dodds PS. Influential networks and public opinion formation. Journal of Consumer Research. 2007;**34**: 441-458

[38] Kitsak M, Gallos LK, Havlin S, Liljeros F, Muchnik L, Stanley HE, et al. Identification of influential spreaders in complex networks. Nature Physics. 2010;**6**:888-893

[39] Pei S, Makse HA. Spreading dynamics in complex networks. Journal of Statistical Mechanics: Theory and Experiment. 2013;**2013**:P12002

[40] Min B, Morone F, Makse HA. Searching for influencers in big-data complex networks. In: Bunde A, Caro J, Karger J, Vogl G, editors. Diffusive Spreading in Nature, Technology and Society. Berlin, Germany: Springer; 2016

[41] Leskovec J, Adamic LA, Huberman BA. The dynamics of viral marketing. ACM Transactions on the Web. 2007;**1**:5

[42] Rogers EM. Diffusion of Innovations. New York, NY, USA: Simon and Schuster; 2010

Chapter 10

# New Variations of the Online $k$-Canadian Traveler Problem: Uncertain Costs at Known Locations

*Davood Shiri and F. Sibel Salman*

## Abstract

In this chapter, we study new variations of the online $k$-Canadian Traveler Problem ($k$-CTP) in which there is an input graph with a given source node O and a destination node D. For a specified set consisting of $k$ edges, the edge costs are unknown (we call these uncertain edges). Costs of the remaining edges are known and given. The objective is to find an online strategy such that the traveling agent finds a route from O to D with minimum total travel cost. The agent learns the cost of an uncertain edge, when she arrives at one of its end-nodes and decides on her travel path based on the discovered cost. We call this problem the online $k$-Canadian Traveler Problem with uncertain edges. We analyze both the single-agent and the multi-agent versions of the problem. We propose a tight lower bound on the competitive ratio of deterministic online strategies together with an optimal online strategy for the single-agent version. We consider the multi-agent version with two different objectives. We suggest lower bounds on the competitive ratio of deterministic online strategies to these two problems.

**Keywords:** multi-agent $k$-CTP, online strategies, deterministic strategies, competitive ratio, undirected graphs

## 1. Introduction

The online $k$-Canadian Traveler Problem ($k$-CTP) is a well-known navigation problem within the field of combinatorial optimization. In the online $k$-CTP, the objective is to reach a destination in a network within minimum travel time under uncertainty of some information. Uncertain information is revealed, while one or more travelers (agents) discover the information during their travels. In the $k$-CTP and its variants studied in the literature, uncertainty is on the locations of blocked edges in the input graph. That is, it is known that there are at most $k$ blocked edges, but their locations are not known. In this study, we consider new variations of the $k$-CTP where a known set of edges have unknown (uncertain) travel times (costs). To the best of our knowledge, this variant of the $k$-CTP with given locations of edges that have unknown traveling costs has not been studied yet in the literature.

Uncertainty in travel times arises in various situations, such as following a disaster or in daily urban traffic systems. After a disaster, uncertainty in travel times arises due to both damage on road segments and traffic congestion on some parts of the road network. We typically know which roads are likely to have damage and to be congested, but the actual travel times can be estimated more accurately when we observe the situation right on the spot. Regarding urban traffic systems, problematic road segments can be detected beforehand since in most current traffic management systems, data indicating locations with high accident frequency are available, but it is difficult to predict the time of occurrence or the intensity of the accident accurately. Also, we usually know where there is a high likelihood of heavy traffic, but travel times show variability. Moreover, nowadays navigation applications indicate which locations have heavy traffic, but the travel times are still not known with certainty, and the situation evolves dynamically as we reach the locations themselves.

In many real-world emergency operations, including response to disasters and daily medical or fire emergencies, operations managers must give dispatching decisions urgently under uncertain travel times. Therefore, it is useful to develop online strategies beforehand. For example, for effective disaster response, these strategies can be adopted before the disaster so that they can be implemented in the shortest time after the disaster. Likewise, when traveling in traffic, in order to reach the desired destination in the shortest time, we need a strategy defined on a network which answers the following questions: when to arrive at the end-node of an uncertain edge to learn its travel cost and when to avoid visiting it; when the travel time of an uncertain edge is learned, whether to take it or change the travel route; and if there exists a route to the destination without any uncertain edges, whether to take it or not. In this chapter, we focus on both developing effective online strategies that answer these questions and analyzing their performances theoretically to reveal their worst-case behavior. We next define our problem and its variants formally.

### 1.1 The online $k$-CTP with uncertain edges

Let $G = (V, E, k)$ denote an undirected graph with O as the source and D as the destination in which the costs of $k$ edges with given locations in the graph are unknown and a traveling agent can only discover their costs when she reaches an end-node of them. The costs of the remaining edges are known and deterministic. We call the edges with unknown costs *uncertain edges* and the edges with known costs *deterministic edges*. The objective is to provide an online strategy such that the traveling agent who is located at O initially receives $G = (V, E, k)$ and the known costs as input and targets to reach D with minimum total travel cost under uncertainty. Since the problem is a new variation of the $k$-CTP, we call this problem the single-agent $k$-CTP with uncertain edges, in short the S-$k$-CTP-U. We also study the multi-agent version of this problem where there are $L$ agents, who are initially located at O. We assume that the agents have the capability to transmit their location and edge cost information to the other agents in real time. We consider the multi-agent version of the problem with two different objectives, where the traveling agents follow an online strategy to ensure that the time when (1) the *first* agent and (2) the *last* agent arrive at D is minimum. We call these problems the M-$k$-CTP-U-f and the M-$k$-CTP-U-l, respectively. In real-life applications mentioned before, e.g., disaster response, the objective of the M-$k$-CTP-U-f is applicable when search-and-rescue teams try to reach a target in the shortest time, whereas the objective of the M-$k$-CTP-U-l is applicable when a convoy of $k$ vehicles delivers aid to a point of distribution.

### 1.2 Competitive analysis

The key concept in analyzing an online strategy is to compare a solution produced by the online strategy with the best possible solution under complete information, which is called the *offline optimum* solution. An offline strategy is to solve the same problem as an online strategy, except that all information about the problem inputs is revealed to an offline strategy from the beginning. An optimal offline strategy is the optimal strategy in the presence of complete input information which produces the offline optimum solution. To analyze the performance of online strategies, *competitive ratio* has been introduced in [1] and used by many researchers. The competitive ratio is the maximum ratio of the cost of the online strategy to the cost of the offline strategy over all instances of the problem. In our problems, the costs of the uncertain edges are known in the offline counterparts. Hence, the offline problems reduce to the shortest path problem.

Next, we discuss related work in the literature. Then, we state our contributions to the defined problems later on in this section.

### 1.3 Previous studies

We focus on studies on the $k$-CTP which are conducted from the online optimization and the competitive analysis perspective, since these are the most related works to our survey. First, we review the literature for the single-agent variants. Next, we discuss the relevant studies on the multi-agent versions.

#### *1.3.1 Single-agent* k*-CTP and variants*

The CTP was defined first in [2]. Papadimitriou and Yannakakis [2] proved that devising an online strategy with a bounded competitive ratio is PSPACE-complete for the CTP. Bar-Noy and Schieber [3] also considered the CTP and its variants. They introduced the $k$-CTP, where an upper bound $k$ on the number of blocked edges is given as input. They showed that for arbitrary $k$, the problem of designing an online strategy that guarantees the minimum travel cost is PSPACE-complete.

Westphal [4] considered the $k$-CTP from the competitive ratio perspective. He showed the lower bounds of $2k + 1$ and $k + 1$ on the competitive ratio of deterministic and randomized online strategies, respectively. He also presented an optimal deterministic online strategy for the $k$-CTP which is called the *backtrack* strategy. Xu et al. [5] also considered the $k$-CTP and presented two online strategies, the *greedy* and the *comparison* strategy, and proved competitive ratios of $2^{k+1} - 1$ and $2k + 1$, respectively, for these strategies. Bender and Westphal [6] presented a randomized online strategy for the $k$-CTP which meets the lower bound of $k + 1$ in special cases. Shiri and Salman [7] modified the strategy given in [7] and proposed an optimal randomized online strategy for the $k$-CTP on O-D edge-disjoint graphs.

#### *1.3.2 Multi-agent* k*-CTP and variants*

A generalization of the $k$-CTP with multiple agents was first considered by Zhang et al. in [8]. They analyzed the multi-agent $k$-CTP in two scenarios, with limited and complete communication. They proposed lower bounds of $2\left\lfloor \frac{k-1}{L_1} \right\rfloor + 1$ and $2\left\lfloor \frac{k}{L} \right\rfloor + 1$ on the competitive ratio of deterministic online strategies for the cases with limited and complete communication, respectively. Note that in the proposed lower bounds $L$ denotes the total number of agents and $L_1$ denotes the number of agents who benefit from complete communication. They also proposed an optimal deterministic online strategy when there are two agents in the graph. Shiri and Salman [9] also investigated the multi-agent $k$-CTP. They provided an updated lower bound on the competitive ratio of deterministic online strategies for the case with limited communication. They also presented a deterministic online strategy which is optimal in both cases with complete and limited communication on O-D edge-disjoint graphs. Randomized online strategies for the multi-agent *k*-CTP are investigated in [10], where lower bounds on the expected competitive ratio together with optimal randomized online strategies on O-D edge-disjoint graphs are proposed for the cases with limited and complete communication.

Xu and Zhang [11] focused on a real-time rescue routing problem from a source node to an emergency spot in the presence of online blockages. They analyzed the problem with the objective to make all the rescuers arrive at the emergency spot with minimum total cost. They studied the problem in two scenarios, without communication and with complete communication. They investigated both of the scenarios on the grid networks and general networks, respectively. They showed that the consideration of both the grid network and the rescuers' communication can significantly improve the rescue efficiency.

### 1.4 Our contributions

In the literature, the common unknown information in the *k*-CTP variants is the locations of the blocked edges in the graph. In fact, in all of the versions of the online *k*-CTP, all of the edges are equally likely to be blocked, and the agents have to explore the blockages in the graph to identify a route from the source node to the destination node with minimum total travel cost. However, in many real-world instances, assuming that all of the edges are equally likely to be congested or blocked ignores valuable information. In other words, there might exist many edges in the graph in which the agent is assured that they are not blocked before she starts her travel. Hence, considering all of the edges to be blocked with equal chance is not a realistic assumption in some of the real-world applications of the *k*-CTP.

As discussed at the beginning of this section, it is possible to identify the potential locations of the blocked edges in the graph in many real-world instances, such as in the urban traffic and post-disaster response. We introduce a new variation of the *k*-CTP with at most *k* number of uncertain edges with given locations and unknown traveling costs. We call this new problem the online *k*-Canadian Traveler Problem with uncertain edges. We consider both single-agent and multi-agent versions of this problem. In the multi-agent version of the problem, we analyze the problem with two different objectives, where the agents aim to ensure the first and the last arrival of the agents at D with minimum travel cost, respectively. The main contributions of our study are detailed below:

1. We introduce new variations of the online *k*-CTP which find applications in real-world problems, namely, the S-*k*-CTP-U, the M-*k*-CTP-U-f, and the M-*k*-CTP-U-l.

2. We provide a tight lower bound on the competitive ratio of deterministic online strategies for the S-*k*-CTP-U and introduce an optimal deterministic online strategy.

3. We derive lower bounds on the competitive ratio of deterministic online strategies for the M-*k*-CTP-U-f and the M-*k*-CTP-U-l.

The rest of this chapter is organized as follows. In Section 2, we describe the assumptions and give preliminaries. In Section 3, we analyze the single-agent version of the problem and provide a tight lower bound and an optimal strategy to this problem. In Section 4, we suggest lower bounds on the competitive ratio for the multi-agent versions of the problem. Finally, we conclude in Section 5.

## 2. Preliminaries

We consider the single-agent and the multi-agent problems defined in Section 1.1 with the following assumptions [1]:

1. The agent(s) are initially located at O. We call this stage the *initial stage* of the problem.

2. If any $k$ edges are removed from the graph, there still exists a path between the source and the destination node. This is a standard assumption in the literature.

3. The cost of the uncertain edges can take any value between 0 and $M$. An uncertain edge with explored cost equal to $M$ would be considered as a blocked edge.

4. Once the cost of an uncertain edge is learned, it remains the same whenever the traveler visits that edge. In other words the cost is not assumed to be time-dependent.

5. We call the time periods in which the cost of a new uncertain edge is identified, *stages* of the problem. That is, there are $k$ stages in the problem, i.e., stage 1 corresponds to the time period starting at the initial stage and ending at the moment before the cost of the first uncertain edge is learned.

We apply the following symbols and definitions to describe our results. We call the O-D paths which contain uncertain edges *uncertain paths* and which do not have uncertain edges *deterministic paths*. Let $D_i$ denote the shortest deterministic path at the $i$th stage and $d_i$ $(i = 1, 2, \dots, k)$ denote its corresponding cost. If there are more than one shortest deterministic path at the $i$th stage, one of them can be selected as $D_i$ arbitrarily. Note that at any stage of the problem there exists at least one deterministic O-D path based on Assumption 2.

We define the *optimistic cost of the O-D path* as the cost of the O-D path after setting the costs of the unvisited uncertain edges on it equal to 0. The *optimistic shortest O-D path* at the $i$th stage of the problem is denoted by $\pi_i$, which corresponds to the shortest O-D path after setting the costs of the remaining uncertain edges equal to 0. We denote its corresponding cost by $p_i$ $(i = 1, 2, \dots, k)$. That is, $\pi_1$ is the optimistic shortest O-D path at the initial stage of the problem. We denote the shortest path after the status of all of the uncertain edges is explored by $\pi_{k+1}$, i.e., $\pi_{k+1}$ is the offline optimum and $p_{k+1}$ is its corresponding cost.

## 3. Single-agent $k$-CTP with uncertain edges

In this section, we analyze the single-agent problem, namely, the S-$k$-CTP-U. We present a lower bound to this problem and prove its tightness by introducing a simple strategy. To suggest a lower bound on the competitive ratio of deterministic strategies, we need to analyze the performance of all of deterministic strategies on a special instance. Below, we propose our lower bound on the S-$k$-CTP-U, by analyzing an instance of O-D edge-disjoint graphs. Note that an O-D edge-disjoint graph is an undirected graph $G$ with a given source node O and a destination node D, such that any two distinct O-D paths in $G$ are edge-disjoint, that is, they do not have a common edge.

Theorem 1.1 For the S-$k$-CTP-U, there is no deterministic online strategy with competitive ratio less than $\min\{d_1/p_1, 2k-1\}$.

**Proof**. Consider the special graph in **Figure 1**. For each of deterministic strategies, we consider the instance when the cost of all of the first $k-1$ visited uncertain edges equals $M$ and the cost of the last visited uncertain edge equals 0. Hence, the cost of the offline shortest path equals $p_1$. For a strategy, we call this instance the *adverse instance*. In the special graph in **Figure 1**, any deterministic strategy corresponds to a permutation which specifies in which order the uncertain paths and $D_1$ (not necessarily all of them) are going to be selected. For each of these strategies, consider the adverse instance. We define $\alpha$ as a binary coefficient which equals 1, if the agent takes $D_1$, and equals 0, if the agent does not take $D_1$ in the strategy. Suppose that the agent has taken $i$ number of uncertain paths before taking $D_1$ when $\alpha$ equals 1. In this case, the competitive ratio of deterministic strategies on the special graph shown in **Figure 1** can be formulated as $\frac{2i(p_1)+\alpha(d_1)+(1-\alpha)p_1}{p_1}$ $(i = 0, 1, 2, ..., k-1)$. Note that in the adverse instance, the agent has to incur a cost equal to $2p_1$ in her first $k-1$ trials at the uncertain paths, since she has to come back to O after finding the uncertain edges blocked. However, since the cost of the $k$th visited uncertain edge equals 0, the agent incurs $p_1$ in her $k$th trial at the uncertain paths and reaches D. Now, we present our proof by considering two cases.

- Case 1. $\frac{d_1}{p_1} \leq 2k-1$. We consider this case for $\alpha = 0$ and $\alpha = 1$ separately.
  - $\alpha = 1$. In this case the competitive ratio of the corresponding strategies can be formulated as $\frac{2i(p_1)+d_1}{p_1}$ $(i = 0, 1, ..., k-1)$. The minimum competitive ratio equals $\frac{d_1}{p_1}$, when $i = 0$, which matches the proposed lower bound of the problem.
  - $\alpha = 0$. In this case the minimum competitive ratio of the corresponding strategies equals $2k-1$, which is greater than or equal to the lower bound of the problem.

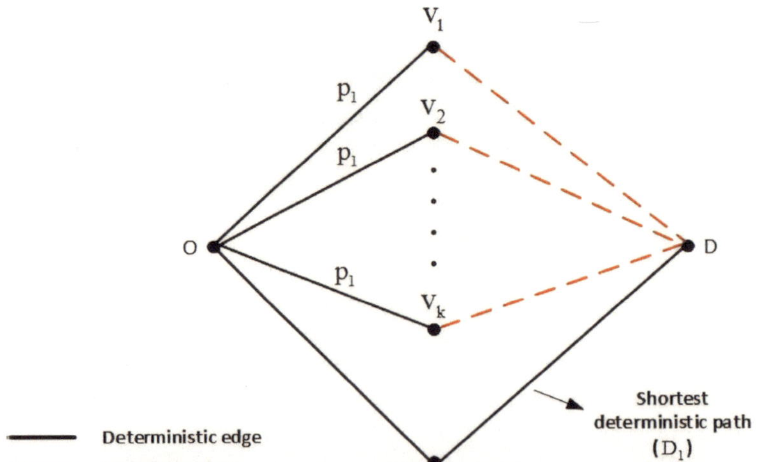


**Figure 1.**
*A special graph.*

- Case 2. $\frac{d_1}{p_1} > 2k - 1$. We also consider this case for $\alpha = 0$ and $\alpha = 1$ separately.
  - $\alpha = 1$. In this case the competitive ratio of the corresponding strategies can be formulated as $\frac{2i(p_1)+d_1}{p_1}$ $(i = 0, 1, \ldots, k-1)$. The minimum competitive ratio equals $\frac{d_1}{p_1}$, when $i = 0$, which is greater than the proposed lower bound of the problem.
  - $\alpha = 0$. In this case the minimum competitive ratio of the corresponding strategies equals $2k - 1$, which matches the lower bound of the problem.

Since we proved that the competitive ratios of all of the deterministic strategies for this special instance are greater than or equal to $\min\{d_1/p_1, 2k-1\}$, the proof is complete.

Now, we introduce a new deterministic strategy which meets the presented lower bound. We call this strategy the *pessimistic strategy* since the agent avoids to explore more than one uncertain edge at each iteration.

### 3.1 Pessimistic strategy

- **Initialization**. Put $i = 0$, where $i$ denotes the iteration number. At each iteration the agent starts her travel from O and explores the cost of one uncertain edge or will reach D without visiting any unvisited uncertain edge. If the agent reaches D, then the strategy ends. Otherwise, she backtracks to O or reaches D without visiting any other unvisited uncertain edge. In the latter case when the agent reaches D, the strategy ends. Note that each iteration corresponds to one of the stages of the problem, because at each iteration the cost of one of the uncertain edges is learned. That is, the first iteration corresponds to stage 1 of the problem. Also note that $p_i$ is nondecreasing in $i$, where $p_i$ is the cost of the optimistic shortest O-D path at the beginning of the $i$th iteration. Let $c_i$ denote the cost of the uncertain edge which is learned at the $i$th iteration. Note that $p_{i+1}$ is computable immediately after the agent observes $c_i$. Let $S$ denote the set of the uncertain edges in the graph.

- **Step 1**. Compute $d_1, p_1$, and $\min\{d_1/p_1, 2k-1\}$. If the minimum is $d_1/p_1$, take $D_1$, otherwise go to step 2.

- **Step 2**. If $(i = k-1)$, then go to step 3; otherwise, put $i = i+1$ and find $\pi_i$. If it does not contain uncertain edges, the agent takes it to reach D. Otherwise, take $\pi_i$ to reach the $i$th visited uncertain edge, observe $c_i$, set the value of the newly visited uncertain edge equal to $c_i$, and remove it from $S$. That is, it is not considered as an uncertain edge hereafter. Next, check the following conditions.

- Condition 1. Check if

$$\frac{2\left(\sum_{j=1}^{i-1} p_j\right) + p_i + c_i}{p_{i+1}} < 2k - 1 \tag{1}$$

and there exists no uncertain edge in the selected path, and proceed to reach D. Otherwise, check condition 2.

- Condition 2. Note that immediately after the agent observes $c_i$, $D_{i+1}$ and $d_{i+1}$ are computable. Check if

$$\frac{2\left(\sum_{i=1}^{i} p_i\right) + d_{i+1}}{p_{i+1}} < 2k - 1, \tag{2}$$

and then go back to O and take $D_{i+1}$. Otherwise, return to O and go to the beginning of step 2.

- **Step 3**. Take $\pi_k$ and observe $c_k$. Then compare

$$A = \frac{2\left(\sum_{i=1}^{k} p_i\right) + p_{k+1}}{p_{k+1}} \tag{3}$$

and

$$B = \frac{2\left(\sum_{i=1}^{k-1} p_i\right) + p_k + c_k}{p_{k+1}}. \tag{4}$$

If $A < B$ return to O and take the shortest path ($\pi_{k+1}$); otherwise, travel through the uncertain edge in the $k$th uncertain path and reach D.

Below we show that our strategy is optimal by using the inequalities which are presented in different steps of the pessimistic strategy.

Theorem 1.2 The pessimistic strategy is optimal for the S-$k$-CTP-U.

**Proof**. Note that if the strategy ends in either step 1 or 2, the competitive ratio would be less than or equal to the lower bound. Hence, we just need to analyze the cases where the strategy ends in step 3. Note that the competitive ratio of the strategy would not exceed $\min\{A, B\}$ in step 3. Thus, it is enough to show that either $A$ or $B$ does not exceed the proposed lower bound of the problem, if the strategy ends in step 3. We consider three different scenarios for $\pi_{k+1}$ to show the optimality of the pessimistic strategy, if the strategy ends in step 3.

- **Scenario 1**. $\pi_{k+1}$ contains the uncertain edge which is visited in the $k$th iteration. In this case, we show that $B$ meets the proposed lower bound of the problem. Since both $\pi_{k+1}$ and $\pi_k$ (i.e., $\pi_{k+1} \geq \pi_k$) contain the $k$th visited uncertain edge, $p_k + c_k$ equals $p_{k+1}$. Hence we can replace $p_k + c_k$ by $p_{k+1}$ in the numerator of $B$. We can also replace $p_i$ values for ($i = 1, 2, ..., k - 1$) by $p_{k+1}$ in the numerator of $B$, since $p_i$ is nondecreasing in $i$. In this case, $B$ would be at most $2k - 1$ which equals the lower bound of the problem.

Here, we note that $\pi_{k+1}$ does not contain the $k$th visited uncertain edge in the next two scenarios.

- **Scenario 2**. $\pi_{k+1}$ contains the uncertain edge which is visited in the $(k - 1)$th iteration. Note that $k \geq 2$ in this scenario, since $\pi_{k+1}$ does not contain the $k$th visited uncertain edge and contains the $(k - 1)$th visited uncertain edge. In this case, we show that $A$ meets the proposed lower bound of the problem. Consider condition 1 in step 2 at the $(k - 1)$th iteration. Since we have assumed that the strategy ends in step 3, we have

$$\frac{2\left(\sum_{i=1}^{k-2} p_i\right) + p_{k-1} + c_{k-1}}{p_k} > 2k - 1. \tag{5}$$

Since, both $\pi_{k+1}$ and $\pi_{k-1}$ (i.e., $\pi_{k+1} \geq \pi_{k-1}$) contain the $(k-1)$th visited uncertain edge, $p_{k-1} + c_{k-1}$ is less than or equal to $p_{k+1}$. Hence, we can replace $p_{k-1} + c_{k-1}$ by $p_{k+1}$ in the numerator above. We can also replace $p_i$ values for $(i = 1, 2, ..., k-2)$ by $p_k$ in the numerator since $p_i$ is nondecreasing in $i$. We obtain $(2k-4)p_k + p_{k+1} > (2k-1)p_k$; hence, $p_{k+1} > 3p_k$.

Now, we replace $p_i$ values for $(i = 1, 2, ..., k)$ by $p_k$ in the numerator of $A$. We obtain

$$A = \frac{2k(p_k) + p_{k+1}}{p_{k+1}}. \tag{6}$$

Now, we can replace $2k(p_k)$ by $\frac{2k}{3}(p_{k+1})$ in the numerator of $A$. In this case, $A$ would be at most $\frac{2k}{3} + 1$ which is less than or equal to the lower bound for $k \geq 2$ since we are comparing $\frac{2k}{3} + 1$ and $\min\{d_1/p_1, 2k-1\}$ for $k \geq 2$. Note that since the strategy ends in step 3, $\min\{d_1/p_1, 2k-1\}$ equals $2k-1$.

- **Scenario 3.** $\pi_{k+1}$ does not contain the uncertain edges which are visited in the $(k-1)$th and the $k$th iterations. In this case, we show that $A$ meets the proposed lower bound of the problem. Note that when $k \leq 2$, $\pi_{k+1} = D_1$ in this scenario. Thus, the strategy ends in step 1 when $k \leq 2$. For $k \geq 3$, consider condition 2 in step 2 at the $(k-2)$th iteration. We have

$$\frac{2\left(\sum_{i=1}^{k-2} p_i\right) + d_{k-1}}{p_{k-1}} > 2k - 1. \tag{7}$$

Since $\pi_{k+1}$ does not contain the uncertain edges which are visited in the $(k-1)$th and the $k$th iterations, $\pi_{k+1}$ is equivalent to $D_{k-1}$. Hence we can replace $d_{k-1}$ by $p_{k+1}$ in the numerator above. We can also replace $p_i$ values for $(i = 1, 2, ..., k-2)$ by $p_{k-1}$ in the numerator since $p_i$ is nondecreasing in $i$. We obtain $(2k-4)p_{k-1} + p_{k+1} > (2k-1)p_{k-1}$. Thus, $p_{k+1} > 3p_{k-1}$. Now, we replace $p_i$ values for $(i = 1, 2, ..., k-1)$ by $p_{k-1}$ in the numerator of $A$. We obtain

$$A = \frac{(2k-2)p_{k-1} + 2p_k + p_{k+1}}{p_{k+1}}. \tag{8}$$

Now, we can replace $(2k-2)p_{k-1}$ by $\frac{2k-2}{3}(p_{k+1})$ in the numerator of $A$. We also replace $p_k$ by $p_{k+1}$, since $p_i$ is nondecreasing in $i$. In this case, $A$ would be at most $\frac{2k-2}{3} + 3$, which is less than or equal to the lower bound for $k \geq 3$.

Since we showed that the competitive ratio of the pessimistic strategy is less than or equal to the lower bound, the proof is complete.

As an illustrative example for the pessimistic strategy, consider the instance given in **Figure 2** which represents a part of the Gulf Coast area of the United States. In **Figure 2**, the nodes represent the cities, and the numbers on the edges denote the edge travel times (per hour) in a post-disaster scenario. The edges (2,6) and (5,6) are the uncertain edges whose costs are not known at the beginning. The traveling agent is initially at node 1 and node 6 is the destination node. Path 1-3-6 is the shortest deterministic path ($D_1$), and path 1-2-6 is the shortest optimistic path ($\pi_1$) at the initial stage, i.e., $d_1 = 11$ and $p_1 = 3$. When step 1 of the pessimistic strategy is implemented, the agent compares $\frac{d_1}{p_1} = \frac{11}{3}$ with

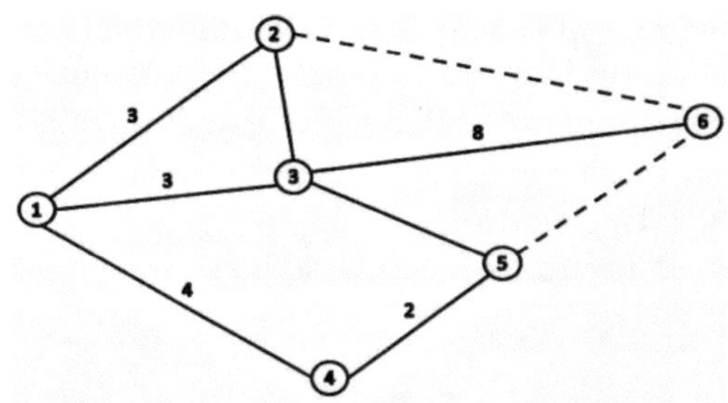


**Figure 2.**
*A scenario from the Gulf Coast area of the United States network with Atlanta as the source node and Wilmington as the destination node.*

$2k-1=3$. Since $\frac{11}{3}>3$, the strategy enters step 2. Next, the agent takes the shortest optimistic path $\pi_1$ and arrives at node 2 after traversing edge (1,2). We assume that the costs of the uncertain edges (2,6) and (5,6) are 3 and 2, respectively. When the agent arrives at node 2, she learns the traveling time of edge (2,6), i.e., $c_1=3$. Then she checks if $\frac{p_1+c_1}{p_2}<2k-1$. Since $\frac{6}{6}<3$, the agent takes edge (2,6) to arrive at node 6 and the strategy ends. Note that the cost of the offline optimum is 6. Therefore, the competitive ratio of the pessimistic strategy is one in the described scenario.

## 4. Multi-agent *k*-CTP with uncertain edges

In this section, we study the M-*k*-CTP-U-f and the M-*k*-CTP-U-l. Note that $L$ denotes the number of agents in the graph in these problems. We assume that there is no distinction between the $L$ agents and all of the agents benefit from complete communication in the sense that they can transmit their locations and explored uncertain edges' cost information to the other agents in real time. By considering an instance of O-D edge-disjoint graphs, we derive lower bounds on the competitive ratio of deterministic online strategies to the M-*k*-CTP-U-f and the M-*k*-CTP-U-l.

Theorem 1.3 For the M-*k*-CTP-U-f and the M-*k*-CTP-U-l, there is no deterministic online strategy with competitive ratio less than $\min\{d_1/p_1, 2(\lfloor\frac{k}{L}\rfloor)+1\}$ and $\min\{d_1/p_1, 2(\lceil\frac{k}{L}\rceil)+1\}$, respectively.

**Proof**. We again consider the special graph in **Figure 1**. In this case, any deterministic strategy corresponds to a permutation which describes in which order the uncertain paths and $D_1$ (not necessarily all of them) are going to be selected by the agents. For all of these strategies, consider the adverse instance which is defined in the proof of Theorem 1.1. Note that the agents will not reach D via uncertain paths unless the costs of all of the uncertain edges are specified. Before we present the rest of our proof, we need to propose the following lemma.

Lemma 1.4 In the adverse instance, the competitive ratio of the strategies in which the arrivals of the agents at D is via the uncertain paths is at least $2\lfloor\frac{k}{L}\rfloor+1$ and $2\lceil\frac{k}{L}\rceil+1$, for the M-*k*-CTP-U-f and the M-*k*-CTP-U-l, respectively.

**Proof**. Note that the agents will not reach D via the uncertain paths unless the costs of all of the uncertain edges are specified since we are considering the adverse instance. Now we present our proof for each claim separately.

- M-$k$-CTP-U-f. In this problem, the agents have to incur a cost of at least $2\left(\left\lfloor\frac{k}{L}\right\rfloor\right)p_1$ to discover the costs of $L\left(\left\lfloor\frac{k}{L}\right\rfloor\right)$ number of uncertain edges and backtrack to O. The agents have to incur $p_1$ to learn the costs of the remaining uncertain edges and deliver at least one of the agents to D. Since the cost of the shortest path is at least $p_1$ in the adverse instance, the competitive ratio of deterministic strategies when none of the agents take $D_1$ would be at least $\frac{2\left(\left\lfloor\frac{k}{L}\right\rfloor\right)p_1+p_1}{p_1}$, which is equal to $2\left\lfloor\frac{k}{L}\right\rfloor+1$.

- M-$k$-CTP-U-l. In this problem, it takes a cost of at least $2\left(\left\lceil\frac{k}{L}\right\rceil\right)p_1$ to explore the costs of all of the $k$ uncertain edges and backtrack the agents to O. It takes at least $p_1$ for all of the agents to take the shortest path and arrive at D. Since the cost of the shortest path is $p_1$ in the adverse instance, the competitive ratio of deterministic strategies when none of the agents take $D_1$ would be at least $\frac{2\left(\left\lceil\frac{k}{L}\right\rceil\right)p_1+p_1}{p_1}$, which is equal to $2\left\lceil\frac{k}{L}\right\rceil+1$.

Note that since we are considering the arrivals of the agents at D via the uncertain paths, the performance of the strategies will not be improved if one or more agents take $D_1$. The proof is complete.

Now, we present the rest of our proof for each problem separately:

- M-$k$-CTP-U-f. We present our proof by considering two cases:
  - Case 1. $\frac{d_1}{p_1}\geq 2\left\lfloor\frac{k}{L}\right\rfloor+1$. In this case, the competitive ratio of the strategies in which the first arrival of the agents at D is via $D_1$ is at least $\frac{d_1}{p_1}$, which is greater than or equal to $\min\left\{d_1/p_1, 2\left(\left\lfloor\frac{k}{L}\right\rfloor\right)+1\right\}$. The competitive ratio of deterministic strategies in which the first arrival of the agents at D is via the uncertain paths would be at least $2\left\lfloor\frac{k}{L}\right\rfloor+1$, which matches the proposed lower bound of the problem.
  - Case 2. $\frac{d_1}{p_1}<2\left\lfloor\frac{k}{L}\right\rfloor+1$. In this case, the competitive ratio of deterministic strategies in which the first arrival of the agents at D is via the uncertain paths would be at least $2\left\lfloor\frac{k}{L}\right\rfloor+1$, which is greater than the proposed lower bound of $\min\left\{d_1/p_1, 2\left(\left\lfloor\frac{k}{L}\right\rfloor\right)+1\right\}$. The competitive ratio of the strategies in which the first arrival of the agents at D is via $D_1$ is at least $\frac{d_1}{p_1}$, which matches the proposed lower bound of the problem.

- M-$k$-CTP-U-l. We present our proof by considering two cases:
  - Case 1. $\frac{d_1}{p_1}\geq 2\left\lceil\frac{k}{L}\right\rceil+1$. In this case, the competitive ratio of the strategies in which the last arrival of the agents at D is via $D_1$ is at least $\frac{d_1}{p_1}$ which is greater than or equal to $\min\left\{d_1/p_1, 2\left(\left\lceil\frac{k}{L}\right\rceil\right)+1\right\}$. The competitive ratio of deterministic strategies in which the last arrival of the agents at D is via the uncertain edges would be at least $2\left\lceil\frac{k}{L}\right\rceil+1$, which matches the proposed lower bound of the problem.
  - Case 2. $\frac{d_1}{p_1}<2\left\lceil\frac{k}{L}\right\rceil+1$. In this case, the competitive ratio of deterministic strategies in which the last arrival of the agents at D is via the uncertain

paths would be at least $2\left\lceil\frac{k}{L}\right\rceil+1$ which is greater than the proposed lower bound of $\min\left\{d_1/p_1, 2\left(\left\lceil\frac{k}{L}\right\rceil\right)+1\right\}$. The competitive ratio of the strategies in which the last arrival of the agents at D is via $D_1$ is at least $\frac{d_1}{p_1}$ which matches the proposed lower bound of the problem.

We just proved that the competitive ratio of deterministic strategies in the adverse instance is not better than $\min\left\{d_1/p_1, 2\left(\left\lfloor\frac{k}{L}\right\rfloor\right)+1\right\}$ and $\min\left\{d_1/p_1, 2\left(\left\lceil\frac{k}{L}\right\rceil\right)+1\right\}$, for the M-$k$-CTP-U-f and the M-$k$-CTP-U-l, respectively. Hence, we conclude that the competitive ratio of the problems cannot be better than the proposed lower bounds.

## 5. Conclusions

We introduced new variants of the online $k$-CTP which find various important real-life applications. In these variants, the locations of the uncertain edges are known, where the traveling costs of these edges are unknown. We investigated both the single-agent and the multi-agent versions of the problem. We proposed a tight lower bound on the competitive ratio of deterministic online strategies and an optimal strategy for the single-agent problem that we call the S-$k$-CTP-U. We derived lower bounds on the competitive ratio of deterministic online strategies for the multi-agent problems called as the M-$k$-CTP-U-f and the M-$k$-CTP-U-l. Providing optimal strategies for the M-$k$-CTP-U-f and the M-$k$-CTP-U-l which match the proposed lower bounds can be considered as a future research direction. Analyzing the problem on special networks such as grid networks is another future research direction for these new variations.

## Author details

Davood Shiri*† and F. Sibel Salman†
Koç University, Istanbul, Turkey

*Address all correspondence to: dshiri@ku.edu.tr

†These authors contributed equally to this work.

## References

[1] Sleator D, Tarjan R. Amortized efficiency of list update and paging rules. Communications of the ACM. 1985;**28**:202-208

[2] Papadimitriou C, Yannakakis M. Shortest paths without a map. Theoretical Computer Science. 1991;**84**: 127-150

[3] Bar-Noy A, Schieber B. The Canadian Traveler Problem. In: SODA '91 Proceedings of the Second Annual ACM-SIAM Symposium on Discrete Algorithms; 1991. pp. 261-270

[4] Westphal S. A note on the $k$-Canadian Traveler Problem. Information Processing Letters. 2008; **106**:87-89

[5] Xu Y, Hu M, Su B, et al. The Canadian Traveler Problem and its competitive analysis. Journal of Combinatorial Optimization. 2009;**18**: 185-205

[6] Bender M, Westphal S. An optimal randomized online algorithm for the $k$-Canadian Traveller Problem on node-disjoint paths. Journal of Combinatorial Optimization. 2015;**30**: 87-96

[7] Shiri D, Salman FS. On the randomized online strategies for the $k$-Canadian Traveler Problem. Journal of Combinatorial Optimization. 2019;**1**: 1-14

[8] Zhang H, Xu Y, Qin L. The $k$-Canadian Travelers Problem with communication. Journal of Combinatorial Optimization. 2013;**26**: 251-265

[9] Shiri D, Salman FS. On the online multi-agent O-D $k$-Canadian Traveler Problem. Journal of Combinatorial Optimization. 2017;**34**: 453-461

[10] Shiri D, Salman FS. Competitive analysis of randomized online strategies for the online $k$-Canadian Traveler Problem. Journal of Combinatorial Optimization. 2019;**37**:848-865

[11] Xu Y, Zhang H. How much the grid network and rescuers' communication can improve the rescue efficiency in worst-case analysis. Journal of Combinatorial Optimization. 2015;**30**: 1062-1076

Chapter 11

# A Geometrical Realisation of Quasi-Cyclic Codes

*Cristina Martinez Ramirez and Alberto Besana*


## Abstract

We study and enumerate cyclic codes which include generalised Reed-Solomon codes as function field codes. This geometrical approach allows to construct longer codes and to get more information on the parameters defining the codes. We provide a closed formula in terms of Stirling numbers for the number of irreducible polynomials and we relate it with other formulas existing in the literature. Further, we study quasi-cyclic codes as orbit codes in the Grassmannian parameterizing constant dimension codes. In addition, we review Horn's algorithm and apply it to construct classical codes by their defining ideals.


**Keywords:** cyclic code, partition, Grassmannian

## 1. Introduction

Function fields are used ubiquitously in algebraic coding theory for their flexibility in constructions and have produced excellent linear codes. Suitable families of function fields, for example good towers of function fields, have been used to construct families of codes with parameters bound better than the asymptotic bound.

Let $q$ a power of a prime number $p$. It is well known, that there exists exactly one finite field with $q$ elements which is isomorphic to the splitting field of the polynomial $x^q - x$ over the prime field $\mathbb{F}_p$. Any other field $F$ of characteristic $p$ contains a copy of $\mathbb{F}_p$. We denote respectively by $\mathbb{A}^n(\mathbb{F}_q)$ and $\mathbb{P}^n(\mathbb{F}_q)$ the affine space and the projective space over $\mathbb{F}_q$. Let $\mathbb{F}_q[x_1, x_2, \dots, x_n]$ be the algebra of polynomials in $n$ variables over $\mathbb{F}_q$.

The encoding of an information word into a $k$-dimensional subspace is usually known as coding for errors and erasures in random network coding [1]. Namely, let $V$ be an $N-$dimensional vector space over $\mathbb{F}_q$, a code for an operator channel with ambient space $V$ is simply a non-empty collection of subspaces of $V$. The collection of subspaces is a code for error correcting errors that happen to send data through an operator channel. The matrix coding the information is parameterised by random variables $a_1, a_2, \dots, a_n$ which constitute the letters of an alphabet. Here the operator channel is an abstraction of the operator encountered in random linear network coding, when neither transmitter nor receiver has knowledge of the channel transfer characteristics. The input and output alphabet for an operator channel is the projective geometry. A good code is capable of correcting error and erasures at the output of the operator channel. Thus in order to construct good codes one need to choose a metric consistent with channel errors and search of a set of vectors with given metric properties as a correcting code. The codes considered here are codes for channels whose errors are consistent with the weighted Hamming metric (WHM).

Let $C$ be a non-singular, projective, irreducible curve defined over $\mathbb{F}_q$, as the vanishing locus of a polynomial $F \in \mathbb{F}_q[x_0, x_1, x_2]$. We define the number $N(q)$ of $\mathbb{F}_q$−rational points on the curve to be

$$N(q) = \ |\{(x_0, x_1, x_2) \in \mathbb{P}^2(\mathbb{F}_q) | F(x_0, x_1, x_2) = 0\}|.$$

It is a polynomial in $q$ with integer coefficients, whenever $q$ is a prime power.

The number of points $\overline{C}(\mathbb{F}_{q^r})$ on $C$ over the extensions $\mathbb{F}_{q^r}$ of $\mathbb{F}_q$ is encoded in an exponential generating series, called the zeta function of $\overline{C}$:

$$Z(C,t) = exp\left(\sum_{r=1}^{\infty} \#\overline{C}(\mathbb{F}_{q^r})\frac{t^r}{r}\right).$$

Garcia and Stichtenoth analysed the asymptotic behaviour of the number of rational places and the genus in towers of function fields, [2]. From Garcia-Stichtenoth's second tower one obtains codes over any field $\mathbb{F}_q$ where $q$ is an even power of a prime [3].

One of the main problems in coding theory is to obtain non-trivial lower bounds of the number $N(F_i)$ of rational places of towers of function fields $\{F_i/\mathbb{F}_q\}_{i=1}^{\infty}$ such that $F_i \subsetneq F_{i+1}$. Suitable families of function fields, for example good towers of function fields, have been used to construct families of codes that beat the Gilbert-Varshamov bound. This paper aims to explore this link for the study and construction of quasi-cyclic codes. For example good codes are obtained for curves of genus 0, they are in fact extended generalised Reed-Solomon codes.

**Notation.** Let $\mathbb{F}_q$ denote the Galois field of $q$ elements and let $(\mathbb{F}_q)^n$ denote the vector spaces of all ordered $n$-tuples over $\mathbb{F}_q$. The Hamming weight of a vector $x$, denoted by $wt(x)$ is then number of non-zero entries in $x$. A linear code $C$ of length $n$ and dimension $k$ over $\mathbb{F}_q$ is a $k$-dimensional subspace of $(\mathbb{F}_q)^n$. Such a code is called $[n, k, d]_q$ code if its minimum Hamming distance is $d$. For $d$ a positive integer, $\alpha = (\alpha_1, \ldots, \alpha_m)$ is a partition of $d$ into $m$ parts if the $\alpha_i$ are positive and decreasing.

## 2. Algebraic geometric codes

Let $\mathbb{F}_q$ be a finite field of $q$ elements, where $q$ is a power of a prime. We consider as an alphabet a set $\mathcal{P} = \{P_1, \ldots, P_N\}$ of $N - \mathbb{F}_q$ rational points lying on a smooth projective curve $C$ of genus $g$ and degree $d$ defined over the field $\mathbb{F}_q$. If $D$ is a divisor on the curve $C$, $\mathcal{L}(D)$ is the linear series attached to this divisor with coefficients in the field.

**Definition 2.1.** *Algebraic Geometric Codes (AGC) are constructed by evaluation of the global sections of a line bundle or a vector bundle on the curve $C$ over $N$ $(N > g)$ distinct rational places $P_1, \ldots, P_N$. Namely, let $F|\mathbb{F}_q$ be the function field of the curve, $\mathcal{D}$ the divisor $P_1 + \cdots + P_N$ and $G$ a divisor of $F|\mathbb{F}_q$ of degree $s \leq N$ such that* $\text{Supp G} \cap \text{Supp D} = \emptyset$. *Then the geometric Goppa code associated with the divisors $D$ and $G$ is defined by*

$$\mathbf{C}(D, G) = \{(x(P_1), \ldots, x(P_n)) | x \in \mathcal{L}(G)\} \subseteq \mathbb{F}_{q^n}.$$

Recall that $\mathbb{F}_{q^n}|\mathbb{F}_q$ is a cyclic Galois extension and it is finitely generated by unique element $\alpha \in \mathbb{F}_{q^n} \backslash \mathbb{F}_q$. $\alpha$ is a primitive element and $\{1, \alpha, \alpha^2, \dots, \alpha^{n-1}\}$ is a basis of the field extension $\mathbb{F}_q \hookrightarrow \mathbb{F}_q(\alpha)$, that is, $\mathbb{F}_{q^n} \cong (\mathbb{F}_q)^n$.

In the sequel, an $[n,k]_q$-code $C$ is a $k$-dimensional subspace of $(\mathbb{F}_q)^n$.

### 2.1 Generalised Reed-Solomon codes as cyclic codes

Another important family of Goppa codes is obtained considering the normal rational curve (NRC) $\mathcal{C}^n$ defined over $\mathbb{F}_q$:

$$\mathcal{C}^n := \{\mathbb{F}_q(1, \alpha, \dots, \alpha^n) : \alpha \in \mathbb{F}_q \cup \{\infty\}\}.$$

Assuming that $(n,p) = 1$ are coprime, the set $\{1, \alpha, \alpha^2, \dots, \alpha^{n-1}\}$ forms a basis of $\mathbb{F}_{q^n}$ over $\mathbb{F}_q$, where $p$ is the characteristic of the field. Thus points in the NRC are in correspondence with $\mathbb{F}_q$−linear combinations of the base vectors up to collineation. The Goppa codes of dimension $n$ defined over $\mathcal{C}^n$ are constructed by evaluating non-zero polynomials of degree less than $n$ over a sequence $\alpha_1, \dots, \alpha_n$ of $n$ distinct elements in $\mathbb{F}_q$, if $k \leq n$, then the map

$$\epsilon : \mathbb{F}_q[x] \rightarrow \mathbb{F}_q^n, \quad f \mapsto (f(\alpha_1, \dots, \alpha_n)) \tag{1}$$

is injective, since the existence of a non-zero polynomial of degree less than $k$ vanishing on all $\alpha_i$ implies $n < k$ by the fundamental theorem of algebra (a non-zero polynomial of degree $r$ with coefficients in a field can have at most $r$ roots). These are just Reed-Solomon codes of parameters $[n,k,d]$ over a finite field $\mathbb{F}_q$, with parity check polynomial $h(x) = \prod_{i=1}^{q}(x - \alpha^i)$, where $\alpha$ is a primitive root of $\mathbb{F}_q$ such that $\alpha^{k+1} = \alpha + 1$. Any codeword $(c_0, c_1, \dots, c_{n-1})$ can be expanded into a $q$-ary $k$ vector with respect to the basis $\{1, \alpha, \dots, \alpha^{k-1}\}$. Construction of generalised Reed-Solomon codes over $\mathbb{F}_q$ only employ elements of $\mathbb{F}_q$, hence their lengths are at most $q+1$. In order to get longer codes, one can make use of elements of an extension of $\mathbb{F}_q$, for instance considering subfield subcodes of Reed-Solomon codes. In this way, one gets cyclic codes. Recall that a linear cyclic code is an ideal in the ring $\mathbb{F}_q[x]/(x^n - 1)$ generated by a polynomial $g(x)$ with roots in the splitting field $\mathbb{F}_q^l$ of $x^n - 1$, where $n|q^l - 1$, ([4]). We shall identify the code with the set of its codewords. A natural question then to ask is how many irreducible polynomials of degree at least 2 are there over the algebraic closure of $\mathbb{F}_q[x]$. Next theorem expresses this number in terms of Stirling numbers.

**Theorem 2.2.** *Assume that $(q,n) = 1$, then the number of polynomials of degree $(n \geq 2)$ decomposable into distinct linear factors over a finite field $\mathbb{F}_q$ of arbitrary characteristic a prime number p, is equal to $\sum_{k=1}^{n}(q)_k$, where $(q)_k$ is the falling factorial polynomial $q \cdot (q-1) \dots (q-k) = \sum_{k=0}^{n} s(n,k) q^k$, where $s(n,k)$ is the Stirling number of the first kind (the number of ways to partition a set of n objects into k non-empty subsets), divided by the order of the affine transformation group of the affine line $\mathbb{A}^1 = \mathbb{P}^1 \backslash \infty$, that is $q^2 - q$.*

*Proof.* We need to count all the polynomials $f_n(x)$ in one variable of degree $n$ fixed. We assume that our polynomial $f_n(x)$ decomposes into linear factors, otherwise we work over $\overline{\mathbb{F}}_q[x]$, where $\overline{\mathbb{F}}_q$ denotes the algebraic closure of the finite field $\mathbb{F}_q$. Since the number of ordered sequences on $q$ symbols is $q!$ and each root is counted with its multiplicity, it follows that the number of monic polynomials with $n-1$

different roots is $q(q-1)(q-2)\dots(q-n+1) := (q-2)_n$. Now we observe that polynomials are invariant by the action of automorphisms of the affine line, so we must divide this number by the order of this group which is $q^2-q$. □

Theorem 2.3. *Given a set of integers $\{0, 1, \dots, n-1\}$ module n, there is a set J of k integers which is a set of roots, that is, there is a polynomial $h(x) = \prod_{j\in J}(x-\alpha^j)$, where $\alpha$ is a generator of $(\mathbb{F}_{p^m})$ for some prime number p and m is the least integer such that $n|p^m-1$. The ideal $h(x)$ generates in $\mathbb{F}_{p^m}[x]/(x^n-1)$ is a cyclic linear code of parameters $(n, k, n-k+1)$.*

*Proof.* Let $m$ be the least integer such that $n$ divides $p^m-1$, then $g.c.d(m, p) = 1$. We define an equivalence relation on the set of integers $\{0, 1, \dots, n-1\}$, by declaring two integers $i$ and $j$ in the range $0 \le i \le n-1$ to be conjugate module $n$

if $p^s i \equiv j \pmod{n}$. This equivalence relation partition the set into cyclotomic cosets. The cyclotomic coset containing $j$, which we will denote by $\Omega_j$, can be described explicitly as the set $\{j, pj, \dots, p^{k-1}j\}$, where $k$ is the least positive integer such that $p^k j \equiv j \pmod{n}$ and $j$ is not necessarily the smallest integer in such coset. Denote by $I_n$ the set consisting of the smallest integers in each cyclotomic coset, then $I_n$ is a set root, that is, it is a set of $k$ integers in arithmetic progression modulo $n$ whose increment is relatively prime to $n$. Let $d = n-k+1$, then the polynomial $\prod_{i\in I_n}(x-\alpha^i)$ defines a cyclic code of parameters $(n, k, d)$. □

As an application of Theorem 2.2, given an integer $n$, we can count the number of cyclic codes of parameters $[n, k]$ for each $0 \le k \le n$ and set of roots $\alpha_1, \dots, \alpha_k$ in the splitting field of $x^n-1$, the corresponding polynomial $g(x) = \prod_{i=1}^{k}(x-\alpha_i)$ generates a linear cyclic code in the ring $\mathbb{F}_q[x]/(x^n-1)$. Thus for each $0 \le k \le n$ there are exactly $(q)_k/(q^2-q)$ cyclic codes.

In the theory of error-correcting codes to a given code $C \subset \mathbb{F}_q^m$, one assigns another important parameter, the minimum distance $d$ which measures how good the decoding is.

**Definition 2.4.** *The distance between vectors $a = (a_1, a_2, \dots, a_n)$ and $b = (b_1, b_2, \dots, b_n)$ in the Weighted Hamming metric (WHM) is defined by a function:*

$$d_{WH}(a, b) = \sum_{i=1}^{n} w_i \overline{d(a_i, b_i)},$$

where $w_i > 0$, $d(a_i, b_i) = 1$ if $a_i \neq b_i$ and $d(a_i, b_i) = 0$ if $a_i = b_i$. The weight of a vector $a$ in the WHM is $wt_{WH}(a) = d_{WH}(a, 0) = \sum_{i:a_i\neq 0} w_i$. The value $w_i$ and vector $w = (w_1, w_2, \dots, w_n)$ are called a weight of position $i$ and a vector of weights of positions respectively.

Geometrically a binary vector $(a_1, \dots, a_n)$ of length $n$ gives the coordinates of a vertex of a unit cube in $n$ dimensions.

**Example 1.** *Consider the Goppa code defined by the rational function $g(x) = \frac{3x^2-5x+5}{x^3-2x^2+x}$ which admits as decomposition into partial fractions the expression $G(x) := \frac{5}{x} - \frac{2}{x-1} + \frac{3}{(x-1)^2}$. The presence of a double factor $(x-1)^2$ corresponds to the existence of an eigenspace E in the vector space $\mathbb{F}_q^n$ of multiplicity 2 and thus an $\alpha$-splitting subspace where the operator $\alpha$ is just the linear operator $A - \lambda I$, with $\lambda$ the eigenvalue associated to E and A is the generator matrix of the code. We recall that an $r$-dimensional W subspace is $\alpha$-splitting if $\alpha^i W = W$ is invariant under the action of any element $\alpha^i$ in the Galois group of the extension $\mathbb{F}_q \rightarrowtail \mathbb{F}_q(\alpha)$.*

## 2.2 Algebraic function field codes

A much greater variety of linear codes is obtained if one uses places of arbitrary degree rather than just places of degree 1. These codes are more naturally described through function field codes. A general viewpoint is that function field codes are certain finite dimensional linear subspaces of an algebraic function field over a finite field as in Goppa's construction.

In the paper [5], the authors introduce another construction where places of arbitrary degree are allowed. The method consists of choosing two divisors $G_1$ and $G_2$ of an algebraic curve over $\mathbb{F}_q$ with $G_1 \leq G_2$. Then $\mathcal{L}(G_1)$ is a subspace of the vector space $\mathcal{L}(G_2)$ over $\mathbb{F}_q$. If we choose a basis of $\mathcal{L}(G_2)$, then the coordinate vectors of the elements of $\mathcal{L}(G_1)$ form a linear code over $\mathbb{F}_q$ of length $n = \dim(\mathcal{L}(G_2))$ and dimension $k = \dim(\mathcal{L}(G_1))$. These are known as function field codes and they provide a general perspective on the construction of algebraic-geometry codes [6].

**Example 2.** *We consider as in* [7] *the Suzuki curve* $\chi$ *defined over* $\mathbb{F}_q$ *by the following equation* $y^q - y = x^{q_0}(x^q - x)$ *with* $q = 2q_0^2 \geq 8$ *and* $q_0 = 2^r$. *This curve has exactly* $q^2 + 1-$*rational places with a single place at infinity* $P_\infty$ *and it is of genus* $g_S = q_0(q-1)$. *We construct a code out of the divisor* $F = mP_\infty$ *and* $Q$ *where* $Q = P_1 + \ldots + P_{q^2}$ *is the sum of the* $q^2-$*rational points and the parameter* $m$ *satisfies the bound* $m > 2g - 2$ *and* $g$ *is the genus of the curve.*

Observe that the geometric Goppa code $\mathbf{C}(F, Q)$ is an $\mathbb{F}_q$-subspace of $(\mathbb{F}_q)^{q^2}$ and its dimension $k$ as an $\mathbb{F}_q-$vector space is the dimension of the code. Geometrically, it corresponds to a point in the Grassmannian $\mathcal{G}_{q^2,k}(\mathbb{F}_q)$. The set of codewords recognised by the code $\mathbf{C}(F, Q)$ admits the following description in terms of monomial ideals in the variables $x, y, z, w$:

$$\left\{x^a y^b z^c w^{d'} \,|\, a, b, c, d' \geq 0, aq + b(q + q_0) + c(q + 2q_0) + d'(q + 2q_0 + 1) \leq d\right\},$$

where $z = x^{2q_0+1}$ and $w = xy^{2q_0} - z^{2q_0}$ are elements in the function field $F_\chi := \mathbb{F}_q(x, y)$ over $\mathbb{F}_q$. Moreover, it is a generating set for the linear series $\mathcal{L}(dP_\infty)$ associated to the divisor $dP_\infty$.

**Theorem 2.5.** *Cyclic codes are function field codes constructed over the curve* $C_{n,m}$ *with affine equation* $y^m + x^n = 1$ *defined over a finite field* $\mathbb{F}_q$ *of* $q$ *elements, where* $q$ *is a power of a prime* $p$ *and* $n, m$ *are integer numbers greater or equal than 2.*

*Proof.* Let us assume $n$ is an integer even number, thus $n = 2^k \cdot s$, with $s$ an integer odd number. We recall that a linear cyclic code is an ideal in the ring $\mathbb{F}_q[x]/(x^n - 1)$ generated by a polynomial $g(x)$ with roots in the splitting field $\mathbb{F}_q^l$ of $x^n - 1$, where $n | q^l - 1$. If we consider the factorisation of the polynomial $x^n - 1$ over $\mathbb{F}_p[x]$, we get $(x^{n/2} - 1)(x^{n/2} + 1) = (x^{n/4} - 1)(x^{n/4} + 1)(x^{n/2} + 1) = \left(x^{n/2^k} - 1\right)\left(x^{n/2^k} + 1\right)\left(x^{n/2^{(k-1)}} - 1\right)\left(x^{n/2^{(k-1)}} + 1\right) \ldots (x^{n/2} + 1)$. We see that the point $P_0 = (\alpha, 0) \in \mathbb{P}\left(\mathbb{F}_q^2\right)$ with $\alpha^{n/2} = p - 1$ is an $\mathbb{F}_{q^2}-$rational place of the affine curve $y^m = (x^{n/2} + 1)$. The other rational places are $P_k = (\beta, 0)$ with $\beta^{n/2^k} = p - 1, \ldots$, $P_2 = (\beta^2, 0)$, $P_1 = (1, 0)$, $P_0 = (-1, 0)$ and the place $P_\infty = (0, \alpha)$ at $\infty$. The cyclic code is realised as the algebraic geometric code associated to the divisors $D = P_0 + P_1 + \ldots + P_k$, $G = \mu P_\infty$ and the parameter $\mu$ satisfies the bound $\mu > 2g - 2$, where $g$ is the genus of the curve $C_{n,m}$. Note that $m$ is the least integer such that $n | p^m - 1$. In particular $\alpha$ is a generator of $(\mathbb{F}_p)^m$.

If $n$ is an integer odd number, by Theorem 2.3, we know there is a set of roots $\{\alpha^j\}_{j\in J}$, such that $\alpha$ is a generator of $(\mathbb{F}_{p^m})$. Now we consider the points $P_j = (\alpha^j, 0)$ with $j \in J$ and the point $P$ at $P_\infty = (0, \alpha) = \infty$, and the cyclic code is realised as the function field code associated to the divisors $D = \sum_{j\in J} P_j$ and $\mu P_\infty$. □

**Remark 2.6.** *The proof given in theorem gives a realisation of cyclic codes as AG codes constructed over the curve with affine model $y^m + x^n = 1$. In particular when $m = n = q + 1$, we cover the codes defined over the Hermitian curve.*

Another important family of cyclic codes is obtained considering the roots of the polynomial $x^n - 1$ over its splitting field. These codes are of great importance in ADN-computing and as they are linear codes, they can be described as function fields. Let $\alpha$ be a primitive element of the underlying vector space over $\mathbb{F}_q$. Since the base field is of characteristic $p$, $x^n - 1$ has $n$ different zeroes. Let $\overline{\mathbb{F}}_q[x]$ be the extension field containing the $n^{th}$ roots of unity $1, \alpha, \ldots, \alpha^{n-1}$, where $\alpha^{n-1} + \alpha^{n-2} + \ldots + \alpha + 1 = 0$. Moreover the set $\{1, \alpha, \ldots, \alpha^{n-1}\}$ constitutes a basis over the prime field $\mathbb{F}_p$, and the field extensions $\mathbb{F}_{p^n} \cong \mathbb{F}_p[x]/(x^{n-1} + \ldots + x + 1)$ are isomorphic.

**Example 3.** *The polynomial $x^2 + x + 1$ over $\mathbb{F}_p[x]$ is irreducible, thus the fields $\mathbb{F}_{p^2} \cong \mathbb{F}_p[x]/(x^2 + x + 1)$ are isomorphic, and the roots $w, w + 1$ correspond to one place of degree 2 in the extension field $\mathbb{F}_p(w)$.*

**Example 4.** *We define the polynomials $f(x) = x^n + a_1x^{n-1} + \cdots + a_n$, with $a_i \in \mathbb{Q}$, and $f(x, t) = f(x) - t$. Then, if $f$ is a separable polynomial, then the Galois group of $f(x, t)$ over $\mathbb{Q}(t)$ is a regular extension with Galois group $S_n$.*

Observe that $\mathbb{Q}(x_1 \ldots, x_n)$ is the function field of an $(n - 1)-$dimensional projective space $\mathbb{P}^{n-1}(\mathbb{Q})$ over $\mathbb{Q}$. Suppose that $z_1, \ldots, z_n$ are the roots of $f$ in a splitting field of $f$ over $\mathbb{Q}$. Each coefficient $a_i$ of $x^i$ in $f$ is symmetric in $z_1, \ldots, z_n$, thus by the theorem on symmetric functions, we can write $a_i$ as a symmetric polynomial in $z_1, \ldots, z_n$ with rational coefficients. On the other side, for a permutation $\sigma \in S_n$, set $E_\sigma = x_1 z_{(\sigma(1))} + \cdots + x_n z_{(\sigma(n))}$ in $\mathbb{Q}(x_1 \ldots, x_n)$ and $f(x) = \prod_\sigma (x - E_\sigma)$, where $\sigma$ runs through all permutations in $S_n$.

**Theorem 2.7.** *(Hilbert) Let $G = S_n$ be acting on $\mathbb{Q}(x_1, \ldots, x_n)$. The field E of $S_n-$ invariants is $\mathbb{Q}(t_1, \ldots, t_n)$, where $t_i$ is the $i^{th}$ symmetric polynomial in $x_1, \ldots, x_n$ and $\mathbb{Q}(x_1, \ldots, x_n)$ has Galois group $S_n$ over E. It is the splitting field of the polynomial $f(x) = x^n - t_1x^{n-1} + \ldots + (-1)^n t_n$.*

Let $F$ be a finite field such that $(char F, n) = 1$. A non-zero polynomial in $\overline{\mathbb{F}}[x, y]$ defines a curve on the plane $\overline{\mathbb{F}}^2$. The elliptic curves are curves of the form $y^2 = f(x)$, where $f(x)$ is a polynomial of degree 3 with coefficients in $\overline{\mathbb{F}}$.

**Proposition 2.8.** *Let $n = rs$ be a factorisation of an integer positive number n into irreducible coprime factors and assume $r < s$, then there is a sequence of field extensions $\mathbb{F}_{q^r} \subset \mathbb{F}_{q^s} \subset \mathbb{F}_{q^n}$.*

*Proof.* Consider the map $T_n : F^n \mapsto F^n$

$$t_j = (-1)^j \sigma_j(x_1, \ldots, x_n),$$

where $\sigma_j$ is the $j$th elementary symmetric function in the variables $x_i$. Thus $\{t_j, j = 1, \cdots n\}$, are the coefficients of the equation:

$$f(z, t_1, \ldots, t_n) = z^n + (-1)t_1 z^{n-1} + \cdots + (-1)^n t_n = (z - x_1)(z - x_2)\cdots(z - x_n).$$

If we apply Theorem 2.7 to the $i^{th}$ elementary symmetric polynomial in the symbols $\alpha, \alpha^q, \alpha^{q^2}, \ldots, \alpha^{q^n}$, we get that the field of $S_n$ invariants of the polynomial

$f(z, t_1, \ldots, t_n)$ contains an extension $\mathbb{F}_{q^n}$ of $\mathbb{F}_q$. Moreover, for any divisor $r$ of $n$, one can consider the field of $S_r$ invariants, and apply Theorem 2.7 to the symbols $\alpha, \alpha^{q^{2s}}, \ldots, \alpha^{q^{rs}}$, where $n = rs$. Then we get an extension $\mathbb{F}_{q^s}$ of $\mathbb{F}_{q^r}$ and all its $\mathbb{F}_q$-subspaces are stable under $Gal(\mathbb{F}_{q^s}/\mathbb{F}_{q^r})$. □

**Example 5.** *Assume $n = q + 1$ and we study again the roots of the polynomial $x^q - 1$ in its splitting field. Let $\xi$ be a non-trivial n-root of unity, for any divisor r of n, one can consider the symbols $\xi^{q^r}, \ldots, \xi^q, \xi$. The field of $S_r$ invariants of the polynomial $f(z, t_1, \ldots, t_r)$ is the set of solutions to the equation:*

$$x^{q^r} + \ldots + x^q + x = a \quad in \quad \mathbb{F}_{q^n}. \tag{2}$$

In $\mathbb{F}_{2^n}$, for any divisor $d$ of $n$, there are exactly $2^{d-1}$ solutions to Eq. (2) if $n/d$ is odd and $2^d$ solutions if $n/d$ is even.

Instead of considering $r, s$ divisors of $n$, we can consider a partition of $n$ into two parts. For example for an integer $0 \leq k \leq n$, we consider the partition $(k, n - k)$ of $n$. Fix two elements $g_1, g_2 \in GL(n, q)$ of rank $k$ and rank $n - k$. These points correspond to linear transformations $T_{g_i} : \mathbb{F}_{q^n} \to \mathbb{F}_{q^n}, i \in 1, 2$. It is well known that the corresponding points $\mathbb{F}_{q^k, q^n} \subset \mathbb{F}_{q^n}$ and $\mathbb{F}_{q^{n-k}, q^n} \subset \mathbb{F}_{q^n}$ in the Grassmannians $\mathcal{G}_{k,n}(\mathbb{F}_q)$ of $k-$dimensional subspaces and the Grassmannian $\mathcal{G}_{n-k}(\mathbb{F}_q)$ of $n - k$ dimensional subspaces respectively are dual subspaces in the underlying vector space $(\mathbb{F}_q)^n$ for the Euclidean inner product. Note that the Hamming weight is preserved under invertible linear transformation. This case is of great interest for applications in coding theory, since the corresponding codes with generator matrices $G_1$ and $G_2$ respectively are dual codes. Namely, given a linear $[n, k]$-code, a parity check matrix for $C$ is an $(n - k) \times n$ matrix $H$ of rank $n - k$ such that $C = \{x \in (\mathbb{F}_q)^n : Hc^T = 0\}$. Then the dual code $C^\perp$ is the linear $[n, n - k]$ code generated by the parity check matrix of $C$. There is a right action of the general linear group $GL(n, \mathbb{F}_q)$ on $\mathcal{G}_{k,n}(\mathbb{F}_q)$:

$$\mathcal{G}_{k,n}(\mathbb{F}_q) \times GL(n, \mathbb{F}_q) \to \mathcal{G}_{k,n}(\mathbb{F}_q) \tag{3}$$

$$(\mathcal{U}, A) \to \mathcal{U}A. \tag{4}$$

One can study the orbits of $\mathcal{G}_{k,n}(\mathbb{F}_q)$ by the action of any subgroup in the general linear group $GL(n, \mathbb{F}_q)$. For example we can study the orbit of any triangle group: the Klein group $\mathbb{Z}_2 \times \mathbb{Z}_2$, the dihedral group, the alternated groups $A_4$ and $A_5$ or the symmetric group $S_n$. Take as $T$ the standard shift operator on $\mathbb{F}_q^n$, a linear code $C$ is said to be quasi-cyclic of index $l$ or $l-$quasi-cyclic if and only if is invariant under $T^l$. If $l = 1$, it is just a cyclic code. The quantity $m := n/l$ is called the co-index of $C$. Namely, if we view a codeword $(c_0, c_1, \ldots, c_{n-1})$ of $C$ as a polynomial $c_0 + c_1 x + \ldots + c_{n-1} x^{n-1} \in \mathbb{F}_q[x]$, then $T(c(x)) = x \cdot c(x) \mod (x^n - 1)$.

**Example 6.** *We study the action of a rotation element on the Grassmannian $\mathcal{G}_{2,4}(\mathbb{F}_q)$ of lines in a 3-dimensional projective space $PG(3, q)$. We apply to any line g a rotation $\tau$ of angle $\alpha = \frac{2\pi}{n}$, represented by the array of vectors $< (1, 0, 0), (0, cos(\alpha), sin(\alpha)), (0, -cos(\alpha), sin(\alpha)) >$. It is easy to see that the orbit code by the composed action $\tau^m$ with m divisor of n is a quasi-cyclic code of index $\frac{m}{n}$.*

In general, we study generalised Grassmannians or more commonly known as flag varieties. Fix a partition $\lambda = (\lambda_1, \ldots, \lambda_r)$ of $n$ and let $\mathcal{F}_\lambda = \mathcal{F}_\lambda(\mathbb{F}_q)$ be the variety of partial flags of $\mathbb{F}_q-$vector spaces

$$\{0\} = E^r \subset E^{r-1} \subset \ldots \subset E^1 \subset E^0 = (\mathbb{F}_q)^n$$

such that $\dim(E^{i-1}/E^i) = \lambda_i$. The group $GL(n, \mathbb{F}_q)$ acts on $\mathcal{F}_\lambda$ in the natural way. Fix an element $X_0 \in \mathcal{F}_\lambda$ and denote by $\mathcal{P}_\lambda$ the stabilizer of $X_0$ in $G$ and by $\mathcal{U}_\lambda$ the subgroup of elements $g \in \mathcal{P}_\lambda$ which induces the identity on $E^i/E^{i+1}$ for all $i = 0, 1, \ldots, n-1$. Put $\mathcal{L}_\lambda = GL_{\lambda_r}(\mathbb{F}_q) \times \ldots, \times GL_{\lambda_1}(\mathbb{F}_q)$, then we have $\mathcal{P}_\lambda = \mathcal{L}_\lambda \times \mathcal{U}_\lambda$.

**Proposition 2.9.** *Let us consider the factorisation of n into irreducible pairwise coprime factors* $n = p_1^{e_1} p_2^{e_2} \cdots p_r^{e_r}$ *with* $e_1 < e_2 < \ldots < e_r$*, and* $\lambda = (e_1, \ldots, e_r)$ *be the partition of exponents. Then there is a flag variety* $\mathcal{F}_\lambda(\mathbb{F}_q)$ *of partial flags of* $\mathbb{F}_q$*—vector spaces:*

$$\{0\} = E^r \subset E^{r-1} \subset \cdots \subset E^1 \subset E^0 = (\mathbb{F}_q)^n,$$

such that $dim(E^{i-1}/E^i) = e^i$.

*Proof* Observe that the result follows trivially for the case in which $n$ is a prime number $e_1 = \cdots = e_r = 1$. If $n = rs$ factorizes into two irreducible prime factors, the result follow as we have seen above, there is a flag $\{0\} = E^r \subset E^s \subset E^0 = (\mathbb{F}_q)^n$ and then by induction in $r$ the result follows. □

Given a cyclic code over $\overline{F}$ of length $n$, its defining set is given by the exponents occurring in $g(x)$, where $g(x)$ is the generator polynomial of the ideal of the code in $F[x]/(x^n - 1)$.

Let $\alpha \in \overline{F}$ be an $n^{th}$ primitive root of unity. The $n^{th}$ cyclotomic polynomial $\Phi_n(x) = \prod_{1<j<n,(j,n)=1} (x - \alpha^j) \in \overline{F}[x]$ is the minimal polynomial of $\alpha$ over $F$. It is monic of degree of the Euler's totient function $\varphi(n)$. It has integer coefficients and it is irreducible over $\mathbb{Q}$. In $\mathbb{Q}[x]$, we have the factorization into irreducible polynomials:

$$x^n - 1 = \prod_{d|n} \Phi_d(x).$$

By Möebius inversion:

$$\Phi_n(x) = \prod_{d|n} (x^d - 1)^{\mu(n/d)}$$

In the case of binary codes where $q = 2$, Bezzateev and Shekhunova [8] have obtained that the number of irreducible normalized polynomials $I_{2^m}(l)$ of degree $l$ over $\mathbb{F}_{2^m}$ satisfy the following equation:

$$I_{2^m}(l) = \frac{1}{l} \sum_{d/l} \mu(d) 2^{m\frac{l}{d}}. \tag{5}$$

where $\mu(d)$ is the Möebius function:

$$\mu(d) = \begin{cases} 1 \ \mathit{if} \ d = 1; \\ (-1)^r \ \mathit{if} \ d \ \mathit{is\ a\ product\ of} \ r \ \mathit{different\ prime\ numbers}; \\ 0 \ \mathit{in\ all\ other\ cases}. \end{cases}$$

Let $g(x)$ equals the least common multiple $lcm\{\Phi_i(x) : \alpha^i \in S\}$, then $S$ is a defining set for $C$. We will describe the defining set by the exponents occurring in $S$ with $S = \{i_1, i_2, \ldots, i_l\}$, where $i_1 < \ldots < i_l$. A parity check matrix for the code $C(S)$ is given by:

$$M(S) = \begin{pmatrix} (\alpha^{i_1})^0 & (\alpha^{i_1})^1 & \cdots & (\alpha^{i_1})^{n-1} \\ (\alpha^{i_2})^0 & (\alpha^{i_2})^1 & \cdots & (\alpha^{i_2})^{n-1} \\ \vdots & & \vdots & \\ (\alpha^{i_r})^0 & (\alpha^{i_r})^1 & \cdots & (\alpha^{i_r})^{n-1}) \end{pmatrix}$$

The code $C \subset F^n$ is obtained as the subfield subcode of $\overline{C}$:

$$C = \{c \in F^n : M(R)c^T = 0\}.$$

Given a triple $(I, J, K)$ of subsets of $\{1, \ldots, n\}$ of the same cardinality $r$, we associate to them partitions $\lambda, \mu$ and $\nu$ as follows. Let $I = \{i_1 < \ldots < i_r\} \subset \{1, \ldots, n\}$, then the corresponding partition is defined as $\lambda = (i_r - r, \ldots, r_1 - 1)$, and respectively for $J, K$. We call the triple $(I, J, K)$ admissible for the Horn problem, if the corresponding triple of partitions $(\lambda, \mu, \gamma)$ occurs as eigenvalues of a triple of Hermitian $r$ by $r$ matrices, with the third one the sum of the first two.

We describe Horn's inductive procedure to produce set of triples $(I, J, K) \subset \{0, 1, \ldots, n\}$.

$$U_r^n = \left\{(I, J, K) | \sum_{i \in I} i + \sum_{j \in J} j = \sum_{k \in K} k + r(r+1)/2\right\},$$

$$T_r^n = \{(I, J, K) \in U_r^n | \ for \ all \ p < r \ and \ all \ (F, G, H) \in T_p^r,$$

$$\sum_{f \in F} i_f + \sum_{g \in G} j_g \leq \sum_{h \in H} k_h + p(p+1)/2\}.$$

**Example 7.** *Let us consider the triple of subsets*

$$(I, J, K) = (\{1, 3, 5\}, \{1, 3, 5\}, \{2, 4, 6\})$$

and the corresponding triple of partitions $(\lambda, \mu, \nu) = ((2, 1, 0), (2, 1, 0)(3, 2, 1))$ arises from the triple of diagonal 3 by 3 matrices with diagonal entries $(2, 1, 0), (1, 0, 2)$ and $(3, 1, 2)$.

**Lemma 2.10.** *For any triple $(I, J, K)$ admissible for the Horn problem, the polynomials defined by $f(x) = \prod_{i \in I}(x - \alpha^i), g(x) = \prod_{j \in J}(x - \alpha^j)$, and $h(x) = \prod_{k \in K}(x - \alpha^k)$ generate a cyclic code of length $n = i_r + j_r + k_r mod p$ and $k = r$, where $r = \ |I| + |J| + |K|$ and $p$ is the characteristic of the field $F$.*

*Proof.* The cyclic code generated by $f(x)$ coincides with the cyclic code generated by $lcm\{m_i(x) : \alpha^i, i \in I\}$ and respectively for $g(x)$ and $h(x)$ the polynomials $lcm\{m_j(x) : \alpha^j, j \in J\}$ and $lcm\{m_k(x) : \alpha^k, k \in K\}$. It is the cyclic code generated by the minimal polynomial of $\alpha^{i_r + j_r + k_r}$. □

**Remark 2.11.** *We see that Horn's algorithm is relevant since some classical code constructions can be seen as ideals in a finite dimensional commutative semi simple algebra over a finite field $\mathbb{F}_q$ with $q = p^r$ elements and $p$ a prime number as in example (3).*

**Lemma 2.12.** *The family of cyclic codes obtained by considering the roots of the polynomial $x^n - 1$ over its splitting fields are indeed AG codes arising from genus 0 curves, and by Riemann-Roch theorem, their parameters satisfy the bound $d \geq n + 1 - k$, where $d$ is the minimum distance.*

*Proof* Let $\overline{\mathbb{F}}_q[x]$ be the extension field containing the $n^{th}$ roots of unity $1, \alpha, \ldots, \alpha^{n-1}$, where $\alpha^{n-1} + \alpha^{n-2} + \ldots + \alpha + 1 = 0$. Moreover the set $\{1, \alpha, \ldots, \alpha^{n-1}\}$

constitutes a basis over the prime field $\mathbb{F}_p$, and the field extensions $\mathbb{F}_{p^n} \cong \mathbb{F}_p[x]/(x^{n-1} + \ldots + x + 1)$ are isomorphic. □

### 2.3 Generating functions of conjugacy classes in a group

The automorphism group of the projective line $\mathbb{P}(\mathbb{F}_q)$ is the projective linear group $PGL(2,q)$. Any finite subgroup $A \subset PGL(2,q)$ defines a $k-$uniform Cayley (sum) hypergraph $\Gamma^k(A)$ whose vertices are the generating $k-$tuples of $A$ and the edges are $k-$element sets $\{x_1, \ldots, x_k\} \in \binom{G}{k}$ represented by random variables $x_1, \ldots, x_k$. In particular, if $f(z)$ is the ordinary generating function that enumerates $A$, that is, number of conjugacy classes in $A$, then $\frac{1}{1-f(z)}$ is the ordinary generating function enumerating sequences of $k$ elements in $A$. If $G$ is an abelian group, then $x_1 + \cdots + x_k \in A$. In general, we will consider $k$-arcs in $\Gamma(A)$ which represent casual connections between the variables. Applications are known in statistics, for example the multinomial experiment consists of $n$ identical independent trials, and there are $k$ possible outcomes (classes, categories or cells) to each trial and the cell counts $n_1, n_2, \ldots, n_k$ are the random variables, the number of observations that fall into each of the $k-$categories.

**Definition 2.13.** *In $PG(n-1,q)$ a $(k;r)-$arc is a set of k points any r of which form a basis for $\mathbb{F}_q^n$, or in other words, $r-1$ of them but not r are collinear.*

Consider the normal rational curve over $\mathbb{F}_q$:

$$\mathcal{V}_1^n := \{\mathbb{F}_q(1, x, x^2, \ldots, x^n) |\ x \in \mathbb{F}_q \cup \{\infty\}\}$$

is a $(q+1)$-arc in the $n$-dimensional projective space $PG(n,q)$.

We see that if $q \le n$, the NRC is a basis of a $q$-dimensional projective subspace, that is, a $PG^q(n,q)$. So we can enumerate how many NRC's are there in a $PG(n,q)$. The answer is $\phi(q;n,q)$, the number of ways of choosing such a set of points in a particular $q$-space. If $q \ge n+2$, the NRC is an example of a $(q+1)$-arc. It contains $q+1$ points, and every set of $n+1$ points are linearly independent.

### 2.4 Conclusion

The problem of considering finite subgroups and conjugacy classes in $PGL(2,q)$ the automorphism group of the projective line can be generalised to that of finite subgroups in $PGL(n,q)$, the collineation group of the normal rational curve.

## Acknowledgements

This research has been partially supported by the COST Action IC1104 and the project ARES (Team for Advanced Research on Information Security and Privacy. Funded by Ministry of Economy and Competitivity).

## Author details

Cristina Martinez Ramirez* and Alberto Besana
Department of Maths, UAB, Barcelona

*Address all correspondence to: cristinamartine@gmail.com

## References

[1] Kötter R, Kschischang FR. Coding for errors and erasures in random network coding. IEEE Transactions on Information Theory. 2008;**54**(8)

[2] Garcia A, Stichtenoth H. On the asymptotic behaviour of some towers of function fields over finite fields. Journal of Number Theory. 1996;**61**(2):248-273

[3] Geil O, Martin S, Martínez-Peas U, Ruano D. Refined analysis of RGHWs of code pairs coming from Garcia-Stichtenoth's second tower. In: Proceedings of 21st Conference on Applications of Computer Algebra

[4] Bezzateev SV, Shekhunova NA. Subclass of cyclic Goppa codes. IEEE Transactions on Information Theory. Nov. 2013;**59**(11)

[5] Niederreiter H, Xing C, Lam KY. A new construction of algebraic-geometry codes. Applicable Algebra in Engineering, Communication and Computing. 1999;**9**:373-381

[6] Hachenberger D, Niederreiter H, Xing C. Function field codes. Applicable Algebra in Engineering, Communication and Computing. 2008;**19**:201-211

[7] Couvreur A, Márquez I, Pellikaan R. A polynomial time attack against algebraic geometry code based public key cryptosystems. arXiv:1401.6025

[8] Bezzateev SV, Shekhunova N. Class of generalized Goppa codes perfect in weighted hamming metric. Des. Codes Criptogr. 2013;**66**:391-399

Chapter 12

# Moments of the Discounted Aggregate Claims with Delay Inter-Occurrence Distribution and Dependence Introduced by a FGM Copula

*Franck Adékambi*


## Abstract

In this chapter, with renewal argument, we derive higher simple moments of the Discounted Compound Delay Renewal Risk Process (DCDRRP) when introducing dependence between the inter-occurrence time and the subsequent claim size. To illustrate our results, we assume that the inter-occurrence time is following a delay-Poisson process and the claim amounts is following a mixture of Exponential distribution, we then provide numerical results for the first two moments. The dependence structure between the inter-occurrence time and the subsequent claim size is defined by a Farlie-Gumbel-Morgenstern copula. Assuming that the claim distribution has finite moments, we obtain a general formula for all the moments of the DCDRRP process.


**Keywords:** compound delay-Poisson process, discounted aggregate claims, moments, FGM copula, constant interest rate

## 1. Introduction

The classical Poisson model is attractive in the sense that the memoryless property of the exponential distribution makes calculations easy. Then the research was extended to ordinary Sparre-Andersen renewal risk models where the inter-claim times have other distributions than the exponential distribution. Dickson and Hipp [1, 2] considered the Erlang-2 distribution, Li and Garrido [3] the Erlang-n distribution, Gerber and Shiu [4] the generalized Erlang-n distribution (a sum of n independent exponential distributions with different scale parameters) and Li and Garrido [5] looked into the Coxian class distributions. One difficulty with these models is that we have to assume that a claim occurs at time 0, which is not the case in usual setting.

Albrecher and Teugels [6] considered modeling dependence with the use of an arbitrary copula. In a similar dependence model to Albrecher and Teugels as well, Asimit and Badescu [7] considered a constant force of interest and heavy tailed claim amounts.

Barges et al. [8] followed the idea of Albrecher and Teugels [6] and supposed that the dependence is introduced by a copula, the Farlie-Gumbel-Morgenstern (GGM) copula, between a claim inter-arrival time and its subsequent claim amount.

Adékambi and Dziwa [9] and Adékambi [10] provide a direct point of extension but assuming that the claim counting process to follow an unknown general distribution in a framework of dependence with random force of interest to calculate the first two moments of the present value of aggregate random cash flows or random dividends.

The discounted aggregate sum has also been applied in many other fields. For example, it can be used in health cost modeling, see Govorun and Latouche [11], Adékambi [12], or in reliability, in civil engineering, see Van Noortwijk and Frangopol [13].

The delayed or modified renewal risk model solves this problem by assuming that the time until the first claim has a different distribution than the rest of the inter-claim times. Not much research has been done for this model at this stage. Among the first works was Willmot [14] where a mixture of a "generalized equilibrium" distribution and an exponential distribution is considered for the distribution of the time until the first claim. Special cases of the model include the stationary renewal risk model and the delayed renewal risk model with the time until the first claim exponentially distributed. Our focus is to extend the work of Bargès et al. [8], Adékambi and Dziwa [9] and Adékambi [10] by allowing the counting process to follow a delay renewal risk process and thus derive a recursive formula of the moments of this subsequent Discounted Compound Delay Poisson Risk Value (DCDPRV).

For example, young performer companies typically have a high growth rate at the beginning, but as they mature their growth rate may decrease with the increasing scarcity of investment opportunities. That makes dividends dependent on the economic climate at the dividend occurrence time. Obviously the distribution of inter-dividends time in times of economic expansion and in times of economic crisis cannot be identically distributed. So it would be appropriate to use a delayed renewal model to model the distribution of the inter-dividend time. A delayed renewal process is just like an ordinary renewal process, except that the first arrival time is allowed to have a different distribution than the other inter-dividends times.

The chapter is organized as follows: In the second section, we present the model of the continuous time discounted compound delay-Poisson risk process that we use and give some notation. In Section 3, we present a general formula for all the moments of the DCDPRV process. A numerical example of the first two moments will then follow in Section 4.

## 2. The model

We use the same model as the one in Bargès et al. [8], where the instantaneous interest rate $\delta$ is constant.

Define our risk model as follows:

i. The number of claims $\{N(t), t \geq 0\}$ and $\{N_d(t), t \geq 0\}$ form, respectively, an ordinary and a delayed renewal process and, for $k \in \mathbb{N} = \{1, 2, 3, ...\}$:

- the positive claim occurrence times are given by $T_k$,
- the positive claim inter-arrival times are given by $\tau_k = T_k - T_{k-1}, k \in \mathbb{N}$, and $T_0 = 0$.
- $(\tau_k)_{k \geq 2} \sim \tau_2$ are independent and identically distributed (i.i.d),

ii. The $k^{th}$ random claim is given by $X_k$, and

- $\{X_k, k \in \mathbb{N}\}$ are independent and identically distributed (i.i.d),
- $\{X_k, \tau_k, k \in \mathbb{N}\}$ are mutually independent; and the higher moments, $\mu_k = E[X_1^k]$ of $X_1$ exist.

iii. The discounted aggregate value at time $t = 0$ of the claims recorded over the period $[0, t]$ yields, respectively, for the ordinary and the delayed renewal case:

$$Z_0(t) = \sum_{k=1}^{N_o(t)} e^{-\delta T_k} X_k, \; Z_d(t) = \sum_{k=1}^{N_d(t)} e^{-\delta T_k} X_k, \tag{1}$$

where $Z_0(t) = Z_d(t) = 0$ if $N_0(t) = N_d(t) = 0$.

### 2.1 The dependence

We introduce a specific structure of dependence based on the Farlie-Gumbel-Morgenstern (FGM) copula. The advantage of using the FGM copula and its generalizations lies in its mathematical manageability. The joint cumulative distribution function (c.d.f.) of $(X_i, \tau_i)$, the $i$th claim and its occurrence time is

$$\begin{aligned} F_{X_i, \tau_i}(x, v) &= C(F_{X_i}(x), F_{\tau_i}(v)) \\ &= F_{X_i}(x) F_{\tau_i}(v) + \boldsymbol{\theta} F_{X_i}(x) F_{\tau_i}(v)\ (1\text{-}F_{X_i}(x))\ (1\text{-}F_{\tau_i}(v)), \end{aligned} \tag{2}$$

for $(x, v) \in \mathbb{R}^+ * \mathbb{R}^+$ and where $F_{X_i}(x)$ and $F_{\tau_i}(v)$ are the marginals of $X_i$ and $\tau_i$ respectively. Recall that the density of the FGM copula is

$$c_\theta^{FGM}\ (u, v) = 1 + \theta(1\text{-}2u)\ (1\text{-}2v), \tag{3}$$

for $(u, v) \in [0, 1] * [0, 1]$ so that the joint probability density function (p.d.f.) of $(X_i, \tau_i)$ is

$$\begin{aligned} f_{X_i, T_i}(x, v) &= c_\theta^{FGM}\ (F_{X_i}(x), \mathrm{F}_{T_i}(v)) f_{X_i}(x) f_{T_i}(v) \\ &= f_{X_i}(x) f_{T_i}(v) + \boldsymbol{\theta} f_{X_i}(x) f_{T_i}(v) (1\text{-}2\mathrm{F}_{X_i}(x))\ (1\text{-}2\mathrm{F}_{T_i}(v)) \end{aligned}, \tag{4}$$

where $f_{X_i}$ and $f_{\tau_i}$ are the p.d.f.'s of $X_i$ and $\tau_i$ respectively.

With these hypotheses, we present in Section 3 recursive formula of the higher moments of this present value risk process, for a constant instantaneous interest rate.

## 3. Recursive expression for higher moments

It is often easier to calculate the moments of the random variable $\{Z_d(t), t \geq 0\}$ than finding its distribution. If the probability generation function of $\{Z_d(t), t \geq 0\}$ or its moment generating function (mgf) exists, it is possible to obtain the corresponding distribution by inversion of its mgf. Since, there is relatively little research devoted to the study of the distribution of the discounted compound renewal sums. We could then think about another technique other than the one proposed by the above authors by studying the moments of $\{Z_d(t), t \geq 0\}$.

### 3.1 Delay renewal case

The mathematical expectation of total claims plays an important role in the determination of the pure premium, in addition to giving a measure of the central tendency of its distribution. The moments centered at the average of order 2, 3 and 4 are the other moments usually considered because they usually give a good indication of the pace of distribution, and these give us respectively a measure of the dispersion of the distribution around its mean, a measure of the asymmetry and flattening of the distribution considered.

Moments, whether simple, joined or conditional, may eventually be used to construct approximations of the distribution of the DCDPRV.

**Theorem 3.1**

The Laplace transform of the $m$th moment of $\{Z_d(t), t \geq 0\}$ is given by:

$$\tilde{\pi}^m_{Z_d}(r) = \left(1 + \frac{\lambda_2}{r + m\delta} + \frac{\lambda_1 - \lambda_2}{r + m\delta + \lambda_1}\right)\tilde{u}_m(r)$$

$$= \lambda_1\left(1 + \frac{\lambda_2}{r + m\delta} + \frac{\lambda_1 - \lambda_2}{r + m\delta + \lambda_1}\right)$$

$$\times \sum_{j=0}^{m-1}\binom{m}{j}\left\{\frac{\left(\mu_{m-j} - \theta\left(\mu'_{m-j} - \mu_{m-j}\right)\right)}{\lambda_1 + m\delta + r} + \frac{2\theta\left(\mu'_{m-j} - \mu_{m-j}\right)}{2\lambda_1 + m\delta + r}\right\}\tilde{\pi}^j_{Z_o}(r) \tag{5}$$

where

$$\tilde{\pi}^m_{Z_d}(r) = \tilde{u}_m(r) + \frac{\lambda_2}{m\delta}\tilde{u}_m(r) \times L_{\tau_1}(m, \delta, r) + \frac{\lambda_1 - \lambda_2}{m\delta + \lambda_1}\tilde{u}_m \times L_{\tau_1}(m\delta + \lambda_1, r). \tag{6}$$

*Proof*

Conditioning on the arrival of the first claim leads to

$$\pi^m_{Z_d}(t) = E[Z^m(t)]$$

$$= E\left[E\left[\left(e^{-\delta s}X_1 + e^{-\delta s}Z_o(t-s)\right)^m | \tau_1 = s\right]\right]$$

$$= \sum_{j=0}^{m-1}\binom{m}{j}\int_0^t f_{\tau_1}(s)e^{-m\delta s}E\left[X^{m-j}|\tau_1 = s\right]\pi^j_{Z_o}(t-s)ds \tag{7}$$

$$+ \int_0^t f_{\tau_1}(s)e^{-m\delta s}\pi^m_{Z_o}(t-s)ds.$$

We have

$$
\begin{aligned}
E\left[X^{m-j}|\tau_1 = s\right] &= \int_0^\infty x^{m-j} f_{X|\tau_1=s}(x)dx \\
&= \int_0^\infty x^{m-j}\{1+\theta(1-2F_X(x))(1-2F_{\tau_1}(s))\} f_X(x)dx \\
&= E\left[X^{m-j}\right] + \theta\int_0^\infty x^{m-j}(1-2F_X(x))(1-2F_{\tau_1}(s)) f_X(x)dx \\
&= E\left[X^{m-j}\right] + \theta\int_0^\infty x^{m-j}(2-2F_X(x))(1-2F_{\tau_1}(s)) f_X(x)dx \\
&\quad - \theta\int_0^\infty x^{m-j}(1-2F_{\tau_1}(s)) f_X(x)dx \\
&= E\left[X^{m-j}\right](1-\theta(1-2F_{\tau_1}(s))) \\
&\quad + \theta(1-2F_{\tau_1}(s))\int_0^\infty (m-j)x^{m-j}(1-F_{\tau_1}(s))dx.
\end{aligned}
\tag{8}
$$

We let,

$$
\begin{aligned}
\mu'_{m-j} = E\left[X'^{m-j}\right] &= \int_0^\infty (m-j)x^{m-j-1}(1-F_X(x))^2 dx \\
&< \int_0^\infty (m-j)x^{m-j-1}(1-F_X(x))dx = E\left[X^{m-j}\right] < \infty
\end{aligned}
\tag{9}
$$

such that the above equation becomes

$$
E\left[X^{m-j}|\tau_1 = s\right] = \mu_{m-j} + \theta(1-2F_{\tau_1}(s))\left(\mu'_{m-j} - \mu_{m-j}\right). \tag{10}
$$

$$
\begin{aligned}
\pi^m_{Z_d}(t) &= E[Z^m(t)] \\
&= E\left[E\left[\left(e^{-\delta s}X_1 + e^{-\delta s}Z_o(t-s)\right)^m|\tau_1 = s\right]\right] \\
&= \sum_{j=0}^{m-1}\binom{m}{j}\int_0^t f_{\tau_1}(s)e^{-m\delta s}\left\{\mu_{m-j} + \theta(1-2F_{\tau_1}(s))\left(\mu'_{m-j} - \mu_{m-j}\right)\right\}\pi^j_{Z_o}(t-s)ds \\
&\quad + \int_0^t f_{\tau_1}(s)e^{-m\delta s}\pi^m_{Z_o}(t-s)ds.
\end{aligned}
$$

Let us $\int_0^t f_{\tau_1}(s)e^{-m\delta s}ds = H_\delta(t)$, $\int_0^t f_{\tau_2}(s)e^{-m\delta s}ds = I_\delta(t)$ then

$$\begin{aligned}
\pi_{Z_d}^m(t) &= \sum_{j=0}^{m-1}\binom{m}{j}\int_0^t f_{\tau_1}(s)e^{-m\delta s}\left\{\mu_{m-j} + \theta(1-2F_{\tau_1}(s))\left(\mu'_{m-j} - \mu_{m-j}\right)\right\}\pi_{Z_o}^j(t-s)ds \\
&\quad + H_{m\delta} * \pi_{Z_o(.)}^m \\
&= u_m + H_{m\delta} * \left\{u_m + I_{m\delta} * \pi_{Z_o(.)}^m\right\} \\
&= u_m + H_{m\delta} * u_m + H_{m\delta} * I_{m\delta} * \pi_{Z_o(.)}^m \\
&= u_m + H_{m\delta} * u_m + H_{m\delta} * I_{m\delta} * \left\{u_m + I_{m\delta} * \pi_{Z_o(.)}^m\right\} \\
&= u_m + H_{m\delta} * u_m + u_m * \sum_{k=1}^{\infty} H_{m\delta} * I_m^{\delta*(k)}(t) = u_m + u_m * \sum_{k=0}^{\infty} H_{m\delta} * I_m^{\delta*(k)}(t) \\
&= u_m + \int_0^t u_m(t-s)e^{-m\delta s}dm_d(s),
\end{aligned} \tag{11}$$

where $u_m(t) = \sum_{j=0}^{m-1}\binom{m}{j}\int_0^t f_{\tau_1}(s)e^{-m\delta s}\left\{\mu_j + \theta(1-2F_{\tau_1}(s))\left(\mu'_j - \mu_j\right)\right\}\pi_{Z_o}^{m-j}$ $(t-s)ds$.

We consider the case where the canonical random variable $\tau_2$ has an Exponential distribution with parameter $\lambda_2 > 0$ and $\tau_1$ has an Exponential distribution with parameter $\lambda_1 > 0$.

That is, we have:

$$f_{\tau_1}(t) = \lambda_1 e^{-\lambda_1 t}, f_{\tau_2}(t) = \lambda_2 e^{-\lambda_2 t}, L_{\tau_1}(\lambda_1, s) = \int_0^\infty e^{-sv}f_{\tau_1}(v)dv = \left(\frac{\lambda_1}{\lambda_1 + s}\right), L_{\tau_2}(\lambda_2, s) = \left(\frac{\lambda_2}{\lambda_2 + s}\right).$$

$$m_d(t) = \lambda_2 t + \frac{\lambda_1 - \lambda_2}{\lambda_1}\left(1 - e^{\lambda_1 t}\right) \tag{12}$$

The $m$th moment of $Z_d(t)$ is then given by,

$$\begin{aligned}
\pi_{Z_d}^m(t) &= u_m + \int_0^t u_m(t-s)e^{-m\delta s}dm_d(s) \\
&= u_m + \lambda_2\int_0^t u_m(t-s)e^{-m\delta s}d(s) + (\lambda_1 - \lambda_2)\int_0^t u_m(t-s)e^{-(m\delta+\lambda_1)s}d(s) \\
&= u_m + \frac{\lambda_2}{m\delta}\int_0^t u_m(t-s)m\delta e^{-m\delta s}d(s) + \frac{\lambda_1 - \lambda_2}{m\delta + \lambda_1}\int_0^t u_m(t-s)(m\delta + \lambda_1)e^{-(m\delta+\lambda_1)s}d(s)
\end{aligned} \tag{13}$$

Taking the Laplace transform of the above equation, we get:

$$\tilde{\pi}_{Z_d}^m(r) = \tilde{u}_m(r) + \frac{\lambda_2}{m\delta}\tilde{u}_m(r) \times L_{\tau_1}(m\delta, r) + \frac{\lambda_1 - \lambda_2}{m\delta + \lambda_1}\tilde{u}_m \times L_{\tau_1}(m\delta + \lambda_1, r) \tag{14}$$

But,

$$
\begin{aligned}
u_m(t) &= \sum_{j=0}^{m-1}\binom{m}{j}\int_0^t f_{\tau_1}(s)e^{-m\delta s}\left\{\mu_j + \theta(1-2F_{\tau_1}(s))\left(\mu_j' - \mu_j\right)\right\}\pi_{Z_o}^{m-j}(t-s)ds \\
&= \sum_{j=0}^{m-1}\binom{m}{j}\int_0^t \lambda_1 e^{-\lambda_1 s}e^{-m\delta s}\left\{\mu_j + \theta(2e^{-\lambda_1 s}-1)\left(\mu_j' - \mu_j\right)\right\}\pi_{Z_o}^{m-j}(t-s)ds \\
&= \frac{\lambda_1\left(\mu_j - \theta\left(\mu_j' - \mu_j\right)\right)}{\lambda_1 + m\delta}\sum_{j=0}^{m-1}\binom{m}{j}\int_0^t (\lambda_1 + m\delta)e^{-(\lambda_1+m\delta)s}\pi_{Z_o}^{m-j}(t-s)ds \\
&\quad + 2\theta\frac{\lambda_1\left(\mu_j' - \mu_j\right)}{2\lambda_1 + m\delta}\sum_{j=0}^{m-1}\binom{m}{j}\int_0^t (2\lambda_1 + m\delta)e^{-(2\lambda_1+m\delta)s}\pi_{Z_o}^{m-j}(t-s)ds
\end{aligned}
\tag{15}
$$

Then the Laplace transform of $u_m(t)$, at $r$, will give:

$$
\tilde{u}_m(r) = \lambda_1\sum_{j=0}^{m-1}\binom{m}{j}\left\{\frac{\left(\mu_j - \theta\left(\mu_j' - \mu_j\right)\right)}{\lambda_1 + m\delta + r} + \frac{2\theta\left(\mu_j' - \mu_j\right)}{2\lambda_1 + m\delta + r}\right\}\tilde{\pi}_{Z_o}^{m-j}(r) \tag{16}
$$

Substituting Eq. (14) into Eq. (13), we have:

$$
\tilde{\pi}_{Z_d}^{m}(r) = \left(1 + \frac{\lambda_2}{r + m\delta} + \frac{\lambda_1 - \lambda_2}{r + m\delta + \lambda_1}\right)\tilde{u}_m(r)
$$

$$
= \lambda_1\left(1 + \frac{\lambda_2}{r + m\delta} + \frac{\lambda_1 - \lambda_2}{r + m\delta + \lambda_1}\right)\sum_{j=0}^{m-1}\binom{m}{j}\left\{\frac{\left(\mu_j - \theta\left(\mu_j' - \mu_j\right)\right)}{\lambda_1 + m\delta + r} + \frac{2\theta\left(\mu_j' - \mu_j\right)}{2\lambda_1 + m\delta + r}\right\}\tilde{\pi}_{Z_o}^{m-j}(r) \tag{17}
$$

Solving the above equation for the ordinary case, where $(\tau_2)_{k\geq 2} \sim \tau_2$, we have:

$$
\begin{aligned}
\tilde{\pi}_{Z_o}^{m}(r) &= \frac{\lambda_2\mu_m}{r(r + \delta m + \lambda_2)} + \frac{\lambda_2}{(r + \delta m + \lambda_2)}\sum_{k=1}^{m-1}C_m^k\,\mu_k\tilde{\pi}_{Z_o}^{(m-k)}(r) \\
&\quad + \theta\left(\mu_m' - \mu_m\right)\frac{\lambda_2(r + \delta m)}{r(r + \delta m + \lambda_2)(r + \delta m + 2\lambda_2)} \\
&\quad + \theta\frac{\lambda_2(r + \delta m)}{(r + \delta m + \lambda_2)(r + \delta m + 2\lambda_2)}\sum_{k=1}^{m-1}C_m^k\left(\mu_k' - \mu_k\right)\tilde{\pi}_{Z_o}^{(m-k)}(r) \\
&\quad + \frac{\lambda_2}{(r + \delta m + \lambda_2)}\tilde{\pi}_{Z_0}^{m}(r)
\end{aligned}
\tag{18}
$$

Rearranging the above equation, we will get

$$\begin{aligned}\tilde{\pi}^m_{Z_o}(r) &= \frac{\lambda_2\mu_m}{r(r+\delta m)} + \frac{\lambda_2}{(r+\delta m)}\sum_{k=1}^{m-1} C^k_m\,\mu_k\tilde{\pi}^{(m-k)}_{Z_o}(r)\\ &+\theta\frac{\lambda_2(\mu'_m-\mu_m)}{r(r+\delta m+2\lambda_2)}+\theta\frac{\lambda_2}{(r+\delta m+2\lambda_2)}\sum_{k=1}^{m-1}C^k_m(\mu'_k-\mu_k)\tilde{\pi}^{(m-k)}_{Z_o}(r)\end{aligned} \tag{19}$$

**Corollary 3.1**

The first moment of $\{Z_d(t), t \geq 0\}$ is given by:

$$\begin{aligned}\pi_{Z_d}(t) &= \left(\theta\lambda_1(\mu'_1-\mu_1)\frac{\lambda_2+\delta}{(\delta+\lambda_1)(\delta+2\lambda_2)}+\frac{\lambda_1(\lambda_2+\delta)}{\delta(\delta+\lambda_1)}\mu_1\right)\\ &+\left(\theta\lambda_1(\mu'_1-\mu_1)\left(\frac{\lambda_1-\lambda_2}{\lambda_1-2\lambda_2}\right)-(\lambda_1-\lambda_2)\mu_1\right)\frac{1}{\delta+\lambda_1}e^{-(\delta+\lambda_1)t}\\ &-\theta\lambda_1\frac{1}{\delta+2\lambda_2}\frac{\lambda_2}{\lambda_1-2\lambda_2}(\mu'_1-\mu_1)e^{-(\delta+2\lambda_2)t}\\ &-2\theta\lambda_1\frac{1}{\delta+2\lambda_1}(\mu'_1-\mu_1)e^{-(\delta+2\lambda_1)t}-\frac{\lambda_2}{\delta}\mu_1e^{-\delta t}\end{aligned} \tag{20}$$

*Proof*:

From Theorem 3.1, we have:

$$\begin{aligned}\tilde{\pi}_{Z_d}(r) &= \frac{\lambda_1\mu_1}{r(r+\delta+\lambda_1)}+\frac{\lambda_1}{(r+\delta+\lambda_1)}\tilde{\pi}_{Z_0}(r)\\ &+\theta(\mu'_1-\mu_1)\frac{\lambda_1(r+\delta)}{r(r+\delta+\lambda_1)(r+\delta+2\lambda_1)}\end{aligned} \tag{21}$$

From Bargès et al. [8], we have

$$\tilde{\pi}_{Z_o}(r) = \frac{\lambda_2\mu_1}{r(r+\delta)}+\theta\frac{\lambda_2(\mu'_1-\mu_1)}{r(r+\delta+2\lambda_2)} \tag{22}$$

Substituting Eq. (22) into Eq. (21), yields

$$\begin{aligned}\tilde{\pi}_{Z_d}(r) &= \frac{\lambda_1\mu_1}{r(r+\delta+\lambda_1)}+\frac{\lambda_1}{(r+\delta+\lambda_1)}\left\{\frac{\lambda_2\mu_1}{r(r+\delta)}+\theta\frac{\lambda_2(\mu'_1-\mu_1)}{r(r+\delta+2\lambda_2)}\right\}\\ &+\theta\lambda_1(\mu'_1-\mu_1)\frac{(r+\delta)}{r(r+\delta+\lambda_1)(r+\delta+2\lambda_1)}\\ &=\left\{\frac{\lambda_1\lambda_2}{r(r+\delta)(r+\delta+\lambda_1)}+\frac{\lambda_1}{r(r+\delta+\lambda_1)}\right\}\mu_1\\ &+\theta\lambda_1(\mu'_1-\mu_1)\left\{\frac{\lambda_2}{r(r+\delta+\lambda_1)(r+\delta+2\lambda_2)}+\frac{r+\delta}{r(r+\delta+\lambda_1)(r+\delta+2\lambda_1)}\right\}\end{aligned} \tag{23}$$

with

$$\frac{\lambda_1}{r(r+\delta+\lambda_1)}=\frac{\lambda_1}{(\delta+\lambda_1)}\cdot\frac{1}{r}-\frac{\lambda_1}{(\delta+\lambda_1)}\cdot\frac{1}{(r+\delta+\lambda_1)} \tag{24}$$

$$\frac{\lambda_1\lambda_2}{r(r+\delta)(r+\delta+\lambda_1)}=\frac{\lambda_1\lambda_2}{\delta(\delta+\lambda_1)}\cdot\frac{1}{r}+\frac{\lambda_2}{(\delta+\lambda_1)}\cdot\frac{1}{(r+\delta+\lambda_1)}-\frac{\lambda_2}{\delta}\cdot\frac{1}{(r+\delta)} \tag{25}$$

$$\begin{aligned}\frac{\lambda_2}{r(r+\delta+\lambda_1)(r+\delta+2\lambda_2)}&=\frac{\lambda_2}{(\delta+\lambda_1)(\delta+2\lambda_2)}\frac{1}{r}\\&+\frac{\lambda_2}{\lambda_1-2\lambda_2}\left[\frac{1}{\delta+\lambda_1}\frac{1}{r+\delta+\lambda_1}-\frac{1}{\delta+2\lambda_2}\frac{1}{r+\delta+2\lambda_2}\right]\end{aligned} \tag{26}$$

$$\begin{aligned}\frac{r+\delta}{r(r+\delta+\lambda_1)(r+\delta+2\lambda_1)}&=\frac{\delta}{(\delta+\lambda_1)(\delta+2\lambda_1)}\frac{1}{r}\\&+\frac{1}{\delta+\lambda_1}\frac{1}{r+\delta+\lambda_1}-\frac{2}{\delta+2\lambda_1}\frac{1}{r+\delta+2\lambda_1}\end{aligned} \tag{27}$$

Substituting Eqs. (24), (25), (26) and (27) into Eq. (23), yields:

$$\begin{aligned}\tilde{\pi}_{Z_d}(r)&=\mu_1\left\{\frac{\lambda_1}{(\delta+\lambda_1)}\cdot\frac{1}{r}-\frac{\lambda_1}{(\delta+\lambda_1)}\cdot\frac{1}{(r+\delta+\lambda_1)}\right\}\\&+\theta(\mu_1'-\mu_1)\left\{\begin{array}{c}\frac{\lambda_2}{(\delta+\lambda_1)(\delta+2\lambda_2)}\frac{1}{r}\\+\frac{\lambda_2}{\lambda_1-2\lambda_2}\left[\frac{1}{\delta+\lambda_1}\frac{1}{r+\delta+\lambda_1}-\frac{1}{\delta+2\lambda_2}\frac{1}{r+\delta+2\lambda_2}\right]\end{array}\right\}\\&+\theta\lambda_1(\mu_1'-\mu_1)\left\{\begin{array}{c}\frac{\delta}{(\delta+\lambda_1)(\delta+2\lambda_1)}\frac{1}{r}\\+\frac{1}{\delta+\lambda_1}\frac{1}{r+\delta+\lambda_1}-\frac{2}{\delta+2\lambda_1}\frac{1}{r+\delta+2\lambda_1}\end{array}\right\}\\&+\mu_1\left\{\frac{\lambda_1\lambda_2}{\delta(\delta+\lambda_1)}\cdot\frac{1}{r}+\frac{\lambda_2}{(\delta+\lambda_1)}\cdot\frac{1}{(r+\delta+\lambda_1)}-\frac{\lambda_2}{\delta}\cdot\frac{1}{(r+\delta)}\right\}\end{aligned} \tag{28}$$

Rearranging the above equation, will give

$$\begin{aligned}\pi_{Z_d}(t)&=\left(\theta\lambda_1(\mu_1'-\mu_1)\frac{\lambda_2+\delta}{(\delta+\lambda_1)(\delta+2\lambda_2)}+\frac{\lambda_1(\lambda_2+\delta)}{\delta(\delta+\lambda_1)}\mu_1\right)\frac{1}{r}\\&+\left(\theta\lambda_1(\mu_1'-\mu_1)\left(\frac{\lambda_1-\lambda_2}{\lambda_1-2\lambda_2}\right)-(\lambda_1-\lambda_2)\mu_1\right)\frac{1}{\delta+\lambda_1}\frac{1}{r+\delta+\lambda_1}\\&-\theta\lambda_1\frac{1}{\delta+2\lambda_2}\frac{\lambda_2}{\lambda_1-2\lambda_2}(\mu_1'-\mu_1)\frac{1}{r+\delta+2\lambda_2}\\&-2\theta\lambda_1\frac{1}{\delta+2\lambda_1}(\mu_1'-\mu_1)\frac{1}{r+\delta+2\lambda_1}-\frac{\lambda_2}{\delta}\mu_1\frac{1}{r+\delta}\end{aligned} \tag{29}$$

**Remark 1**

If $\lambda_1 = \lambda_2$ then Eq. (29) becomes

$$\begin{aligned}\tilde{\pi}_{Z_d}(r) &= \left(\theta\lambda(\mu_1' - \mu_1)\left(\frac{1}{\delta+2\lambda}\right) + \frac{\lambda}{\delta}\mu_1\right)\frac{1}{r} \\ &\quad - \frac{1}{\delta+2\lambda}\theta\lambda(\mu_1' - \mu_1)\frac{1}{r+\delta+2\lambda} - \frac{\lambda}{\delta}\mu_1\frac{1}{(r+\delta)} \\ &= \frac{\lambda}{\delta}\mu_1\left[\frac{1}{r} - \frac{1}{r+\delta}\right] + \theta\lambda(\mu_1' - \mu_1)\left(\frac{1}{\delta+2\lambda}\right)\left[\frac{1}{r} - \frac{1}{r+\delta+2\lambda}\right] \\ &= \frac{\lambda\mu_1}{r(r+\delta)} + \frac{\lambda\theta(\mu_1' - \mu_1)}{r(r+\delta+2\lambda)},\end{aligned} \tag{30}$$

which is exactly the result of Bargès et al. [8].

The inverse of the Laplace transform in Eq. (29) will give

$$\begin{aligned}\pi_{Z_d}(t) &= \left(\theta\lambda_1(\mu_1' - \mu_1)\frac{\lambda_2+\delta}{(\delta+\lambda_1)(\delta+2\lambda_2)} + \frac{\lambda_1\lambda_2+\lambda_1\delta}{\delta(\delta+\lambda_1)}\mu_1\right) \\ &\quad + \left(\theta\lambda_1(\mu_1' - \mu_1)\left(\frac{\lambda_2}{\lambda_1 - 2\lambda_2} + 1\right) + (\lambda_2 - \lambda_1)\mu_1\right)\frac{1}{\delta+\lambda_1}e^{-(\delta+\lambda_1)t} \\ &\quad - \theta\lambda_1\frac{1}{\delta+2\lambda_2}\frac{\lambda_2}{\lambda_1 - 2\lambda_2}(\mu_1' - \mu_1)e^{-(\delta+2\lambda_2)t} \\ &\quad - 2\theta\lambda_1\frac{1}{\delta+2\lambda_1}(\mu_1' - \mu_1)e^{-(\delta+2\lambda_1)t} - \frac{\lambda_2}{\delta}\mu_1 e^{-\delta t}\end{aligned} \tag{31}$$

**Remarks 2**

If $\theta = 0$ and $\lambda_1 \neq \lambda_2$ then

$$\begin{aligned}\pi_{Z_d}(t) &= \left[\frac{\lambda_1}{\delta}\left(\frac{\lambda_2+\delta}{\delta+\lambda_1}\right) - \frac{\lambda_2}{\delta}e^{-\delta t}\right]\mu_1 + \left(\frac{\lambda_2 - \lambda_1}{\delta+\lambda_1}\right)\mu_1 e^{-(\delta+\lambda_1)t} \\ &= \lambda_2\left(\frac{1-e^{-\delta t}}{\delta}\right)\mu_1 + (\lambda_1 - \lambda_2)\left(\frac{1-e^{-(\delta+\lambda_1)t}}{\delta+\lambda_1}\right)\mu_1 \\ &= \left\{\lambda_2\bar{a}_{t|\delta} + (\lambda_1 - \lambda_2)\bar{a}_{t|\delta+\lambda_1}\right\}\mu_1\end{aligned} \tag{32}$$

which is exactly the result of Léveillé et al. [15].

If $\lambda_1 = \lambda_2$ and $\theta \neq 0$ then

$$\begin{aligned}\pi_{Z_o}(t) &= \left(\theta\lambda(\mu_1' - \mu_1)\frac{1}{(\delta+2\lambda_2)} + \frac{\lambda}{\delta}\mu_1\right) \\ &\quad - \theta\lambda\frac{1}{\delta+2\lambda}(\mu_1' - \mu_1)e^{-(\delta+2\lambda)t} - \frac{\lambda}{\delta}\mu_1 e^{-\delta t} \\ &= \frac{\lambda}{\delta}(1-e^{-\delta t})\mu_1 + \theta\lambda\left(\frac{1-e^{-(\delta+2\lambda)t}}{\delta+2\lambda}\right)(\mu_1' - \mu_1)\end{aligned} \tag{33}$$

which is exactly the result of Bargès et al. [8].

If $\lambda_1 = \lambda_2$ and $\theta = 0$ then

$$\pi_{Z_o}(t) = \frac{\lambda}{\delta}(1-e^{-\delta t})\mu_1 = \lambda\bar{a}_{t|\delta}\mu_1, \tag{34}$$

which is exactly the result of Léveillé et al. [15].

**Corollary 3.2**
The second moment of $\{Z_d(t), t \geq 0\}$ is given by the following development:
The result in Theorem 3.1 when $n = 2$ gives:

$$\tilde{\pi}^2_{Z_d}(r) = \frac{2\lambda_1(\lambda_1 - \lambda_2)}{r(r+2\delta+\lambda_1)(r+2\delta+2\lambda_1)}\left(\frac{1}{r}\mu_2 + 2\mu_1\tilde{\pi}_{Z_o}(r)\right)$$
$$+\frac{\lambda_1}{(r+2\delta+\lambda_1)}\left(\frac{(r+2\delta)(r+2\delta+2\lambda_2)}{\lambda_2(r+2\delta+2\lambda_1)}+1\right)\tilde{\pi}^2_{Z_0}(r), \tag{35}$$

From Bargès et al. [8], we have.

$$\tilde{\pi}_{Z_o}(r) = \frac{\lambda_2\mu_1}{r(r+\delta)} + \theta\frac{\lambda_2(\mu_1' - \mu_1)}{r(r+\delta+2\lambda_2)} \tag{36}$$

and

$$\tilde{\pi}^2_{Z_o}(r) = \frac{\lambda_2\mu_2}{r(r+2\delta)} + \theta\frac{\lambda_2(\mu_2' - \mu_2)}{r(r+2\delta+2\lambda_2)}$$
$$+\frac{2\lambda_2^2\mu_1^2}{r(r+\delta)(r+2\delta)} + \frac{2\theta\lambda_2^2\mu_1(\mu_1'-\mu_1)}{r(r+\delta+2\lambda_2)(r+2\delta)} + \frac{2\theta\lambda_2^2\mu_1(\mu_1'-\mu_1)}{r(r+2\delta+2\lambda_2)(r+\delta)}$$
$$+\frac{2\theta^2\lambda_2^2(\mu_1'-\mu_1)^2}{r(r+\delta+2\lambda_2)(r+2\delta+2\lambda_2)} \tag{37}$$

Substituting Eqs. (39) and (40) into Eq. (38), yields:

$$\tilde{\pi}^2_{Z_d}(r) = \frac{2\lambda_1(\lambda_1-\lambda_2)}{r(r+2\delta+\lambda_1)(r+2\delta+2\lambda_1)}\left(\frac{1}{r}\mu_2 + 2\mu_1\left(\frac{\lambda_2\mu_1}{r(r+\delta)} + \theta\frac{\lambda_2(\mu_1'-\mu_1)}{r(r+\delta+2\lambda_2)}\right)\right)$$
$$+\frac{\lambda_1}{(r+2\delta+\lambda_1)}\left(\frac{(r+2\delta)(r+2\delta+2\lambda_2)}{\lambda_2(r+2\delta+2\lambda_1)}+1\right)$$
$$\times\left\{\begin{array}{l}\dfrac{\lambda_2\mu_2}{r(r+2\delta)} + \theta\dfrac{\lambda_2(\mu_2'-\mu_2)}{r(r+2\delta+2\lambda_2)} \\ +\dfrac{2\lambda_2^2\mu_1^2}{r(r+\delta)(r+2\delta)} + \dfrac{2\theta\lambda_2^2\mu_1(\mu_1'-\mu_1)}{r(r+\delta+2\lambda_2)(r+2\delta)} + \dfrac{2\theta\lambda_2^2\mu_1(\mu_1'-\mu_1)}{r(r+2\delta+2\lambda_2)(r+\delta)} \\ +\dfrac{2\theta^2\lambda_2^2(\mu_1'-\mu_1)^2}{r(r+\delta+2\lambda_2)(r+2\delta+2\lambda_2)}\end{array}\right\}, \tag{38}$$

and rearranging Eq. (38), will give:

$$\tilde{\pi}^2_{Z_d}(r) = \frac{\lambda_1\mu_2}{r(r+2\delta+\lambda_1)} + \frac{2\lambda_1\lambda_2\mu_1^2}{r(r+\delta)(r+2\delta+\lambda_1)} + \frac{2\theta\lambda_1\lambda_2\mu_1(\mu_1'-\mu_1)}{r(r+\delta+2\lambda_2)(r+2\delta+\lambda_1)}$$
$$+\theta\lambda_1(\mu_2'-\mu_2)\frac{(r+2\delta)}{r(r+2\delta+\lambda_1)(r+2\delta+2\lambda_1)}$$
$$+2\theta\lambda_1\lambda_2\mu_1(\mu_1'-\mu_1)\frac{(r+2\delta)}{r(r+\delta)(r+2\delta+\lambda_1)(r+2\delta+2\lambda_1)}$$
$$+2\theta^2\lambda_1\lambda_2(\mu_1'-\mu_1)^2\frac{(r+2\delta)}{r(r+\delta+2\lambda_2)(r+2\delta+\lambda_1)(r+2\delta+2\lambda_1)} + \lambda_1\lambda_2\mu_2\frac{1}{r(r+2\delta)(r+2\delta+\lambda_1)}$$
$$+\theta\lambda_1\lambda_2(\mu_2'-\mu_2)\frac{1}{r(r+2\delta+2\lambda_2)(r+2\delta+\lambda_1)} + 2\lambda_1\lambda_2^2\mu_1^2\frac{1}{r(r+\delta)(r+2\delta)(r+2\delta+\lambda_1)}$$
$$+2\theta\lambda_1\lambda_2^2\mu_1(\mu_1'-\mu_1)\frac{1}{r(r+\delta+2\lambda_2)(r+2\delta)(r+2\delta+\lambda_1)}$$
$$+2\theta\lambda_2^2\lambda_1\mu_1(\mu_1'-\mu_1)\frac{1}{r(r+2\delta+2\lambda_2)(r+\delta)(r+2\delta+\lambda_1)}$$
$$+2\theta^2\lambda_2^2\lambda_1(\mu_1'-\mu_1)^2\frac{1}{r(r+\delta+2\lambda_2)(r+2\delta+2\lambda_2)(r+2\delta+\lambda_1)}, \tag{39}$$

which can be simplified to

$$\tilde{\pi}^2_{Z_d}(r) = \frac{\gamma_0}{r} + \frac{\gamma_1}{r+\delta} + \frac{\gamma_2}{r+2\delta} + \frac{\gamma_3}{r+2\delta+\lambda_1} + \frac{\gamma_4}{r+\delta+2\lambda_2} + \frac{\gamma_5}{r+2\delta+2\lambda_1} + \frac{\gamma_6}{r+2\delta+2\lambda_2}, \tag{40}$$

with,

$$\gamma_0 = \left\{ \begin{array}{l} \dfrac{\lambda_1\mu_2}{(2\delta+\lambda_1)} + \dfrac{2\lambda_1\lambda_2\mu_1^2}{\delta(2\delta+\lambda_1)} + \dfrac{2\theta\lambda_1\lambda_2\mu_1(\mu_1'-\mu_1)}{(2\delta+\lambda_1)(\delta+2\lambda_2)} + \dfrac{\theta\delta\lambda_1(\mu_2'-\mu_2)}{(2\delta+\lambda_1)(\delta+\lambda_1)} \\ + \dfrac{2\theta\lambda_1\lambda_2\mu_1(\mu_1'-\mu_1)}{(2\delta+\lambda_1)(\delta+\lambda_1)} + \dfrac{2\delta\theta^2\lambda_1\lambda_2(\mu_1'-\mu_1)^2}{(2\delta+\lambda_1)(\delta+\lambda_1)(\delta+2\lambda_2)} + \dfrac{\lambda_1\lambda_2\mu_2}{2\delta(2\delta+\lambda_1)} \\ + \dfrac{\theta\lambda_1\lambda_2(\mu_2'-\mu_2)}{2(\delta+\lambda_2)(2\delta+\lambda_1)} + \dfrac{\lambda_1\lambda_2^2\mu_1^2}{\delta^2(2\delta+\lambda_1)} + \dfrac{\theta\lambda_1\lambda_2^2\mu_1(\mu_1'-\mu_1)}{\delta(\delta+2\lambda_2)(2\delta+\lambda_1)} \\ + \dfrac{\theta\lambda_1\lambda_2^2\mu_1(\mu_1'-\mu_1)}{\delta(\delta+\lambda_2)(2\delta+\lambda_1)} + \dfrac{\theta^2\lambda_1\lambda_2^2(\mu_1'-\mu_1)^2}{2(\delta+2\lambda_2)(\delta+\lambda_2)(2\delta+\lambda_1)} \end{array} \right\} \tag{41}$$

$$\gamma_1 = \left\{ -\frac{2\lambda_1\lambda_2\mu_1^2}{\delta(\delta+\lambda_1)} - \frac{2\theta\lambda_1\lambda_2\mu_1(\mu_1'-\mu_1)}{(\delta+\lambda_1)(\delta+2\lambda_1)} - \frac{2\lambda_1\lambda_2^2\mu_1^2}{\delta^2(\delta+\lambda_1)} - \frac{2\theta\lambda_1\lambda_2^2\mu_1(\mu_1'-\mu_1)}{\delta(\delta+2\lambda_2)(\delta+\lambda_1)} \right\} \tag{42}$$

$$\gamma_2 = \left\{ -\frac{\lambda_2\mu_2}{2\delta} + \frac{\lambda_2^2\mu_1^2}{\delta^2} + \frac{\theta\lambda_2^2\mu_1(\mu_1' - \mu_1)}{\delta(\delta - 2\lambda_2)} \right\} \tag{43}$$

$$\gamma_3 = \left\{ \begin{array}{l} -\frac{\lambda_1\mu_2}{2\delta+\lambda_1} + \frac{2\lambda_1\lambda_2\mu_1^2}{(2\delta+\lambda_1)(\delta+\lambda_1)} + \frac{2\theta\lambda_1\lambda_2\mu_1(\mu_1' - \mu_1)}{(2\delta+\lambda_1)(\lambda_1+\delta-2\lambda_2)} + \frac{\theta\lambda_1(\mu_2' - \mu_2)}{2\delta+\lambda_1} \\ -\frac{2\theta\lambda_1\lambda_2\mu_1(\mu_1' - \mu_1)}{(2\delta+\lambda_1)(\delta+\lambda_1)} - \frac{2\theta^2\lambda_1\lambda_2(\mu_1' - \mu_1)^2}{(2\delta+\lambda_1)(\lambda_1+\delta-2\lambda_2)} + \frac{\lambda_2\mu_2}{2\delta+\lambda_1} \\ +\frac{\theta\lambda_1\lambda_2(\mu_2' - \mu_2)}{(2\delta+\lambda_1)(\lambda_1-2\lambda_2)} - \frac{2\lambda_2^2\mu_1^2}{(2\delta+\lambda_1)(\delta+\lambda_1)} - \frac{2\theta\lambda_2^2\mu_1(\mu_1' - \mu_1)}{(2\delta+\lambda_1)(\lambda_1+\delta-2\lambda_2)} \\ +\frac{2\theta\lambda_1\lambda_2^2\mu_1(\mu_1' - \mu_1)}{(2\lambda_2-\lambda_1)(\delta+\lambda_1)(2\delta+\lambda_1)} + \frac{2\theta^2\lambda_1\lambda_2^2(\mu_1' - \mu_1)^2}{(2\delta+\lambda_1)(\lambda_1+\delta-2\lambda_2)(2\lambda_2-\lambda_1)} \end{array} \right\} \tag{44}$$

$$\gamma_4 = \left\{ \begin{array}{l} \frac{-2\theta\lambda_1\lambda_2\mu_1(\mu_1' - \mu_1)}{(\delta+2\lambda_2)(\lambda_1+\delta-2\lambda_2)} + \frac{2\theta^2\lambda_1\lambda_2(\mu_1' - \mu_1)^2(2\lambda_2-\delta)}{(\delta+2\lambda_2)(\lambda_1+\delta-2\lambda_2)(2\lambda_1-2\lambda_2+\delta)} \\ -\frac{2\theta\lambda_1\lambda_2^2\mu_1(\mu_1' - \mu_1)}{(\delta+2\lambda_2)(\delta-2\lambda_2)(\lambda_1+\delta-2\lambda_2)} - \frac{2\theta^2\lambda_1\lambda_2^2(\mu_1' - \mu_1)^2}{\delta(\delta+2\lambda_2)(\lambda_1+\delta-2\lambda_2)} \end{array} \right\} \tag{45}$$

$$\gamma_5 = \left\{ -\frac{\theta\lambda_1(\mu_2' - \mu_2)}{\delta+\lambda_1} + \frac{2\theta\lambda_1\lambda_2\mu_1(\mu_1' - \mu_1)}{(\delta+\lambda_1)(\delta+2\lambda_1)} - \frac{2\theta^2\lambda_1\lambda_2(\mu_1' - \mu_1)^2}{(\delta+\lambda_1)(2\lambda_1-\delta-2\lambda_2)} \right\} \tag{46}$$

$$\gamma_6 = -\frac{\theta\lambda_1\lambda_2(\mu_2' - \mu_2)}{2(\delta+\lambda_2)(\lambda_1-2\lambda_2)} + \frac{\theta\lambda_1\lambda_2^2\mu_1(\mu_1' - \mu_1)}{(\delta+\lambda_2)(\delta+2\lambda_2)(\lambda_1-2\lambda_2)} + \frac{\theta^2\lambda_1\lambda_2^2(\mu_1' - \mu_1)^2}{\delta(\delta+\lambda_2)(\lambda_1-2\lambda_2)} \tag{47}$$

**Remark 2**
When

$$\lambda_1 = \lambda_2$$

$$\begin{aligned} \tilde{\pi}^2_{Z_d}(r) = {} & \frac{2\lambda^2\mu_2}{r(r+\delta)(r+2\delta)} + \frac{2\theta\lambda^2\mu_1(\mu_1' - \mu_1)}{r(r+\delta+2\lambda)(r+2\delta)} \\ & + \frac{\lambda\mu_2}{r(r+2\delta)} + \frac{\theta\lambda(\mu_2' - \mu_2)}{r(r+2\delta+2\lambda)} \\ & \frac{2\theta\lambda^2\mu_1(\mu_1' - \mu_1)}{r(r+\delta)(r+2\delta+2\lambda)} + \frac{2\theta^2\lambda^2(\mu_1' - \mu_1)^2}{r(r+\delta+2\lambda)(r+2\delta+2\lambda)}, \end{aligned} \tag{48}$$

which is exactly the result of Bargès et al. [8].

The Laplace transform in Eq. (49), is a combination of terms of the form:

$$\tilde{g}(r) = \frac{1}{r(\alpha_1 + r)(\alpha_2 + r)...(\alpha_n + r)}, \tag{49}$$

with $g$ a function defined for all non-negative real numbers. As described in the proof of Theorem 1.1 in Baeumer [16], each of these terms can be expressed as a combinations of partial fraction such as $\tilde{g}(r) = \gamma_0 \frac{1}{r} + \gamma_1 \frac{1}{\alpha_1 + r} + ... + \gamma_n \frac{1}{\alpha_n + r}$ where.

$\gamma_0 = \frac{1}{\alpha_1 ... \alpha_n}$, for $i = 1, ..., n$, $\gamma_i = -\frac{1}{\alpha_i} \prod_{j=1, j\neq i} \frac{1}{\alpha_j - \alpha_i}$.

Since the inverse Laplace transform of $\frac{1}{\alpha_i + r}$ is $e^{-\alpha_i t}$, it is easy to invert $\tilde{g}$ and obtain

$$g(t) = \gamma_0 + \gamma_1 e^{-\alpha_1 t} + \gamma_2 e^{-\alpha_2 t} + ... + \gamma_n e^{-\alpha_n t}. \tag{50}$$

Using Eq. (49) in Eq. (53), it results that

$$\pi^2_{Z_d}(t) = \gamma_0 + \gamma_1 e^{-\delta t} + \gamma_2 e^{-2\delta t} + \gamma_3 e^{-(2\delta+\lambda_1)t} + \gamma_4 e^{-(\delta+2\lambda_2)t} + \gamma_5 e^{-(2\delta+2\lambda_1)t} + \gamma_6 e^{-(2\delta+2\lambda_2)t}, t \geq 0, \tag{51}$$

where $(\gamma_i)_{i\in\{0,1,2,...,6\}}$ are given by equation Eq. (50).

**Remarks**

If $\theta = 0$ then

$$\begin{aligned} \gamma_0^{\theta=0} &= \frac{\lambda_1\mu_2}{2\delta+\lambda_1} + \frac{2\lambda_1\lambda_2\mu_1^2}{\delta(2\delta+\lambda_1)} + \frac{\lambda_1\lambda_2\mu_2}{2\delta(2\delta+\lambda_1)} + \frac{\lambda_1\lambda_2^2\mu_1^2}{\delta^2(2\delta+\lambda_1)} \\ &= \frac{\lambda_1}{2\delta+\lambda_1}\left(\frac{2\delta+\lambda_2}{2\delta}\right)\mu_2 + \frac{\lambda_1\lambda_2}{\delta(2\delta+\lambda_1)}\left(\frac{2\delta+\lambda_2}{\delta}\right)\mu_1^2 \\ &= \frac{\lambda_1(2\delta+\lambda_2)}{2\delta(2\delta+\lambda_1)}\left\{\mu_2 + \frac{2\lambda_2\mu_1^2}{\delta}\right\}, \end{aligned} \tag{52}$$

$$\gamma_1^{\theta=0} = -\frac{2\lambda_1\lambda_2\mu_1^2}{\delta(\delta+\lambda_1)} - \frac{2\lambda_1\lambda_2^2\mu_1^2}{\delta^2(\delta+\lambda_1)} = -\frac{2\lambda_1\lambda_2\mu_1^2}{\delta(\delta+\lambda_1)}\left(1+\frac{\lambda_2}{\delta}\right), \gamma_2^{\theta=0} = \frac{\lambda_2^2\mu_1^2}{\delta^2} - \frac{\lambda_2\mu_2}{2\delta} \tag{53}$$

$$\begin{aligned} \gamma_3^{\theta=0} &= -\frac{\lambda_1\mu_2}{2\delta+\lambda_1} + \frac{2\lambda_1\lambda_2\mu_1^2}{(2\delta+\lambda_1)(\delta+\lambda_1)} + \frac{\lambda_2\mu_2}{2\delta+\lambda_1} - \frac{2\lambda_2^2\mu_1^2}{(2\delta+\lambda_1)(\delta+\lambda_1)} \\ &= \frac{(\lambda_2-\lambda_1)\mu_2}{2\delta+\lambda_1} + \frac{2\lambda_2(\lambda_1-\lambda_2)\mu_1^2}{(2\delta+\lambda_1)(\delta+\lambda_1)}, \end{aligned} \tag{54}$$

$$\gamma_4^{\theta=0} = \gamma_5^{\theta=0} = \gamma_6^{\theta=0}. \tag{55}$$

Then,

$$
\begin{aligned}
\pi^2_{Z_d}(t) &= \frac{\lambda_1(2\delta+\lambda_2)}{2\delta(2\delta+\lambda_1)}\left\{\mu_2+\frac{2\lambda_2\mu_1^2}{\delta}\right\}-\frac{2\lambda_1\lambda_2\mu_1^2}{\delta(\delta+\lambda_1)}\left(1+\frac{\lambda_2}{\delta}\right)e^{-\delta t}\\
&+\left(\frac{\lambda_2^2\mu_1^2}{\delta^2}-\frac{\lambda_2\mu_2}{2\delta}\right)e^{-2\delta t}+\left(\frac{1}{2\delta+\lambda_1}(\lambda_2-\lambda_1)\mu_2+\frac{2\lambda_2(\lambda_1-\lambda_2)\mu_1^2}{(2\delta+\lambda_1)(\delta+\lambda_1)}\right)e^{-(2\delta+\lambda_1)t}\\
&=\left(\frac{\lambda_1(2\delta+\lambda_2)}{2\delta(2\delta+\lambda_1)}-\frac{\lambda_2}{2\delta}e^{-2\delta t}+\frac{(\lambda_2-\lambda_1)}{2\delta+\lambda_1}e^{-(2\delta+\lambda_1)t}\right)\mu_2\\
&+\left(\frac{\lambda_2\lambda_1(2\delta+\lambda_2)}{\delta^2(2\delta+\lambda_1)}-\frac{2\lambda_1\lambda_2}{\delta(\delta+\lambda_1)}\left(1+\frac{\lambda_2}{\delta}\right)e^{-\delta t}+\frac{\lambda_2^2}{\delta^2}e^{-2\delta t}+\frac{2\lambda_2(\lambda_1-\lambda_2)}{(2\delta+\lambda_1)(\delta+\lambda_1)}e^{-(2\delta+\lambda_1)t}\right)\mu_1^2\\
&=\left(\frac{\lambda_1(2\delta+\lambda_2)}{2\delta(2\delta+\lambda_1)}-\frac{\lambda_2}{2\delta}\left(1-2\delta\overline{a}_{t|2\delta}\right)+\frac{(\lambda_2-\lambda_1)}{2\delta+\lambda_1}\left(1-(2\delta+\lambda_1)\overline{a}_{t|2\delta+\lambda_1}\right)\right)\mu_2\\
&+\left(\frac{\lambda_2\lambda_1(2\delta+\lambda_2)}{\delta^2(2\delta+\lambda_1)}-\frac{2\lambda_1\lambda_2}{\delta(\delta+\lambda_1)}\left(1+\frac{\lambda_2}{\delta}\right)e^{-\delta t}+2\frac{\lambda_2^2}{\delta^2}e^{-2\delta t}-\frac{\lambda_2^2}{\delta^2}e^{-2\delta t}\right)\mu_1^2\\
&+\left\{\frac{2\lambda_2(\lambda_1-\lambda_2)}{\delta(\delta+\lambda_1)}-\frac{2\lambda_2(\lambda_1-\lambda_2)}{\delta(2\delta+\lambda_1)}\right\}\mu_1^2e^{-(2\delta+\lambda_1)t}\\
&=\left(\lambda_2\overline{a}_{t|2\delta}+(\lambda_1-\lambda_2)\overline{a}_{t|2\delta+\lambda_1}\right)\mu_2\\
&+\left(\frac{\lambda_2\lambda_1(2\delta+\lambda_2)}{\delta^2(2\delta+\lambda_1)}-\frac{2\lambda_1\lambda_2}{\delta(\delta+\lambda_1)}\left(1+\frac{\lambda_2}{\delta}\right)e^{-\delta t}+2\frac{\lambda_2^2}{\delta^2}e^{-2\delta t}-\frac{\lambda_2^2}{\delta^2}\left(1-2\delta\overline{a}_{t|2\delta}\right)\right)\mu_1^2\\
&+\frac{2\lambda_2(\lambda_1-\lambda_2)}{\delta(\delta+\lambda_1)}e^{-(2\delta+\lambda_1)t}\mu_1^2-\frac{2\lambda_2(\lambda_1-\lambda_2)\left(1-(2\delta+\lambda_1)\overline{a}_{t|2\delta+\lambda_1}\right)\mu_1^2}{\delta(2\delta+\lambda_1)}\\
&=\left(\lambda_2\overline{a}_{t|2\delta}+(\lambda_1-\lambda_2)\overline{a}_{t|2\delta+\lambda_1}\right)\mu_2+\frac{2\lambda_2}{\delta}\left(\lambda_2\overline{a}_{t|2\delta}+(\lambda_1-\lambda_2)\overline{a}_{t|2\delta+\lambda_1}\right)\mu_1^2\\
&-e^{\delta t}\left\{\frac{2\lambda_1\lambda_2}{\delta(\delta+\lambda_1)}\left(1+\frac{\lambda_2}{\delta}\right)-2\frac{\lambda_2^2}{\delta^2}e^{-\delta t}+\frac{2\lambda_2(\lambda_1-\lambda_2)}{\delta(\delta+\lambda_1)}e^{-(\delta+\lambda_1)t}\right\}
\end{aligned}
\tag{56}
$$

$$
\begin{aligned}
\pi^2_{Z_d}(t) &= \left(\lambda_2\overline{a}_{t|2\delta}+(\lambda_1-\lambda_2)\overline{a}_{t|2\delta+\lambda_1}\right)\mu_2+\frac{2\lambda_2}{\delta}\left(\lambda_2\overline{a}_{t|2\delta}+(\lambda_1-\lambda_2)\overline{a}_{t|2\delta+\lambda_1}\right)\mu_1^2\\
&-e^{\delta t}\left\{\frac{2\lambda_1\lambda_2}{\delta(\delta+\lambda_1)}\left(1+\frac{\lambda_2}{\delta}\right)-2\frac{\lambda_2^2}{\delta^2}e^{-\delta t}+\frac{2\lambda_2(\lambda_1-\lambda_2)}{\delta(\delta+\lambda_1)}e^{-(\delta+\lambda_1)t}\right\}\\
&=\left(\lambda_2\overline{a}_{t|2\delta}+(\lambda_1-\lambda_2)\overline{a}_{t|2\delta+\lambda_1}\right)\mu_2+\frac{2\lambda_2}{\delta}\left(\lambda_2\overline{a}_{t|2\delta}+(\lambda_1-\lambda_2)\overline{a}_{t|2\delta+\lambda_1}\right)\mu_1^2\\
&-e^{\delta t}\left\{\frac{2\lambda_1\lambda_2}{\delta(\delta+\lambda_1)}\left(1+\frac{\lambda_2}{\delta}\right)-2\frac{\lambda_2^2}{\delta^2}\left(1-\delta\overline{a}_{t|2\delta}\right)+\frac{2\lambda_2(\lambda_1-\lambda_2)}{\delta(\delta+\lambda_1)}\left(1-(\delta+\lambda_1)\overline{a}_{t|\delta+\lambda_1}\right)\right\}\\
&=\left(\lambda_2\overline{a}_{t|2\delta}+(\lambda_1-\lambda_2)\overline{a}_{t|2\delta+\lambda_1}\right)\mu_2+\frac{2\lambda_2}{\delta}\left(\lambda_2\overline{a}_{t|2\delta}+(\lambda_1-\lambda_2)\overline{a}_{t|2\delta+\lambda_1}\right)\mu_1^2\\
&-\frac{2\lambda_2}{\delta}e^{\delta t}\left\{\left(\lambda_2\overline{a}_{t|\delta}+(\lambda_1-\lambda_2)\overline{a}_{t|\delta+\lambda_1}\right)\right\}\\
&=\left(\lambda_2\overline{a}_{t|2\delta}+(\lambda_1-\lambda_2)\overline{a}_{t|2\delta+\lambda_1}\right)\mu_2\\
&+\frac{2\lambda_2}{\delta}\left(\lambda_2\overline{a}_{t|2\delta}+(\lambda_1-\lambda_2)\overline{a}_{t|2\delta+\lambda_1}-e^{\delta t}\left\{\left(\lambda_2\overline{a}_{t|\delta}+(\lambda_1-\lambda_2)\overline{a}_{t|\delta+\lambda_1}\right)\right\}\right)\mu_1^2.
\end{aligned}
\tag{57}
$$

To finally have:

$$
\begin{aligned}
\pi^2_{Z_d}(t) &= \left(\lambda_2\overline{a}_{t|2\delta}+(\lambda_1-\lambda_2)\overline{a}_{t|2\delta+\lambda_1}\right)\mu_2\\
&+\frac{2\lambda_2}{\delta}\left(\lambda_2\overline{a}_{t|2\delta}+(\lambda_1-\lambda_2)\overline{a}_{t|2\delta+\lambda_1}-e^{\delta t}\left\{\left(\lambda_2\overline{a}_{t|\delta}+(\lambda_1-\lambda_2)\overline{a}_{t|\delta+\lambda_1}\right)\right\}\right)\mu_1^2,
\end{aligned}
\tag{58}
$$

which is exactly the result of Léveillé et al.[15].
If $\lambda_1 = \lambda_2$ then

$$\gamma_0^{\lambda_1=\lambda_2} = \frac{\lambda\mu_2}{2\delta} + \frac{\lambda^2\mu_1^2}{\delta^2} + \left(\frac{\theta\lambda^2}{\delta(\delta+\lambda)} + \frac{\theta\lambda^2}{\delta(\delta+2\lambda)}\right)\mu_1(\mu_1'-\mu_1) + \frac{\theta\lambda(\mu_2'-\mu_2)}{2(\delta+\lambda)} + \frac{\theta^2\lambda^2(\mu_1'-\mu_1)^2}{(\delta+\lambda)(\delta+2\lambda)}, \tag{59}$$

$$\gamma_1^{\lambda_1=\lambda_2} = -\frac{2\lambda^2\mu_1^2}{\delta^2} - \frac{2\theta\lambda^2\mu_1(\mu_1'-\mu_1)}{\delta(\delta+2\lambda)}, \gamma_2^{\lambda_1=\lambda_2} = -\frac{\lambda\mu_2}{2\delta} + \frac{\lambda^2\mu_1^2}{\delta^2} + \frac{\theta\lambda^2\mu_1(\mu_1'-\mu_1)}{\delta(\delta-2\lambda)}, \gamma_3^{\lambda_1=\lambda_2} = 0 \tag{60}$$

$$\gamma_4^{\lambda_1=\lambda_2} = -\frac{2\theta\lambda^2\mu_1(\mu_1'-\mu_1)}{(\delta+2\lambda)(\delta-2\lambda)} - \frac{2\theta^2\lambda^2(\mu_1'-\mu_1)^2}{\delta(\delta+2\lambda)}, \tag{61}$$

$$\gamma_5^{\lambda_1=\lambda_2} = -\frac{\theta\lambda(\mu_2'-\mu_2)}{\delta+\lambda} + \frac{2\theta\lambda^2\mu_1(\mu_1'-\mu_1)}{(\delta+\lambda)(\delta+2\lambda)} + \frac{2\theta^2\lambda^2(\mu_1'-\mu_1)^2}{\delta(\delta+\lambda)}, \tag{62}$$

$$\gamma_6^{\lambda_1=\lambda_2} = \frac{\theta\lambda(\mu_2'-\mu_2)}{2(\delta+\lambda)} - \frac{\theta\lambda^2\mu_1(\mu_1'-\mu_1)}{(\delta+\lambda)(\delta+2\lambda)} - \frac{\theta^2\lambda^2(\mu_1'-\mu_1)^2}{\delta(\delta+\lambda)}. \tag{63}$$

Then,

$$\begin{aligned}\pi_{Z_d}^2(t) = {} & \lambda\mu_2\left(\frac{1}{\delta} - \frac{e^{-\delta t}}{2\delta}\right) + \theta\lambda(\mu_2'-\mu_2)\left(\frac{1}{2\lambda+2\delta} - \frac{e^{-(2\lambda+2\delta)t}}{2\lambda+2\delta}\right) \\ & + 2\lambda^2\mu_1^2\left(\frac{1}{2\delta^2} - \frac{e^{-\delta t}}{\delta^2} + \frac{e^{-2\delta t}}{2\delta^2}\right) \\ & + 2\theta\lambda^2\mu_1(\mu_1'-\mu_1)\left(\frac{1}{2\delta(2\lambda+\delta)} - \frac{e^{-(2\lambda+\delta)t}}{(2\lambda+\delta)(-2\lambda+\delta)} + \frac{e^{-2\delta t}}{2\delta(-2\lambda+\delta)}\right) \\ & + 2\theta\lambda^2\mu_1(\mu_1'-\mu_1)\left(\frac{1}{\delta(2\lambda+2\delta)} - \frac{e^{-\delta t}}{\delta(2\lambda+\delta)} + \frac{e^{-(2\lambda+2\delta)t}}{(2\lambda+2\delta)(2\lambda+\delta)}\right) \\ & + 2\theta^2\lambda^2(\mu_1'-\mu_1)^2\left(\frac{1}{(2\lambda+\delta)(2\lambda+2\delta)} - \frac{e^{-(2\lambda+\delta)t}}{\delta(2\lambda+\delta)} + \frac{e^{-(2\lambda+2\delta)t}}{\delta(2\lambda+2\delta)}\right)\end{aligned} \tag{64}$$

which is exactly the result of Bargès et al. [8].
If $\lambda_1 = \lambda_2$ and $\theta = 0$ then

$$\pi_{Z_o}^2(t) = \lambda\bar{a}_{t|2\delta}\mu_2 + (\lambda\bar{a}_{t|2\delta}\mu_1)^2, \tag{65}$$

which is exactly the result of Léveillé et al. [15].
**Remark 3.1**
Noting for $i = 1, 2, ..., m, j = 1, 2, ..., m, p = 0, 1$ and $k \in \mathbb{N} - \{0\}$

$$\eta_k(i,j,p) = \frac{\binom{i}{j}\theta^p\lambda^k(E[X^j])^{1-p}\left(E\left[X^{\prime j}\right] - E[X^j]\right)^p}{(r + p\times 2\lambda + i\delta)^k}. \tag{66}$$

We can rewrite $\tilde{\pi}_{Z_o}(r)$ and $\tilde{\pi}^2_{Z_o}(r)$ as

$$\tilde{\pi}_{Z_o}(r) = \frac{1}{r}[\eta_k(1,1,0) + \eta_k(1,1,1)], \tag{67}$$

$$\begin{aligned}\tilde{\pi}^2_{Z_o}(r) &= \frac{1}{r}[\eta_k(2,2,0) + \eta_k(2,2,1) + (\eta_k(2,1,0) + \eta_k(2,1,1))(\eta_k(1,1,0) + \eta_k(1,1,1))] \\ &= \frac{1}{r}\begin{bmatrix}\eta_k(2,2,0) + \eta_k(2,2,1) + \eta_k(2,1,0)\eta_k(1,1,0) + \eta_k(2,1,0)\eta_k(1,1,1) \\ + \eta_k(2,1,1)\eta_k(1,1,0) + \eta_k(2,1,1)\eta_k(1,1,1)\end{bmatrix}\end{aligned} \tag{68}$$

The term $\tilde{\pi}^m_{Z_o}(r)$ can also be expressed using

$$\tilde{\pi}^m_{Z_o}(r) = \frac{1}{r}\sum_{i=1}^{m} (i_1,j_1,p_1)...(i_n,j_n,p_n) \in \wp_{m,n}\eta_k(i_n,j_n,p_n) \times ... \times \eta_k(i_1,j_1,p_1), \tag{69}$$

where

$$\wp_{m,n} = \begin{Bmatrix}(i_1,j_1,p_1),...,(i_n,j_n,p_n) : i_1 = m, i_1 + ... + i_n = m-1+n, i_1 > ... > i_n, \\ j_1 = m-1+n, j_1 + ... + j_n = m, j_1 > ... > j_n, p \in \{0,1\}\end{Bmatrix}, \tag{70}$$

and

$$\begin{aligned}\tilde{\pi}^m_{Z_d}(r) = &\frac{2\lambda_1(\lambda_1-\lambda_2)}{r(r+m\delta+\lambda_1)(r+m\delta+2\lambda_1)}\left(\frac{1}{r}\mu_m + \sum_{k=1}^{m-1}C^k_m E\left[X^k_1\right]\tilde{\pi}^{(m-k)}_{Z_o}(r)\right) \\ &+\frac{\lambda_1}{(r+m\delta+\lambda_1)}\left(\frac{(r+m\delta)(r+m\delta+2\lambda_2)}{\lambda_2(r+m\delta+2\lambda_1)}+1\right)\tilde{\pi}^m_{Z_0}(r),\end{aligned} \tag{71}$$

## 4. Application

### 4.1 First two moments

For the numerical illustration, suppose that $X \sim pExp\left(\beta_1 = \frac{1}{80}\right) + (1-p)\,Exp\,\left(\beta_2 = \frac{1}{200}\right)$, the inter-claim time distribution parameters $\lambda_1 = 2; 4$ and $\lambda_2 = 1$, the interest rate $\delta = 3\%$ (**Tables 1–4**). We use three different values for the copula parameter $\theta = -1; 0; 1$, $p = 1/3$ and fix the time $t = 1; 10; 100$. The $m$th moment of $X$ is

$$\mu_m = p\frac{m!}{\beta_1^m} + (1-p)\frac{m!}{\beta_2^m} \text{ and } \mu'_m = \int_0^\infty mx^{m-1}(1-F_X(x))^2dx = \mu_m = p\frac{m!}{(2\beta_1)^m} + (1-p)\frac{m!}{(2\beta_2)^m}. \tag{72}$$

| $\theta$ | $t = 1$ | $t = 10$ | $t = 100$ |
|---|---|---|---|
| −1 | 482.3375 | 4450 | 16,428 |
| 0 | 438.1057 | 4407.1 | 16,385 |
| 1 | 393.874 | 4364.2 | 16,342 |

**Table 1.**
$E[Z_d(t)]$ *for* $\lambda_1 = 1, \lambda_2 = 10, \delta = 3\%$.

| θ | t = 1 | t = 10 | t = 100 |
|---|---|---|---|
| −1 | 597.1633 | 4578.7 | 16,557 |
| 0 | 554.5237 | 4535.5 | 16,513 |
| 1 | 511.8841 | 4492.3 | 16,470 |

**Table 2.**
$E[Z_d(t)]$ *for* $\lambda_1 = 5, \lambda_2 = 10, \delta = 3\%$.

| θ | t = 1 | t = 10 | t = 100 |
|---|---|---|---|
| −1 | 33900.85 | 29646.71 | 7928.481 |
| 0 | 23972.65 | 20976.12 | 5711.548 |
| 1 | 359.8795 | 1036.223 | 1542.412 |

**Table 3.**
$Std[Z_d(t)]$ *for* $\lambda_1 = 1, \lambda_2 = 10, \delta = 3\%$.

| θ | t = 1 | t = 10 | t = 100 |
|---|---|---|---|
| −1 | 34027.1 | 29756.44 | 7954.606 |
| 0 | 24061.85 | 21053.72 | 5731.564 |
| 1 | 366.1484 | 1035.491 | 1545.671 |

**Table 4.**
$Std[Z_d(t)]$ *for* $\lambda_1 = 5, \lambda_2 = 10, \delta = 3\%$.

## 4.2 Premium calculation

From the results in Section 4.1, we can compute the premium related to the risk of an insurance portfolio represented by $G(t)$, depending on the premium calculation principles adopted by the insurance company. The loaded premium $Z_d(t)$ consists in the sum of the pure premium $E[Z_d(t)]$, the expected value of the costs related to the portfolio, and a loading for the risk $M(t)$ as

$$G(t) = E[Z_d(t)] + M(t) \tag{73}$$

The loading for the risk differs according to the premium calculation principles.

### *4.2.1 The expected value principle*

Denote by $\theta > 0$ the safety loading. The expected value principle defines the loaded premium as:

$$G(t) = E[Z_d(t)] + \theta E[Z_d(t)], \tag{74}$$

where $M(t) = \theta E[Z_d(t)]$.

### *4.2.2 The variance principle*

Denote by $\theta > 0$ the safety loading. The variance principle defines the loaded premium as:

$$G(t) = E[Z_d(t)] + \theta Var[Z_d(t)], \tag{75}$$

where $M(t) = \theta Var[Z_d(t)]$.

### *4.2.3 The standard deviation principle*

Denote by $\theta > 0$ the safety loading. The standard deviation principle defines the loaded premium as:

$$G(t) = E[Z_d(t)] + \theta\sqrt{Var[Z_d(t)]}, \tag{76}$$

where $M(t) = \theta\sqrt{Var[Z_d(t)]}$.

### *4.2.4 The quantile principle*

The standard deviation principle defines the loaded premium as:

$$G(t) = F^{-1}_{Z_d(t)}(1 - \varepsilon), \tag{77}$$

where $\varepsilon$ is smallest (for example: $\varepsilon = 0.5\%, 1\%, 2.5\%, 5\%$).
In this case, the safety loading $M(t)$ is given by

$$M(t) = F^{-1}_{Z_d(t)}(1 - \varepsilon) - E[Z_d(t)] \tag{78}$$

The principles of standard deviation and variance only require partial information on the distribution of the random variable, $Z_d(t)$, i.e., its expectation and its variance.

Often, the actuary only has this information for different reasons (time constraints, information ...).

If the actuary has more information about the random variable, $Z_d(t)$ i.e., he knows the form of $F_{Z_d(t)}$, then he can apply the quantile principle.

But he does not know much about $F_{Z_d(t)}$, then he can approximate the distribution of $Z_d(t)$ using the matching moments technique.

## 5. Conclusion

We have derived exact expressions for all the moments of the DCDPRV process using renewal arguments, again disproving the popular belief that renewal techniques cannot be applied in the presence of economic factors. Our results, for the DCDPRV process, are consistent: (i) with the results of Léveillé et al. [15] for $\theta = 0, \lambda_1 \neq \lambda_2$ and for $\theta = 0, \lambda_1 = \lambda_2$, (ii) with the results of Bargès et al. [8] for $\theta \neq 0, \lambda_1 = \lambda_2$.

Within this framework, further research is needed to get exact expressions (or approximations) of certain functional of the $\{Z_d(t), t \geq 0\}$ process, as stop-loss premiums and ruin probabilities.

Our models have applications in reinsurance, house insurance and car insurance. They can also be used in evaluation of health programs, finance, and other areas.

For example, consider the case of a male currently aged 25 who is starting a defined contribution (DC) pension plan and is planning to retire in, say, 40 years at the age of 65. He anticipates that when he reaches that age he will convert his accumulated pension fund into a life annuity in order to hedge his own longevity risk and avoid outliving his own financial resources. The value of his retirement income will depend not only on the value of his pension fund, but also on the price of annuities at the time. Other things being equal, this means that his retirement income prospects will be affected by the distribution on future annuity value: the greater the dispersion of that distribution, the riskier his retirement income will be. For the assessment of the accumulated pension fund and its variability our models can be used. We can suppose that this man makes a deposit to a bank account, and that the time between successive deposits follows a renewal process and the force of interest is stochastic. Our model allows us to calculate the accumulated pension fund and its variability at the age of 65.

Another possible application is in reliability, to model the net present value of aggregate equipment failures costs until its total breakdown. A piece of equipment is deemed to be beyond repair when the repair time exceeds a predetermined gap. Of course, another possible definition of total breakdown is when the cost of repair exceeds a predetermined gap. But, since the cost of repair is defined per unit time, the two definitions are somewhat equivalent.

## Author details

Franck Adékambi
School of Economics, University of Johannesburg, South Africa

*Address all correspondence to: fadekambi@uj.ac.za

## References

[1] Dickson DCM, Hipp C. Ruin probabilities for Erlang(2) risk process. Insurance: Mathematics & Economics. 1998;**22**:251-262

[2] Dickson DCM, Hipp C. On the time to ruin for Erlang(2) risk process. Insurance: Mathematics & Economics. 2001;**29**:333-344

[3] Li S, Garrido J. On ruin for the Erlang (n) risk process. Insurance: Mathematics & Economics. 2004;**34**: 391-408

[4] Gerber H, Shiu E. The time value of ruin in a sparre Andersen model. North American Actuarial Journal. 2005;**9**: 49-69

[5] Li S, Garrido J. On a general class of renewal risk process: Analysis of the Gerber-Shiu function. Advances in Applied Probability. 2005;**37**:836-856

[6] Albrecher Hö, Teugels JL. Exponential behaviour in the presence of dependence in risk theory. Journal of Applied Probability. March 2006;**43**(1): 257-273

[7] Asimit AV, Badescu AL. Extremes on the discounted aggregate claims in a time dependent risk model. Scandinavian Actuarial Journal. 2010;**2**: 93-104

[8] Bargès M, Cossette H, Loisel S, Marceau E. On the moments of aggregate discounted claims with dependence introduced by a FGM copula. ASTIN Bulletin: The Journal of the IAA. 2011;**41**(1):215-238

[9] Adékambi F, Dziwa S. Moment of the discounted compound renewal cash flows with dependence: The use of Farlie-Gumbel-Morgenstern copula. In: Proceedings of the 58th Annual Conference of the South African Statistical Association for 2016; Cape Town, South Africa: University of Cape Town. 2016. pp. 1-8

[10] Adékambi F. Second moment of the discounted aggregate renewal cash flow with dependence. In: Proceedings of the 58th Annual Conference of the South African Statistical Association for 2016; Cape Town, South Africa: University of Cape Town. 2017. pp. 1-8

[11] Govorun M, Latouche G. Modeling the effect of health: Phase-type approach. European Actuarial Journal. 2014;**4**(1):197-218

[12] Adékambi F. Moments of phase-type aging modeling for health dependent costs. Advances in Decision Sciences. 2019;**23**(2):1-28

[13] Van Noortwijk J, Frangopol D. Two probabilistic life-cycle maintenance models for deteriorating civil infrastructures. Probabilistic Engineering Mechanics. 2004;**19**(4): 345-359

[14] Willmot GE. A note on a class of delayed renewal risk processes. Insurance: Mathematics & Economics. 2004;**34**:251-257

[15] Léveillé G, Garrido J. Moments of compound renewal sums with discounted claims. Insurance: Mathematics & Economics. 2001;**28**: 217-231

[16] Baeumer B. On the inversion of the convolution and Laplace transform. Transactions of the American Mathematical Society. 2003;**355**(3): 1201-1212

Chapter 13

# Modelling the Information-Psychological Impact in Social Networks

*Igor Goncharov, Nikita Goncharov, Pavel Parinov, Sergey Kochedykov and Alexander Dushkin*

## Abstract

The paper considers the objects, subjects, purposes, tools, methods and implementation of information-psychological impact (IPI). It suggests a cellular automata model of the diffusion process of information-psychological impact in social networks, the hierarchy of the changes in the states of the subjects of information-psychological impact and the chart of transitions from state to state used in the cellular automaton algorithm. The suggested cellular automaton takes into account the effect of forgetting the information-psychological impact, as well as social and psychological parameters and probabilistic characteristics of the subjects of the social network. It therefore allows for the modelling of the diffusion of the information-psychological impact in the social network. The model can be used to determine the number of subjects affected by the information-psychological impact and the possibility of successful diffusion of the impact. The modelling of the suggested algorithm was performed. The results of the modelling are analysed in the paper.

**Keywords:** social network, information-psychological impact, negative information-psychological impact, positive information-psychological impact, cellular automata, diffusion, social and psychological parameters

## 1. Introduction

Information-psychological impact (IPI) is the informational influence on people's minds, which alters their perception of the reality, behavioural functions and, in some cases, even the functioning of their inner organs and body systems [1–3]. Information-psychological impact (IPI) may affect individuals, groups of people, communities and the whole society. IPI can be either positive or negative, depending on the intended purpose. Positive IPI is used for medical treatment purposes, rehabilitation, improvement of behavioural patterns and creative purposes. It can also be used to unite people for a good cause. Negative IPI is used for manipulating—directly or indirectly—individuals, groups of people or the whole society into actions that violate either their own interests or interests of others. Negative IPI may cause emotional, psychological and social tensions, deterioration of moral standards and behavioural norms, as well as moral and political disorientation. This, in turn, leads to dramatic changes in individual, group and public conscience and alterations in the moral, political, social and psychological environment within the society [1–3].

Information-psychological impact is implemented by means of various tools and techniques. At the moment, negative information-psychological impacts are more common. They influence individuals, groups of people or the society by means of telecommunication systems, mass media and social networks. Negative IPIs are used to control the society, force certain opinions on various issues, recruit members to religious cults and terrorist groups and to alter people's mental state. Among the examples of such IPIs are colour revolutions, the so-called "death groups" on social networking sites, as well as active recruitment campaigns to terrorist groups, which are based on films or video games aimed primarily at young people.

It is thus very important to model IPIs in social networks in order to analyse and select the most effective methods of using positive IPIs and combating negative IPIs [1, 2, 4].

## 2. Overview of previous studies

Social networks are usually represented as graphs with multiple vertices (agents) and edges representing the links between the agents. Agents represent various subjects of the network, from individuals to large groups, organisations and communities. Links denote the relationships between the agents, such as information exchange, social relations and communication [4–9]. The process of IPI can be divided into two stages: diffusion of the IPI and alteration of the agents' opinions. Gubanov et al. [4] consider various models of social networks and divide the tasks connected with studying IPI into following groups: modelling of the informational influence, modelling of the information management and modelling of the information confrontation.

Models of the informational influence are used to study the behaviour of the subject affected by IPI. The influence may be intentional or unintentional. Social influence becomes obvious during communication or in case of comparison. Models of the informational influence are used for information management, as they help the managing subject to determine the kind of informational influence that will make the controlled subject behave in the desired way. The information management model, in turn, is used to model information confrontation, that is, the interaction of several subjects with conflicting interests who apply their informational influence to the same controlled subject [4]. There are a number of approaches to modelling the influence.

1. The Independent Cascade and Linear Threshold Models [4, 8, 10–13]. In these models, the subject (a vertice of the graph) can be either active or inactive. The state may only change from active to inactive, not the other way round. The agent becomes active depending on the selected threshold. The threshold can be uniform for all agents or may be randomly selected according to a probabilistic distribution. These models do not take into account groups, game interaction between the subjects, individual activity of the subjects or incomplete awareness of the subjects.

2. Network autocorrelation models. In these models, the opinion and behaviour of the subject are affected by the opinion of the neighbouring subjects and represent the reaction of the subject to the IPI. The authors [14–19] consider a determined time-digital linear process, where opinions (properties) of the

subject are presented as vector $y_t$ and change under the influence of other subjects according to the so-called influence matrix $W$: $y_{t+1} = Wy_t$.

3. Ising models [20, 21]. Invented for studying the phenomenon of ferromagnetism, Ising model is often used to verify the results of numerical modelling. When studying the diffusion of IPI in social networks, the model helps to describe the changes in the behaviour of a large social group caused by the nearest neighbours. The influence of the nearest neighbours plays a key role, and the willingness of the group to accept a new idea serves as the analogue of the temperature.

4. Influence models based on Markov chains. Such models employ corresponding mathematical tools to represent the activities of every subject and the group as a whole. They are used to analyse social dynamics and determine the patterns of the group behaviour. The authors [22–24] consider the similarity of opinions of the subject, the authors [7, 24] focus on the time over which the opinions become similar, and the authors [7, 25] study the conditions under which a uniform final opinion is formed.

All the above-mentioned models represent the rules of interaction between the subjects or groups of subjects. However, they either do not at all represent the specifics and characteristics of the network influence and the interaction process or do this inadequately.

When a social network is considered as a set of agents [4, 26–28], we assume that every agent has a certain degree of influence on the other agents. It is therefore necessary to determine a small group of agents with the maximal level of influence, that is, to solve the influence maximisation problem [4, 10, 29]. These agents can be used as key nods for influencing other subjects of the social network or to monitor the social network in order to reveal the presence of IPI. The influence maximisation problem has been considered in papers focusing on the following issues.

- Viral marketing [29], where a social network is represented by a Markov chain with each agent A having his own value that depends on the profit from sales to other agents influenced by agent A.

- Influence maximisation in the models of innovations' diffusion [10]. They include a set of active agents, and at a certain point in time, a new active agent can activate his neighbours with a set probability.

- Voting process modelling [9], where every agent can, at any stage, change his opinion by accidentally voting for one of his neighbours and adapting their opinion. The agent is more likely to adapt the opinion supported by the majority of his neighbours.

Besides analysing the influence, management and confrontation, there is also a problem of diffusion of information-psychological impact in the information space [5]. Information may spread in the following directions [5, 30]: from a subject to another subject, from a subject to a group, from the information production centre to an individual subject or a group.

The authors [5, 26–28, 31] suggest a multi-agent model of information diffusion. The model takes into account the growth of the number of agents over time. Agents may appear themselves, produce new agents, disappear from the subjects' neighbourhood or receive links from other agents.

In [5, 30], the life cycle of the information flow is represented by information diffusion models based on cellular automata. In these models, each cell of the automaton can have various states, such as "influence taken," "influence not relevant," or "influence rejected." The information spreads according to probabilistic rules. The observed states of the objects alter simultaneously in discrete time intervals following the constant local probabilistic rules. The rules themselves depend on the state of variables describing the nearest neighbours of the agent or on the state of the subject itself. For instance, the authors [8, 32] present a model of word-of-mouth information transfer considering strong and weak links between the subjects.

In order to analyse the information diffusion process, the authors [6, 33, 34] compare information diffusion to virus transmission using infiltration and contamination models such as SIR model and SIRS model.

Runkov [35] compares the structure of social networks and neural networks. Individual users are viewed as neurons. Using the information about the users' activities, the neural network may forecast the kind of news they will be interested in [35, 36] also suggests using neural networks to forecast the behaviour of the subject of IPI and their recruitability to certain assignments, as well as to assess their reliability using the data available in the social network.

From the information security perspective, it is vital to identify IPI as soon as possible. For this purpose, the authors [4, 37] suggest monitoring the states of a small group of nods in the network using graph models. The problem is to determine the set of nods to be monitored. Deviations from the standard dynamics of transmission of some information messages may serve as an indicator of information-psychological impact. In order to analyse the dynamics of the information spread and determine the channels caused by external factors, wavelet analysis can be used [5, 15, 17].

Dodonov and Lande [5] introduce the term information reservation for an isolated area of the information space and suggest certain modification to information diffusion models in order to model the dynamics of information flows in information reservations. Information reservations are information areas subject to constant information-psychological impact. They can be used for information and psychological control over the society.

We should say, however, that all the suggested models do not fully consider social and psychological factors, such as the psychological state of the subjects during IPI diffusion in social networks. IPI diffusion process depends on the probabilistic characteristics of the subjects of the social network and the links between them. It is, therefore, interesting to study IPI diffusion taking into account social and psychological factors and the psychological state of the subjects of the social network.

The aim of this paper is to model the process of IPI diffusion in social networks considering social and psychological factors and the psychological state of the subjects of the social network. This can be done using a cellular automaton model, as cellular automata can most adequately represent the process of IPI diffusion in a social network and the changes in the opinions of its subjects caused by their immediate neighbours, taking into account social and psychological factors.

## 3. Materials and methods

When modelling and analysing the process of IPI diffusion, we regarded the social network as a two-dimensional cellular automaton. A two-dimensional cellular automaton is a set of finite automata (subjects of the social network) allocated on the reference frame and marked with integer coordinates $(i,j)$. Each automaton can have certain properties and be in one of the states $S_{i,j} \in \{S_1, S_2, .., S_k\}$. The state of a finite automaton $(i,j)$ at a certain moment in time $t+1$ is determined as follows Eq. (1):

$$S_{i,j}(t+1) = F\left(S_{i,j}(t), N(i,j), t\right), \tag{1}$$

where $F$ is the rule for the transition of state of the automaton; $N(i,j)$ is the point neighbourhood $(i,j)$ and $t$ is a step on the axis of time.

In the cellular automaton model, each cell changes its state while interacting with a limited number of other cells, normally adjacent ones with the same edge or vertex. Such models allow for a simultaneous change of the state of all cells following the general principle of the cellular automaton. Therefore, it is easy to see the connection between the processes occurring on the micro level and the processes of spatial interaction between the elements.

Due to the simplicity of their implementation and the ability to describe complex processes, cellular automata are widely used for the modelling of systems, which consist of a large number of nonlineary interacting particles (fluid and gas dynamics in various environments, fires, traffic, and so on), as well as for representing collective phenomena, such as turbulence, arrangement and chaos.

### 3.1 Suggested models of IPI in social networks

Given below are the models we suggest for describing the process of information-psychological impact diffusion in social networks.

1. Information interaction within the social network is presented as a two-dimensional cellular automaton, whose grid is a two-dimensional array, where each cell is numbered with an ordered pair $(i,j)$. Each cell is a subject of the social network. The nearest neighbours of each cell are considered the cells that have a common vertex with the one observed (Moore neighbourhood). Thus, each cell has eight nearest neighbours. To eliminate the tip effect, the grid of the cellular automaton is topologically twisted into a torus [5, 30, 38], that is, the first line is considered to be the continuation of the last one, and the last one precedes the first one. The same applies to the columns [5, 30, 38–40].

2. The informational interaction in the social network is presented as a cellular automaton, whose grid is a free-scale network generated by a Barabási-Albert algorithm.

Each cell may be in one of the following states: highly positive, neutral (mild negative or positive attitude) or highly negative. Depending on its state and social and psychological characteristics, a cell may or may not spread the information (by influencing the neighbouring cells) [5, 30, 38]. The state and behaviour of cells change according to the set of rules for the suggested model. These rules take into account social and psychological factors as well as the psychological state of the subjects of the social network.

A state transition graph is presented in **Figure 1**. $S_0$ is the initial state; $S_1$ is the subject that does not spread the information $I$ and his negative opinion (negative feedback); $S_2$ is the subject that does not spread the information $I$ and his positive opinion (positive feedback); $S_3$ is the subject that spreads the information $I$ together

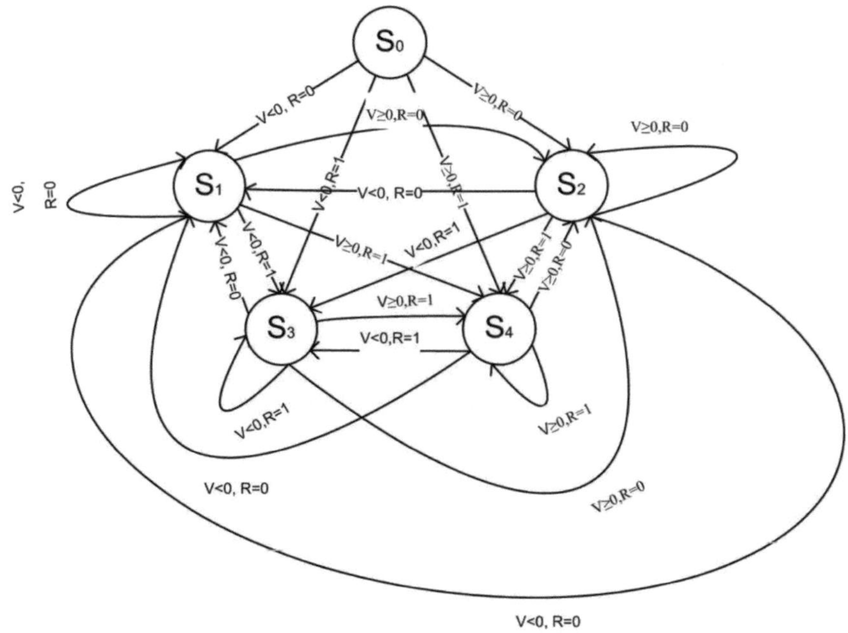


**Figure 1.**
*State transition graph.*

with his negative opinion (negative feedback); $S_4$ is the subject that spreads the information $I$ together with his positive opinion (positive feedback).

Each subject $P_k$ of the social network is interested in a certain number of topics $T^k = \{T^k_m\}$ and is indifferent to other topics. Subject $P_k$ has the following social and psychological parameters [41].

1. Initial personal opinion $V_k$ about the information presented in the IPI, which depends on individual psychological characteristics, education, moral principles, environment and so on. This parameter is evaluated by the experts using Harrington scale, according to which values $V_k$ can be interpreted as follows [41]:

    - $[-1; -0.64)$ interval—highly negative opinion that motivates the subject to spread the information $I$ together with the negative opinion (negative feedback);
    - $[-0.64; 0)$ interval—mild negative opinion that does not motivate the subject to spread the information $I$;
    - $[0; 0.64)$ interval—mild positive opinion that does not motivate the subject to spread the information $I$;
    - $[0.64; 1)$ interval—highly positive opinion that motivates the subject to spread the information $I$ together with the positive opinion (positive feedback).

2. Level of trust ${TR_{kj}}^{T_l}$ to the $j$-th user concerning the topic $T$. This parameter influences the attitude of subject $P_k$ to the information presented in the IPI,

received from $j$-th source. The set of ${TR_{kj}}^T$ forms a "trust matrix" $TR^{T_I}$ for the topic $T_I$. The TRTI matrix should not necessarily be symmetric.

3. Communication skills $O_k$. This parameter is evaluated using various psychological tests, such as Ryakhovsky's test for communication skills. Let $O_k$ = {Bad; Average; Good} [41, 42].

4. Information transfer coefficient $G_k$, showing the force of influence transmitted by subject $P_k$ to the neighbouring subjects.

5. Level of perception $C_k$, showing how much subject $P_k$ relies on his own opinion within the topic $T_I$.

In order to evaluate the current (at a specific time interval $t = t + 1$) opinion $V_k^{t+1}$ about the information presented in the IPI, the following relations are suggested [41]:

$$V_k^{t+1} = \begin{cases} 1, \textit{whenever } X \geq 1, \\ X, \quad \textit{whenever } -1 < X < 1, \; X = V_k^t + C_k^{T_I} \cdot CSP_k^{t+1} \\ -1, \textit{whenever } X \leq -1, \end{cases} \tag{2}$$

$$CSP_k^{t+1} = \frac{\sum_{i=1}^{N} F_i TR_{ki}^{TI}}{N}, \quad F_i = G_i V_i^t,$$

where $CSP_k^{t+1}$ is an "integral social force," denoting the degree of influence on the opinion of subject $P_k$ about the information in the IPI received from the subject $P_k$ is interacting with; $N$ is the number of subjects interacting with subject $P_k$; $F_i$ is the force of IPI with which the $i$-th subject influences subject $P_k$ and $V_i^t$ is the opinion of the $i$-th subject.

Whether subject $P_k$ will spread the IPI with the force $F$ depends on his opinion $V_k$ and his communication skills $O_k$. To evaluate the coefficient of the information transfer by subject $P_k$ at a specific time interval $(t + 1)$, the following formula is used [41]:

$$R_k^{t+1} = \begin{cases} 0, \textit{if } O_k = "\textit{bad}" \text{и } V_k \in [-0,64; 0,64); \\ \qquad 1, \textit{else}. \end{cases} \tag{3}$$

The subject affected by the IPI in the social network develops his own opinion about the received information, which depends on his individual parameters and the force of the IPI. The opinion can be positive or negative and may change over time under the influence of other factors. Depending on his opinion about the information and his communication skills, the subject may or may not spread the received IPI [43–46].

The effectiveness of the IPI can be defined by the following relation Eq. (4):

$$P = \frac{N_{S_2} + N_{S_4}}{N}, \tag{4}$$

where $N_{S_2}$ is the number of subjects in state $S_2$, $N_{S_4}$ is the number of subjects in state $S_4$ and $N$ is the total number of subjects.

Users of the social network may be subject to various kinds of IPI aimed at different groups of people. IPIs may also differ by their purpose and the effectiveness of implementation. IPI in social networks may also be used to influence specific public officers.

Using the results of the IPI modelling, we can perform a comprehensive assessment of the general level of information and psychological security and suggest practical recommendations on how to eliminate the negative effect of the information-psychological influence. The assessment can be based on the methodology for calculating the security indices in the military, political, economic and other spheres developed by the PIR Center [48, 49]. This means that the index of general information and psychological security (IGIPS) is calculated according to the following formula:

$$IIPS = \frac{G_0}{H}\left[f_1(1-\beta_1)+f_2(1-\beta_2)+\dots+f_H(1-\beta_H)\right]+ \\ +\frac{G_{tar}}{K}\left[\mathrm{h}_1(1-\gamma_1)+h_2(1-\gamma_2)+\dots+h_K(1-\gamma_K)\right]\chi_i, \tag{5}$$

where $G_o$ is the coefficient of the degree of IPI on the social network; $H$ is the number of IPIs; $f_i$ is the coefficient of the importance of the $i$-th IPI; $\beta_i$ is the probability of using the $i$-th IPI in the social network determined by the Eq. (4); $G_{tar}$ is the coefficient of the degree of the IPI on the specific management system; $K$ is the number of public officers that may be subject to the IPI; $h_i$ is the coefficient of the importance of the $i$-th; $\gamma_i$ is the probability of effective implementation of the IPI used to influence the $i$-th public officer and $\chi_i$ is the coefficient of importance of the $i$-th management object.

$G_o$, $G_{tar}$, $f_i$, $h_i$ and $\chi_i$ are determined by means of an expert survey. The probability of effective implementation of the IPI $\gamma_i$ used to influence the $i$-th public officer is calculated using Eq. (6):

$$\gamma_i = \frac{S}{D}, \tag{6}$$

where $S$ is the number of wrong decisions made and $D$ is the total number of decisions made after the IPI.

The probability of the IPI being aimed at a specific public officer is calculated using Eq. (7) [2]:

$$P = 1-(1-\mathrm{a}_i)(1-\mathrm{b}_j)\dots(\mathrm{g}_s), \tag{7}$$

where $\mathrm{a}_i$, $\mathrm{b}_j$, ..., $\mathrm{g}_s$ are informational factors determined by the expert survey that indicate that the IPI is aimed at a certain public officer.

The suggested method of assessing the IGIPS has the following advantages. It registers the increase in the degree of the IPI on the social network in good time. It registers the connection between the IPI on public officials and the decisions they make. It allows for calculating the index of information and psychological security and developing a strategy to decrease negative IPIs.

### 3.2 Modelling algorithm

**Figure 2** presents a flow chart of the algorithm for modelling IPI. During the initial stage, main parameters of the social network's subjects are determined. The trust matrix is formed, and the communication skills of the subjects, their

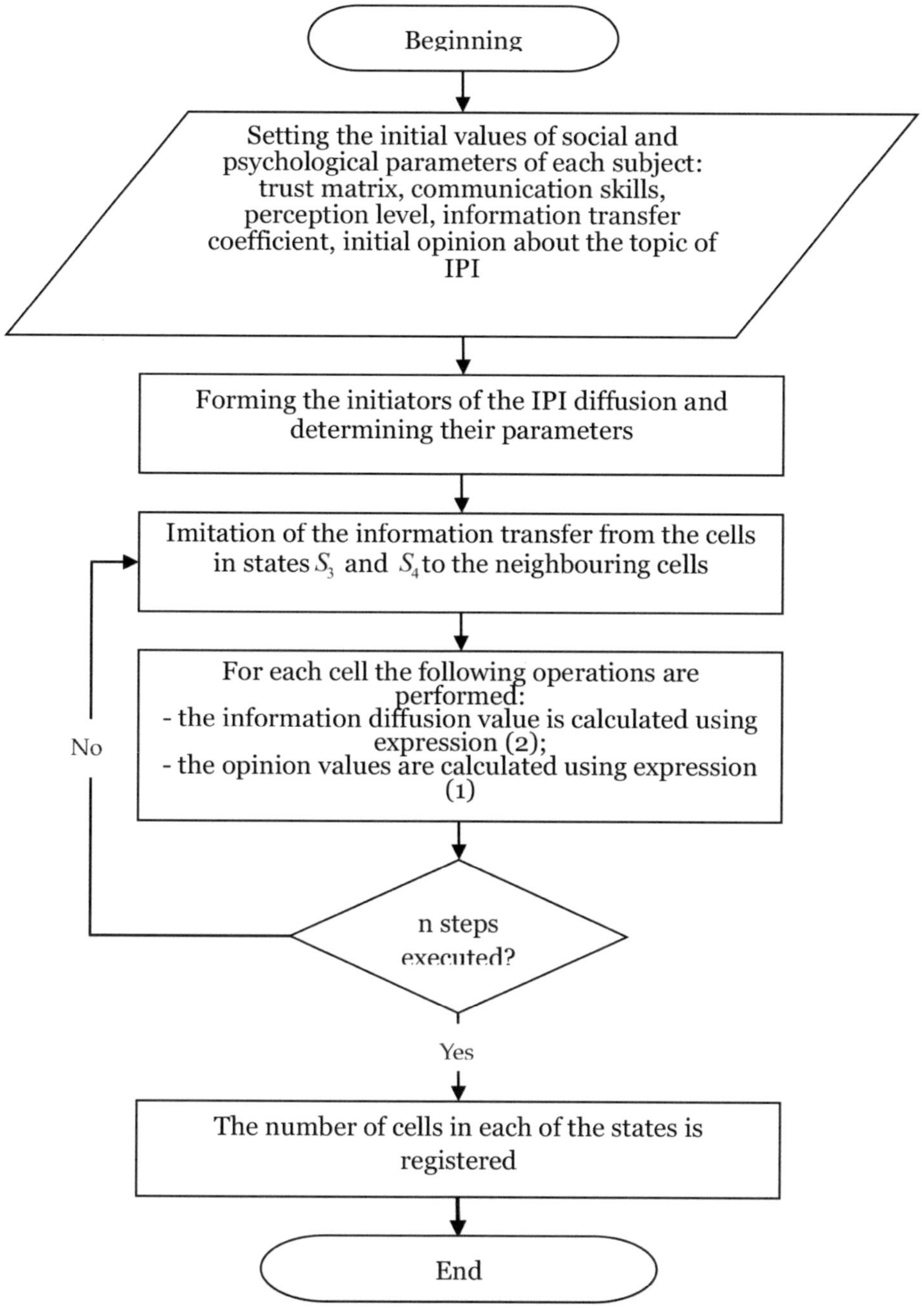


**Figure 2.**
*Flow chart of the algorithm for modelling IPI.*

perception level, information transfer coefficient and the initial opinion about the given issue are determined [43–47].

During the first stage, which corresponds to the origin on the time axis ($t = 0$), the whole grid consists of cells in state $S_0$, except for certain cells that initiate the diffusion of the IPI together with their positive opinion about the information.

The second stage involves information diffusion and exchange of opinions between the subjects along the time axis $t = t + 1$. The information diffusion is calculated using Eq. (3), and the opinions are calculated using Eq. (2). Cells with the value of information diffusion equal 1 spread the information to the neighbouring cells.

A cell may change its state receiving influence $F_i$ from the neighbouring cells whose information transfer value equals 1. When the influence is received, the current values of opinion $V_k$ and information diffusion $R_k$ are calculated.

## 4. Experiments and discussion

### 4.1 Two-dimensional array implementation

The suggested algorithm was implemented on a 100 × 100 grid. The automaton was tested in the following way: the initial values were distributed following the normal distribution law; 10 random initiators of the IPI and 2 opponents were selected out of all the subjects; the automaton was tested 100 times, each test run including 1000 steps; average number of subjects in each of the states was determined. The initial personal opinion of subject $V_k$ about the information was distributed according to the normal distribution rule within the intervals $[-1; -0.5]$, $[-0.5; 0.5]$, $[0.5; 1]$. Trust level $TR_{kj}{}^{T_I}$ was distributed according to the normal distribution rule within the interval $[0; 1]$ or $[-1; 1]$. **Figures 3–5** demonstrate the functioning of the automaton.

**Figure 4** demonstrates the functioning of the automaton, when $V_k \in [-0,5; 0,5]$, that is, most subjects are neutral to the IPI. **Figure 5** demonstrates the functioning of the automaton, when $V_k \in [-1; -0,5]$, that is, most subjects are negative to the IPI. **Figure 6** demonstrates the functioning of the automaton, when $V_k \in [0,5; 1]$, that is, most subjects are positive to the IPI. Figures "a" demonstrate the functioning of the automaton, when $TR_{kj}{}^{T_I} \in [0; 1]$, that is, the subjects adopt opinions of other subjects. Figures "b" demonstrate the functioning of the automaton, when $TR_{kj}{}^{T_I} \in [-1; 1]$, that is, the subject has the opposite opinion to the one imposed by the IPI.

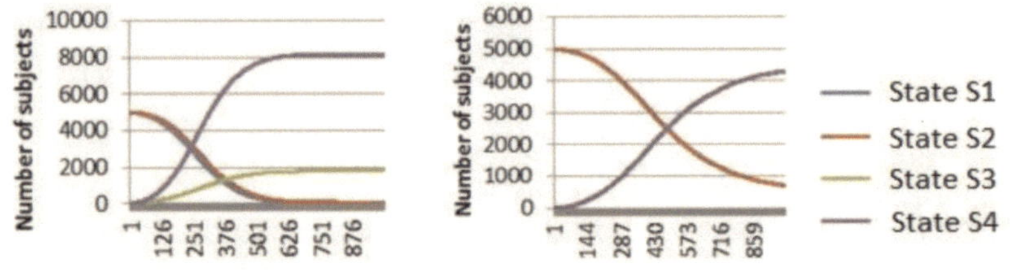


**Figure 3.**
*Distribution of cells according to the discrete time whenever* $V_k \in [-0,5; 0,5]$.

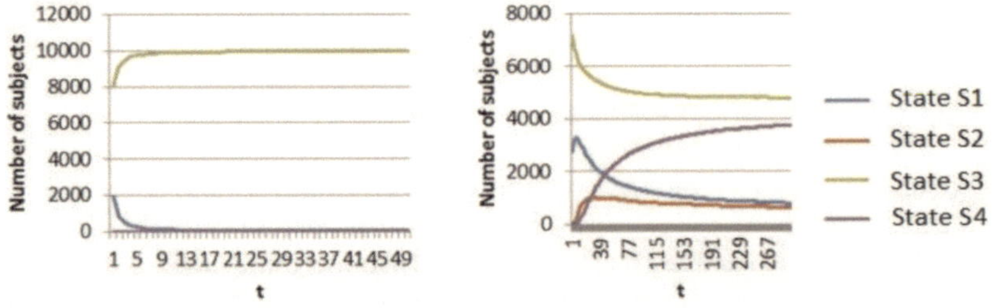


**Figure 4.**
*Distribution of cells according to the discrete time whenever* $V_k \in [-1; -0,5]$.

*DOI: http://dx.doi.org/10.5772/intechopen.88252*

### 4.2 Barabási-Albert model implementation

The suggested algorithm was implemented using a random scale-free network generated by Barabási-Albert algorithm. The network consisted of 1000 nods. The results are given in **Figure 6**. The automaton was tested in the following way: the initial values were distributed following the normal distribution law; 90 random initiators of the IPI and 10 opponents were selected out of all the subjects; the automaton was tested 100 times, each test run including 300 steps; average number of subjects in each of the states was determined. The initial personal opinion of the subject $V_k$ about the information was distributed according to the normal distribution rule within the intervals $[-1; -0.5]$, $[-0.5; 0.5]$, $[0.5; 1]$. Trust level ${TR_{kj}}^{T_I}$ was distributed according to the normal distribution rule within the interval $[0; 1]$ or $[-1; 1]$. **Figures 7–9** demonstrate the functioning of the automaton.

**Figure 7** demonstrates the functioning of the automaton, when $V_k \in [-0, 5; 0, 5]$, that is, most subjects are neutral to the IPI. **Figure 8** demonstrates the functioning of the automaton, when $V_k \in [-1; -0, 5]$, that is, most subjects are negative to the IPI. **Figure 9** demonstrates the functioning of the automaton, when $V_k \in [0, 5; 1]$, that is, most subjects are positive to the IPI. Figures "a" demonstrate the functioning of the automaton, when ${TR_{kj}}^{T_I} \in [0; 1]$, that is, the subjects adopt opinions of

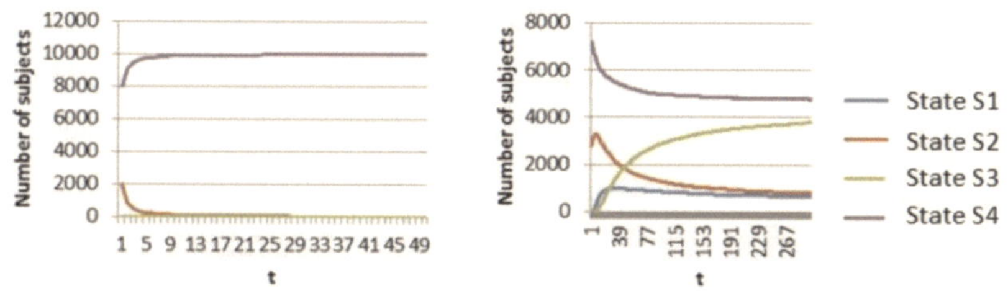


**Figure 5.**
*Distribution of cells according to the discrete time whenever $V_k \in [0, 5; 1]$.*

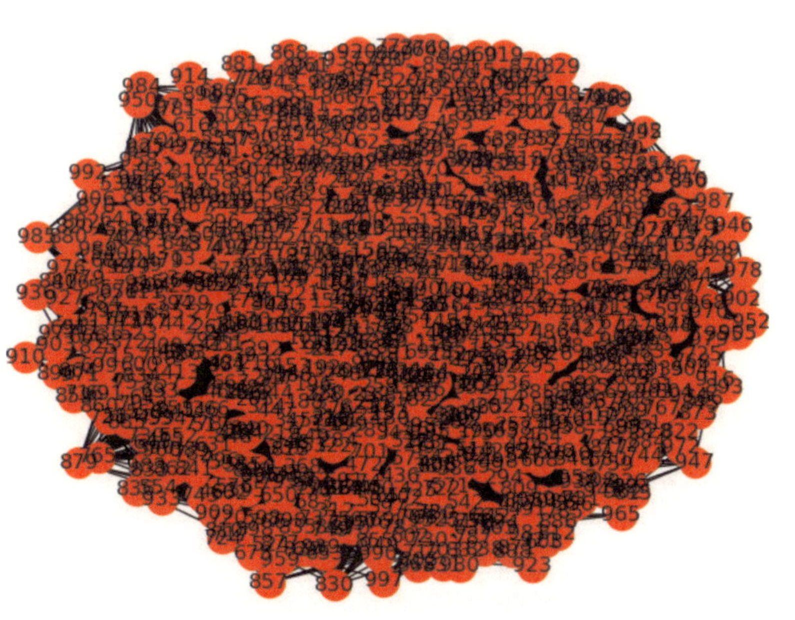

**Figure 6.**
*Random scale-free network generated by Barabási-Albert model.*

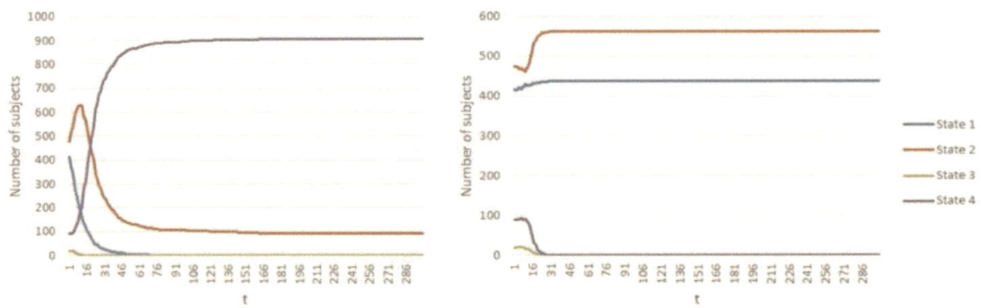


**Figure 7.**
*Distribution of cells according to the discrete time whenever* $V_k \in [-0,5; 0,5]$.

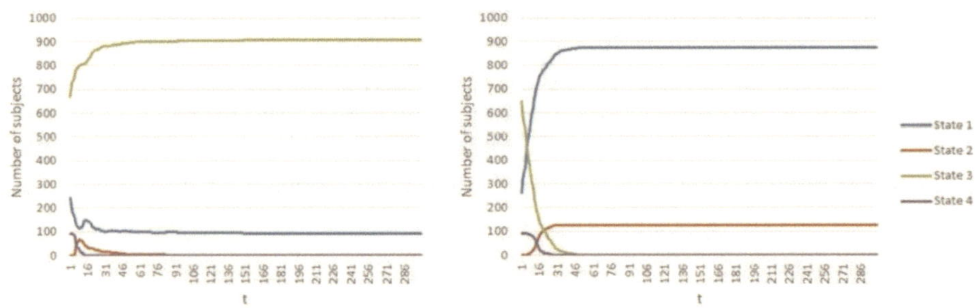


**Figure 8.**
*Distribution of cells according to the discrete time whenever* $V_k \in [-1; -0,5]$.

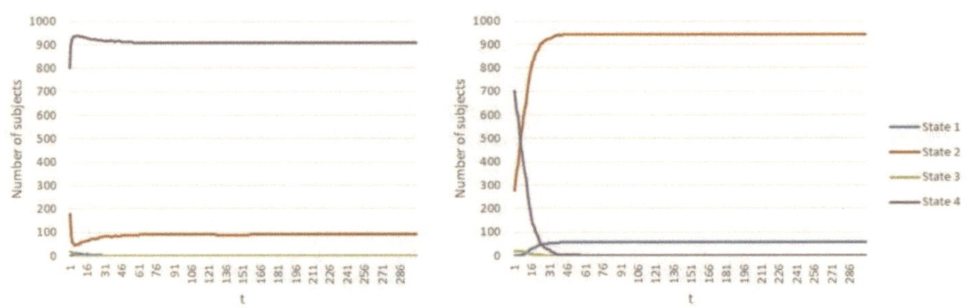


**Figure 9.**
*Distribution of cells according to the discrete time whenever* $V_k \in [0,5; 1]$.

other subjects. Figures "b" demonstrate the functioning of the automaton, when $TR_{kj}{}^{T_I} \in [-1; 1]$, that is, the subject has the opposite opinion to the one imposed by the IPI.

### 4.3 Discussion

Analysis of **Figures 3–9** shows that

- the character of the IPI diffusion within the social network is practically exponential;
- when the subjects are neutral to the IPI (**Figures 3a** and **7a**), just a small number of initiators can successfully perform the IPI;
- when the subjects are negative or positive to the IPI (**Figures 4a, 5a, 8a,** and **9a**), the IPI does not influence their state;

when the subjects do not trust each other and change their opinions to the opposite ones (**Figures 3–5b** and 7–**9b**), the number of subjects in states S3 and S4 is similar, irrespective of their initial state.

The results obtained using the suggested models agree with the results presented in Refs. [4, 5, 30]. These works consider the information diffusion, which is an individual case of IPI diffusion in social networks. As opposed to Refs. [4, 5, 30, 39, 40], the suggested model is not based on the probabilistic characteristics of the subjects of the social network but takes into account the social and psychological parameters of the subjects and their psychological state during IPI diffusion in social networks.

## 5. Conclusion

The paper suggests a model for describing the diffusion process of information-psychological impact in social networks based on cellular automata. Cellular automata models can change the states of a large number of cells over a minimal period of time, which is very useful for the modelling of the process of information-psychological impact diffusion in social networks. The suggested models can thus represent the process of IPI diffusion in a social network and the corresponding changes in the opinions of its subjects caused by their immediate neighbours, taking into account social and psychological factors.

## Author details

Igor Goncharov[1*], Nikita Goncharov[1], Pavel Parinov[1], Sergey Kochedykov[2] and Alexander Dushkin[3]

1 JSC "NGO" Infosecurity, Voronezh, Russia

2 Voronezh Institute of the Federal Penitentiary Service of Russia, Voronezh, Russia

3 National Research University of Electronic Technology, Moscow, Russia

*Address all correspondence to: goncharov@infobez.org

## References

[1] Barishpolets V. Field of application of information-psychological impact. Information Technologies. 2014;**6**(1): 52-79

[2] Barishpolets V. Information-psychological security: Main principles. Information Technologies. 2013;**2**(3): 62-104

[3] The Doctrine of Information Security of the Russian Federation, approved by Presidential Decree of December 5, 2016, No. 646

[4] Gubanov D, Nobikov D, Chkhartishvili A. Models of influence in social networks. Management in Social and Economic Systems. 2008;**27**:205-281

[5] Dodonov A, Lande D. Information reservations modeling. In: Proceedings of the International Scientific and Technological Conference OSTIS; Minsk; 2016; 1. pp. 253-256

[6] Bailey N. The Mathematical Theory of Infectious Diseases and its Applications. New York: Hafner Press; 1975

[7] Golub B, Jackson M. Naïve learning in social networks and the wisdom of crowds. American Economic Journal Microeconomics. 2010;**2**(1):112-149

[8] Granovetter M. Threshold models of collective behavior. American Journal of Sociology. 1978;**83**(6):1420-1443

[9] Even-Dar E, Shapira A. A note on maximizing the spread of influence in social networks. Internet and Network Economics. In: Proceedings Third International Workshop, WINE 2007. San Diego, CA, USA, December 12-14, 2007

[10] Kempe D, Kleinberg J, Tardos E. Maximizing the spread of influence through a social network. In: Proceedings of the 9-th ACM SIGKDD International Conference on Knowledge Discovery and Data Mining; 2003. pp. 137-146

[11] Masuda N, Kawamura Y, Kori H. Analysis of relative influence of nodes in directed networks. Physical Review. 2009;**80**:046114

[12] Rolfe M. Social Networks and Threshold Models of Collective Behavior. Preprint. Chicago: University of Chicago; 2004

[13] Valente T. Network Models of the Diffusion of Innovations. Cresskill, NJ: Hampton Press; 1995

[14] Goncharov I, Demianenko N, Ya M. Formalization of the process of information and psychological impact. In: Vestnik VGU, System Analysis and Information Technologies. Vol. 2. Voronezh: Publishing House VSU; 2012. p. 41

[15] Goncharov I, Demianenko N, Mishina Y. Modelling the information and psychological impact using artificial neural networks. In: Proceedings of the 12th International Scientific Conference “Informatics: Problems, Methodology, and Technologies”; Voronezh; 2013

[16] Goncharov I, Demianenko N, Khachumov A, Nozdrachev S. Analysis of the capabilities and classification of technological tools used for creating a channel for information and psychological impact. In: Proceedings of the Russian Scientific and Technological Conference; Voronezh; 2009. pp. 168-174

[17] Goncharov I, Mishina Y. Presenting the state of the subjects and objects of the process of information and psychological impact by means of the wavelet transform. In: Proceedings of the International Scientific Conference

"Technologies and Security of the Penal System 2013"; Voronezh Institute of the Federal Penitentiary Service; 2013

[18] Freidkin NE. Structural cohesion and equivalence explanations of social homogeneity. Sociological Methods & Research. 1984;**12**:235-261

[19] Leenders R. The specification of weight structures in network autocorrelation models of social influence. 2002

[20] Breer V. Stochastic models of social networks. Large-Scale Systems Control. 2009;**27**:169-204

[21] Tarnow E. Like water and vapor—Conformity and independence in the large group. Available from: http://cogprints.org/4274/1/LargeGroupOrderTarnow.pdf [Accessed: 10 June 2019]

[22] Beger RL. A necessary and sufficient conditions for reaching a consensus using De Groot's method. Journal of American Statistical Assotiation. 1981; **76**:415-419

[23] Chatterjee S, Seneta E. Toward consensus: Some convergence theorems on repeated averaging. Journal of Applied Probability. 1977;**14**:159-164

[24] De Marzo P, Vayanos D, Zweibel J. Persuasion bias, social influence and unidimensional opinions. The Quarterly Journal of Economics. 2003;**118**(3): 909-968

[25] Hesgelman R, Krause U. Opinion dynamics and bounded confidence models, analysis and simulation. Journal of Artificial Societies and Social Simulation. 2002;**5**:3

[26] Kazakov V, Karpenko I. Multi-agent systems and the concepts of the Web extension. Information society: education, science, culture, and technologies of the future. In: Proceedings of the 28th Conference "Internet and Modern Society: IMS"; 23–25 June 2015; St. Petersburg; pp. 260-268

[27] Okhapkina E, Voronova L. Developing the ontology fragment for the multi-agent moderation system of users' posts. Vestnik of NGU. 2014;**3**:60-67

[28] Kazakov V, Karpenko I. Social networks and multi-agent systems. Available from: http://conf.ict.nsc.ru/files/conferences/dicr2014/fulltext/245478/248602/kazakov.pdf [Accessed: 10 June 2019]

[29] Domingos P, Richardson M. Mining the network value of customers. In: Proceedings of the Seventh International Conference on Knowledge Discovery and Data Mining; 2002. pp. 57-66

[30] Lande D. Information diffusion model. Information technologies and security. Information security management. Institute for Information Transmission Problems. 2007. 10: pp. 51-67

[31] Zaitsev I. Multi-Agent Systems in Modelling Socio-Economic Relations: Studying the Behaviour and Verifying the Properties Using Markov Chains. Novosibirsk; 2014. p. 142

[32] Goldenberg J, Libai B, Muller E. Talk of the network: A complex systems look at the underlying process of word-of-mouth. Marketing Letters. 2001;**2**: 11-34

[33] Romualdo P, Alessandro V. Epidemic spreading in scale-free networks. Physical Review Letters. 2001;**14**(86):3200-3203

[34] Wu F, Huberman B, Adamic L, TylerR J. Information flow in social groups. Statistical and Theoretical Physics. 2004;**337**:327-335

[35] Ya R. Applying neural networks in social media communications and

corresponding internet sites. International journal of experimental education. 2015;**12**:676-677

[36] Neural Networks and Social Engineering, Ideas and Prospects. Available from: http://ai-news.ru/2017/07/nejronnye_seti_i_socialnaya_inzhene riya_idei_i_perspektivy.html [Accessed: 10 June 2019]

[37] Leskovec J, Krause A, Guestrin C, Faloutsos C, Vanbriesen J, Glance N. Cost-effective outbreak detection in networks. In: Proceedings of the 13-th ACM SIGKDD International Conference on Knowledge Discovery and Data Mining; 2007. pp. 420-429

[38] Smidt Y, Lobodina O. Some approaches to modeling the spatial diffusion of innovations. Spatial Economics. 2015;**2**:103-115

[39] Lomakin S, Fedotov A. The analysis of information transfer model in the network of cellular automaton. Vestnik NGU: Information Technologies. 2014; **3**:86-99

[40] Nizhegorodtsev R, Sekerin V, Lisafyev S. The cellular automata models in a principle of innovation diffusion. Issues of New Economy. 2012; **3**(23):39-43

[41] Azhmukhamedov I, Machueva D, Zholobov D. Modelling the information interaction process in social systems. Fundamental Research. 2017;**5**:18-39

[42] Greben N. Psychological Tests for Professionals. Minsk: Modern School; 2007. p. 496

[43] Goncharov N, Goncharov I, Parinov P, Dushkin A, Maximova M. Modeling of information processes for modern information system security assessment. 2019. pp. 1758-1763. Doi: 10.1109/EIConRus.2019.8656828

[44] Goncharov I, Parinov P, Sirota A. Modeling the Processes of Information and Psychological Impact in Social Networks Vestnik VGU, System Analysis and Information Technologies. Vol. 2. Voronezh: Publishing House VSU; 2018. pp. 93-104

[45] Goncharov I, Goncharov N, Parinov P, Kochedykov S, Dushkin A. Probabilistic analysis of the influence of staff qualification and information-psychological conditions on the level of systems information. In: Security, Probabilistic Modeling in System Engineering, Andrey Kostogryzov. Rijeka: Intech Open; 2018. DOI: 10.5772/intechopen.75079

[46] Goncharov N, Goncharov I, Parinov P, Dushkin A. Simulation and evaluation of conflict interactions in information systems. Journal of Physics: Conference Series. 2019;**1203**:012063. DOI: 10.1088/1742-6596/1203/1/012063

[47] Markov A, Barabanov A, Tsirlov V. Periodic Monitoring and Recovery of Resources in Information Systems, Probabilistic Modeling in System Engineering, Andrey Kostogryzov. Rijeka: IntechOpen; 2018. DOI: 10.5772/intechopen.75232

[48] iSi Methodology. Available from: http://www.pircenter.org/static/isi-methodology

[49] Kefeli I, Yusupov M, editors. Information-Psychological and Cognitive Security. St. Petersburg: Metropolis; 2017. 300 p

Chapter 14

# Combinatorial Cosmology

*Martin Tamm*

## Abstract

In this chapter, a combinatorial model for cosmology is analyzed. We consider each universe as a path in a graph, and the set of all such paths can be made into a finite probability space. We can then consider the probabilities for different kinds of behavior and under certain circumstances argue that a scenario where the behavior of the entropy is monotonic, either increasing or decreasing, should be much more likely than a scenario where the behavior is symmetric with respect to time. In this way we can attempt to construct a model for a multiverse which is completely time symmetric but where the individual universes tend to be time asymmetric, i.e., have an arrow of time. One of the main points with this approach is that this kind of broken symmetry can be studied in very small models using exact mathematical methods from, e.g., combinatorics. Even if the amount of computations needed increases very rapidly with the size of the model, we can still hope for valuable information about what properties more realistic models should have. Some suggestions for further research are pointed out.

**Keywords:** cosmology, multiverse, graph theory, entropy, time's arrow

## 1. Introduction

Applications of combinatorics have in recent years invaded many new areas of research. Still, cosmology is probably not the first such area which comes to your mind. Traditional cosmology is usually based on differential geometry and general relativity, often in combination with various ideas from fundamental physics and high-precision astronomical measurements. However, it is very much at the heart of cosmology that any model that we study must be based on rather drastic simplifications. In fact, when the object of study in a sense contains everything, finding the right way to discard nonessential information becomes a fundamental problem. From this point of view, the combinatorial approach is just one of several possible ways to proceed. For a discussion of this question from a more general point of view, see [1].

Different problems may of course call for different kinds of simplifications. As a rather extreme example, I will in this chapter discuss the long open problem of time's arrow, where it can be argued that the best method of attack may be to discard almost everything we know about the universe, just to uncover the underlying combinatorial skeleton. In other words, we should forget almost everything we know about ordinary physics and instead consider all the possible states that a universe could be in as the nodes of a huge graph. Each possible universe then becomes a path in this graph, and our mission becomes to try to decide what kinds of paths are the most common ones.

The ambition here is not to claim any kind of final solution to "the riddle of time." Rather, the ambition is to give a new angle to a well-known problem. And also to show that from this point of view, it may even make sense to study models which are ridiculously small in comparison with the real universe.

After a short introduction to the problem of time's arrow in Section 2, the basic structure of the combinatorial multiverse is presented in Section 3. But in order to study the asymmetry of time, we also need the concept of entropy which is introduced in Section 4, and in Section 5 we then turn to the dynamics. Both the entropy and the dynamics are treated in a very simplified way. However, the essential point here is to try to explain how the time-asymmetric development of the entropy that we observe in our universe can arise from time-symmetric dynamical laws and boundary conditions. In Section 6, the object is to show how standard methods from combinatorics can be used to make computations in the combinatorial multiverse. In Sections 7 and 8, we then consider some very simple probabilistic assumptions which turn the combinatorial multiverse into a probability space and discuss the consequences for time asymmetry.

In the simple model discussed so far, it is not difficult to obtain similar results by heuristic reasoning. However, the approach here should mainly be considered as a preparation for more complicated models, where the same combinatorial methods could be used, but heuristics would be difficult to apply.

Thus, this should more be considered as a starting point for further research than as an endpoint. In Section 9, I will therefore take a step in this direction by suggesting one such possible generalization (out of many), which could be used to obtain a stronger kind of time asymmetry. Finally, some conclusions are then discussed in Section 10.

Many of the ideas presented here have appeared before, e.g., in [2, 3], and in particular [4], although from a somewhat different angle.

## 2. The arrow of time

The term "time's arrow" was coined by Eddington [5] and refers to the fact that macroscopic time is asymmetric. In fact, we all know that the future is very different from the past. For instance, how does it come that we can remember yesterday but we cannot remember tomorrow? This can also be expressed by saying that we all agree that there is a well-defined direction from the past toward the future. Ever since the time of Ludwig Boltzmann, it has been clear that this has something to do with the growth of entropy and the second law of thermodynamics, although it may still not be quite obvious exactly what the connection is.

What is mysterious about time's arrow is that somehow the macroscopic laws that we observe must emerge from the underlying microscopic laws of motion, and these are in general considered to be essentially time invariant. So how can asymmetric macroscopic laws arise from symmetric microscopic ones?

Few questions in physics have generated such a variety of completely different answers (see, e.g., Barbour [6], Halliwell et al. [7], Zeh [8]), and the problem is still wide open. But, as has repeatedly been pointed out by Price [9], most such tentative answers seem to contain some (more or less hidden) asymmetry from the beginning, either in the boundary conditions or in the dynamical laws.

Here I will advocate a different viewpoint. We can consider the set of all possible universes as a probability space, a "multiverse," and this probability space will be completely time symmetric in the sense that reversing the direction of time would generate the same probability space. But for observers, like ourselves, who are by necessity confined to our own universe, it can still be that the symmetry appears to be broken in the sense that in one direction of time the entropy is increasing and in the other it is decreasing. Another way to express this would be to say that all the universes in the multiverse would share the same endpoints, the Big Bang and the Big Crunch. But only half of them would have the same Big Bang as we have. In the other half, our Big Bang would instead be the Big Crunch.

## 3. The combinatorial multiverse

This is not the place to try to describe all possible combinatorial models for cosmology. Rather, I have chosen to just discuss the simple case of a closed, finite universe. Many cosmologists these days support open models, and it is of course possible to apply combinatorial methods to them too. However, since such models tend to be infinite, they may be considerably more complicated from a probabilistic point of view.

To model the set of all universes in the simplest possible way, let us for each moment of time between the endpoints $-T_0$ and $T_0$ (i.e., the Big Bang and the Big Crunch) consider the finite set of all possible "states" of a universe. To make everything extremely simple, let us suppose that time is discrete in the sense that we only consider it at a finite number of points as follows:

$$-T_0 \quad -T_0+1 \quad \cdots \quad -1 \quad 0 \quad 1 \quad \cdots \quad T_0-1 \quad T_0 \tag{1}$$

Thus, we can measure time just by counting the number of time intervals, which means that time can be viewed as integer valued. At the endpoints $-T_0$ and $T_0$, there will just be one unique state (with zero volume), but in between, there will be many states for each $t$. All such states will be the nodes of an enormous graph, and a universe will then be just a path in this graph with the property that there is exactly one state for each moment of time. The dynamics of the model can then be specified by choosing at certain collection of edges between adjacent moments of time, say $t$ and $t$+1, which correspond to those time developments which are possible. A quite schematic picture is displayed in **Figure 1**.

*Remark* 1. For the readers taking interest in the underlying physics: the word "state" is not referring to quantum states as they are usually interpreted. A better way of thinking of them is to say that they represent "distinguishable configurations." This is in fact a kind of semiclassical approximation (see Tamm [2]).

The important point here is that a given state can lead to different states in the future. This is very much what actually happens when, say, a particle decays: whether or not this happens may, according to the multiverse interpretation, lead to very different futures within a rather short time. And there is no contradiction

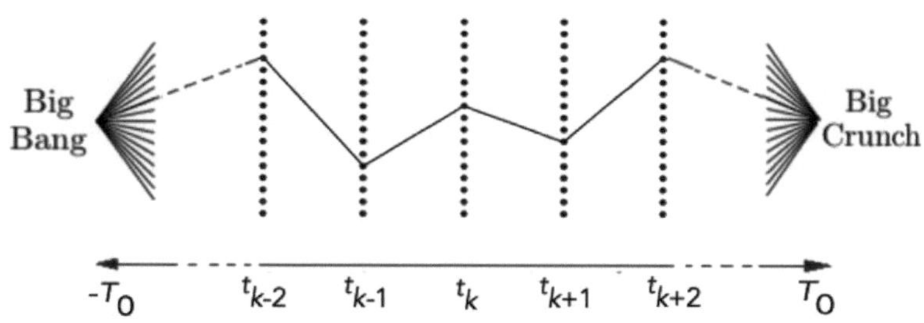


**Figure 1.**
*One universe in the combinatorial multiverse [3].*

between this and the fact that the development of the underlying wave function for the whole universe is unique.

Summarizing:

**Definition 1.** A *universe* $U$ is a chain of states (one state $U_t$ for each moment of time $t$), with the property that the transition between adjacent states is always possible.

**Definition 2.** A *multiverse* $M$ is the set of all possible universes $U$ in the sense of Definition 1 together with a probability measure on this set.

It may of course be said that quantum mechanics should allow for transitions between all kinds of states, although the probability for most such transitions may be extremely small. In this extremely simplified treatment, I will assume that for a given state at a given moment of time $t$, the dynamical laws will only permit transitions to a very limited number of states at the previous and next moments, which will make the probabilistic part of the investigation particularly simple. However, modifications are called for near the endpoints (the Big Bang and the Big Crunch); see Section 5.

As it stands, the model presented so far is too simple to generate any results. In fact, there are no observable differences at all between the states, which mean that there are no measurable variables which could be related to the (so far non-specified) dynamics.

There are of course many different variables which we can choose to enrich this structure, and which ones to choose must depend on what properties we want to explain. For explaining the second law of thermodynamics, the obvious choice is the entropy.

## 4. Entropy

According to Boltzmann, the total entropy of a certain macro-state at a certain time is given by

$$S = k_B \ln \Omega, \tag{2}$$

or inversely

$$\Omega = W^S, \quad \text{with} \quad W = e^{1/k_B}, \tag{3}$$

where $\Omega$ denotes the number of corresponding micro-states and $k_B$ is Boltzmann's constant.

This formula was from the beginning derived for simple cases, like an ideal gas. Nevertheless, it does represent a kind of universal truth in statistical mechanics: the number of possible micro-states corresponding to a given macro-state grows exponentially with the entropy. Although there are many complications when one tries to consider the entropy of the universe as a whole, I will still take it as the starting point for the discussion that the entropy (at a given time $t$) is an exponential function of the total entropy as in (3). A more difficult question is if and how the constant $W$ may vary with time, but for the purpose of the present paper, I will simply let it be constant.

One may of course argue that this can only be true when the universe is still quite ordered and the entropy is very far from reaching its maximum. But this is certainly what the situation is like in our universe today, and according to the computations in [10, 11], it would take an almost incredibly long time to reach such a state of maximal entropy. Thus, it will in the following be taken for granted that this time is much longer than the life-span of our universe.

## 5. The dynamics

The next step is to construct a model for the dynamics. The idea, which essentially goes back to Boltzmann (see [12]), is that any given macro-state at any given time is extremely likely to develop into a state with higher entropy at the next moment of time, simply because there are so many more states with higher entropy than with lower entropy (compare with (3)). The problem with this in the present situation, however, is that this way of thinking in fact presupposes a preferred direction of time. Otherwise, given that the dynamical laws are time symmetric, why can we not similarly argue that the entropy should also grow when we go backward in time? (compare [9]).

There have been many attempts to avoid this problem by looking for defects in the symmetries. But my conclusion here is that we must actually accept Boltzmann's argument in both directions of time and hence we are led to the following:

**Principle 1**. At every moment of time $t$ and for every state with entropy $S$, there are very many "accessible states" with higher entropy, both at the previous moment of time $t-1$ and at the next one $t+1$. On the other hand, the chance for finding such accessible states with lower entropy, both at times $t-1$ and $t+1$, is extremely small.

This principle also implies a shift of perspective in the search for time's arrow. Rather than trying to find the reason for the asymmetry, we must concentrate on understanding why we cannot observe the symmetric structure of the multiverse as a whole.

As still one more simplification, let us assume that the entropy can only change by $\pm 1$ during each unit of time. This assumption, however, has to be modified near the endpoints (BB and BC) for the following reason: it is a very important aspect of this approach to assume that physics during the first and last moments is very different from the rest of the time, since at these moments quantum phenomena can be expected to become global. To model this in a simple way, we can split the life-span of our multiverse up into three parts:

$$[-T_0, -T_1] \cup [-T_1, T_1] \cup [T_1, T_0]. \tag{4}$$

Here the first and last parts may be called "the extreme phases," which are characterized by the property that transition between very different states can be possible. During the "normal phase" in between on the other hand, physics is supposed to behave more or less as we are used to.

## 6. Modeling the dynamics

To construct a miniature multiverse for computational purposes, one can proceed as follows: first of all, in the very small multiverses studied here, the extreme phases will only last for one single unit of time. Also, for ease of notation, let us put $T_1 = m$, so that the moments of time can in this context be denoted as

$$-m-1, \ -m, \ -m+1, \ \dots, m-1, m, m+1. \tag{5}$$

The dynamics is specified by randomly choosing for each state at time $t$ with entropy $S$, $K$ edges to states at time $t+1$ with entropy $S+1$, and similarly $K$ edges to states at time $t-1$ with entropy $S+1$ (with obvious modifications at the endpoints). In this section, again to make everything as simple as possible, $K$ will be set equal to 2. These random choices are in practice carried out by the random number generator in, e.g., Mathematica or MATLAB. But once these are chosen, they specify a model for the dynamics for the miniature multiverse, and we are faced with the problem of computing the number of paths of different kinds. It should be observed that if $K \ll W$, then only a small fraction of all states will be connected to states with lower entropy at the next or previous moment, in spite of the fact that all states are connected to several states with higher entropy, just as in the Principle 1 in Section 5.

As an illustration, a schematic picture of the set of the possible states in the case of a very small multiverse with only 5 moments of time between the Big Bang and the Big Crunch and with $W = 4$ is shown in **Figure 2**.

Note that due to the way we have set up the dynamics, the entropy can grow with at most one unit during each unit of time. This means that if we start from an ordered state with $S = 0$ at one end of the normal phase, then only values of $S$ less than or equal to four can occur during the life-span of the corresponding universe. This means that the part of the multiverse graph displayed in **Figure 2** is sufficient for computing the number of all possible universes with zero entropy at one end. To actually carry out the computation, we can proceed as follows: it is easy to compute the number of paths with monotonically increasing entropy. According to the above assumptions, each state with entropy $S$ is connected to exactly two states with entropy $S + 1$, both at the next and at the previous moment (with an obvious restriction to just one side at times $-m$ and $m$). This clearly implies that for each unit of time, the number of paths doubles: from the state with $S = 0$ at time $-m$, there are precisely two edges to states with $S = 1$ at time $-m + 1$, and for each of these, there will also be precisely two edges to states with $S = 2$ at time $-m + 2$ which gives in total four paths. At the next step, there will then be eight paths to states with $S = 3$ at time $-m + 3$ and so on.

In the case $m = 2$, we obtain $2^4 = 16$ such universes, since there are in this case four unit intervals of time. For $m = 3$, we get in an analogous way $2^6 = 64$ universes since there are in this case six unit intervals of time.

One has to work harder to compute the number of paths with zero entropy at both ends, at least if we want exact results and not just heuristic ones. The number of such universes must be considered as a statistical variable which depends on the random choices of the edges which defines the dynamics. In fact, the most

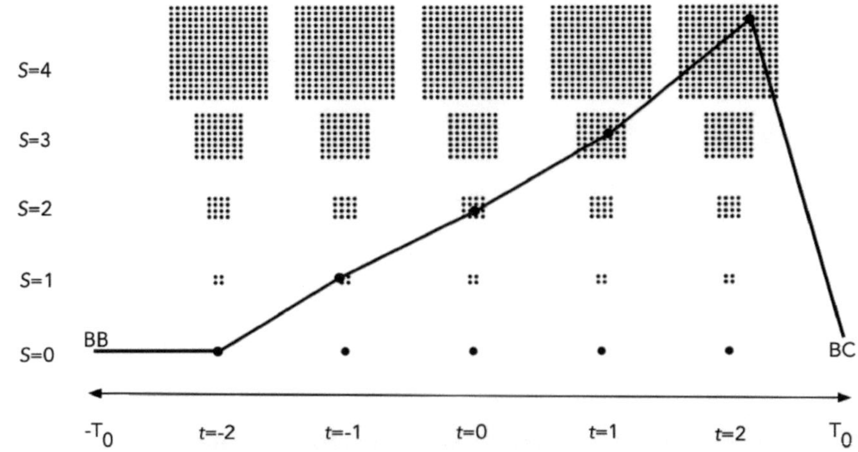


**Figure 2.**
*A schematic picture of a universe in a very small multiverse with only five moments of time between the endpoints (i.e., $m = 2$). In this case, the universe has a monotonically increasing entropy [4].*

significant variable will be the average number of universes when we consider many graphs at the same time.

The basic combinatorial tool for making these computations is the adjacency matrix (see [13] or [14]) of the graph.

Thus, recall that a (directed) path $[v_1, v_2, \dots, v_m]$ from $v_1$ to $v_m$ is a sequence of nodes such that for each $j = 1, 2, \dots, m - 1$, the pair $v_j, v_{j+1}$ belongs to the set of (directed) edges of the graph. A lot of work has been done in combinatorics to calculate the number of paths of a given length between two nodes. In general, this is a hard problem or at least a time-consuming one. But for graphs with the special time-related properties in this chapter, the task may be somewhat easier. In particular, in this case all paths joining two given nodes all have the same length.

**Definition 3.** The *adjacency matrix* of the (directed) graph $G$ with nodes $v_1, v_2, \dots, v_m$ is the $m \times m$-matrix $A = (a_{ij})$, where $a_{ij} = 1$ if the pair $v_i, v_j$ determines a (directed) edge in $G$ and $a_{ij} = 0$ otherwise.

The reason why this matrix is useful to us lies in the following classical result:

**Theorem 1.** The element at position $ij$ in the k$^{\text{th}}$ power of the adjacency matrix, $A^k$, equals the number of paths of length $k$ starting at $v_i$ and ending at $v_j$.

*Remark* 2. The fact that I have chosen to work with directed graphs here should not be confused with some kind of preferred direction of time. It would in fact be possible to work with two-sided paths as well. This would however introduce more elements different from zero in the adjacency matrix and hence slow down the computations. In other words, the choice to work with directed graphs is just for technical reasons. In fact, when considering the universes in this chapter, the number of directed paths from $t = -m$ to $t = m$ is precisely the same as the number of nondirected path between $t = -m$ and $t = m$.

When considering powers of the adjacency matrix below, everything we need to know about paths starting with $S = 0$ at $-m$ can be obtained from the first row of $A^{2m}$. Thus, this can all essentially be done by simple linear algebra. Although simple in principle, the size of $A$ grows very fast with the size of the model, i.e., primarily with $m$ and $W$.

In view of our simple choice for the dynamics and in particular of the fact the entropy can only change by $\pm 1$ at each step during the normal phase, it suffices to consider nodes in the graph with $S \le t + m$.

Starting from $S = 0$ at time $-m - 2$, we observe that at time $t = -1$, only states which have $S \le 1$ have to be considered, which gives 1 + 4 = 5 states. In the same way, we get for $t = 0$, 1 + 4 + 16 = 21 states; for $t = 1$, 1 + 4 + 16 + 64 = 85 states; and finally for $t = 2$, 1 + 4 + 16 + 64 + 256 = 341 states.

The adjacency matrix can now be written as a block matrix in the following way:

$$A = \begin{pmatrix} & B_{12} & & & \\ & & B_{23} & & \\ & & & B_{34} & \\ & & & & B_{45} \\ & & & & \end{pmatrix}. \tag{6}$$

Here empty blocks should be understood as containing just zeros. Each of the five-block rows/columns correspond to a moment of time, i.e., to $-2, -1, 0, 1, 2$ as in **Figure 2**. Inside each such row/column, the states are ordered according to entropy: the first element is the unique state with $S = 0$. Then (if $t \ge -1$) the four elements with $S = 1$ follow, thereafter (if $t \ge 0$) the 16 elements with $S = 2$, and so on. The internal order between all the elements with equal $S$ and $t$ is not at all important in the following.

With this setup and the random dynamics introduced earlier, each $B$-matrix contains all the information about the edges from all the states at one moment of time to the states at the next one. For example, $B_{12}$ contains the information about all edges from the single state with $S = 0$ at time $t = -2$ to the five states with $S \leq 1$ when $t = -1$. In the same way, $B_{23}$ gives a complete description of the edges from the 5 states with $S \leq 1$ at time $t = -1$ to the 21 states with $S \leq 2$ when $t = 0$.

The number of rows and columns in the $B$-matrices are now given as follows:

$$B_{12} : 1 \times 5, \qquad B_{23} : 5 \times 21, \qquad B_{34} : 21 \times 85, \qquad B_{45} : 85 \times 341. \tag{7}$$

For the quadratic adjacency matrix $A$, this gives the format $453 \times 453$. The matrices $B_{k,k+1}$ can also be described as block matrices in the following way: $B_{12} = (0|0101)$ (the first element is always a 0 and among the other four, two randomly chosen elements will be one instead of zero). For the following matrix, we obtain (with certain random choices of ones as before)

$$B_{23} = \left(\begin{array}{c|c|c} & C_1 & \\ \hline C_2 & & C_3 \end{array}\right) =$$

$$= \left(\begin{array}{c|cccc|cccccccccccccccc} 0 & 1 & 0 & 1 & 0 & 0 & 0 & 0 & 0 & 0 & 0 & 0 & 0 & 0 & 0 & 0 & 0 & 0 & 0 & 0 & 0 \\ \hline 1 & 0 & 0 & 0 & 0 & 0 & 0 & 0 & 0 & 1 & 0 & 0 & 1 & 0 & 0 & 0 & 0 & 0 & 0 & 0 & 0 \\ 0 & 0 & 0 & 0 & 0 & 0 & 0 & 0 & 0 & 0 & 1 & 0 & 0 & 0 & 0 & 0 & 0 & 0 & 0 & 0 & 1 \\ 1 & 0 & 0 & 0 & 0 & 0 & 0 & 0 & 0 & 0 & 0 & 0 & 0 & 0 & 0 & 0 & 0 & 1 & 1 & 0 & 0 \\ 0 & 0 & 0 & 0 & 0 & 1 & 0 & 0 & 0 & 1 & 0 & 0 & 0 & 0 & 0 & 0 & 0 & 0 & 0 & 0 & 0 \end{array}\right). \tag{8}$$

Both $C_1$ and $C_3$ have rows containing only zeros, except for two randomly chosen positions where there are ones instead (these are the edges which connect to states with higher entropy one unit of time later), and $C_2$ is a column of zeros with two randomly chosen ones instead (these are the edges which connect to states with lower entropy one unit of time later).

The structures of $B_{34}$ and $B_{45}$ are similar:

$$B_{34} = \left(\begin{array}{c|c|c|c} & D_1 & & \\ \hline D_2 & & D_3 & \\ \hline & D_4 & & D_5 \end{array}\right), \qquad B_{45} = \left(\begin{array}{c|c|c|c|c} & E_1 & & & \\ \hline E_2 & & E_3 & & \\ \hline & E_4 & & E_5 & \\ \hline & & E_6 & & E_7 \end{array}\right), \tag{9}$$

where now all $D$:s and $E$:s with odd indices have rows with two randomly chosen ones and those with even indices have columns with two randomly chosen ones.

## 7. Modeling the combinatorial multiverse as a probability space

Now when we have specified the dynamics of the model, i.e., decided which paths (universes) can occur, it is time to attribute to each such path its probability weight so that the multiverse becomes a probability space. Following the tradition in statistical mechanics, I will frequently make use of un-normalized probabilities. This means that summing up all (un-normalized) probabilities will give the "state sum," which in general is not equal to one. To obtain the usual probabilities, one has to divide by the state sum. This may seem unnatural at first but turns out to be very practical in situations where only the relative sizes of the probabilities are needed.

As for the normal phase, the choice will, to start with, be the simplest possible one: each path is either possible or not, corresponding to the probability weights 1 and 0. During the extreme phases, this assumption is no longer reasonable. Again the model will be extremely simplified, but still it is based on physical intuition and, most importantly, completely time symmetric. Assume that the only types of edges having a non-neglectable chance of occurring during the extreme phase $[-m-1, -m]$ are of the following two kinds: The first scenario is that the universe passes through the extreme phase into a state of zero entropy. The other scenario is that it passes into a state with high entropy (equal to $2m$). Universes of one of these two types will be given the (un-normalized) probability 1 or $p$, respectively. Here $p > 0$ should be thought of as a very small number, at least when the size of the model becomes large. During the other extreme phase $[m, m+1]$, near the Big Crunch, we make the completely symmetric assumption.

*Remark* 3. These assumptions may perhaps seem somewhat arbitrary. And to a certain extent, this may be so. However, they do represent the following viewpoint of what may happen at the full cosmological scale: we may think of the Big Bang and the Big Crunch as states of complete order with zero volume and entropy. Such states can very well be metastable, very much like an oversaturated gas at a temperature below the point of condensation. If no disturbance takes place, such metastable states can very well continue to exist for a substantial period of time. In particular, a low-entropy state can have a very good chance of surviving the intense but extremely short extreme phase. On the other hand, if a sufficiently large disturbance occurs, then the metastable state may almost immediately decay into a very disordered state of high entropy.

It is not my intension to further argue in favor of this viewpoint here. The main thing in this chapter is to show that completely symmetric boundary conditions at the endpoints may give rise to a broken time symmetry.

The multiverse now splits up into four different kinds of paths:

- LL: The entropy is low (=0) at both ends ($-m$ and $m$).
- LH: The entropy is 0 at $-m$ and $2m$ at $m$.
- HL: The entropy is $2m$ at $-m$ and 0 at $m$.
- HH: The entropy is high ($= 2m$) at both ends ($-m$ and $m$).

If we now denote by $N_{LL}, N_{LH}, N_{HL}$ and $N_{HH}$ the number of paths of the indicated kinds, then with the above assumptions we also get the corresponding probability weights for the corresponding types as

$$P_{LL} = N_{LL}, \quad P_{LH} = pN_{LH}, \quad P_{HL} = pN_{HL}, \quad P_{HH} = p^2 N_{HH}. \tag{10}$$

We can now consider the following two types of broken time symmetry:

**Definition 4.** A multiverse is said to exhibit a *weak* broken time symmetry if

$$P_{LL} \ll P_{LH} + P_{HL}. \tag{11}$$

**Definition 5.** A multiverse is said to exhibit a *strong* broken time symmetry if

$$P_{LL} + P_{HH} \ll P_{LH} + P_{HL}. \tag{12}$$

Both these definitions should of course be made more precise when applied to specific models for the multiverse, e.g., by showing that the corresponding limits

$$\lim \frac{P_{LL}}{P_{LH}+P_{HL}} \quad \text{and} \quad \lim \frac{P_{LL}+P_{HH}}{P_{LH}+P_{HL}} \tag{13}$$

equal zero when certain parameters tend to infinity in some well-defined way. However, it is worthwhile at this stage to note their implications for cosmology.

The strong broken symmetry in Definition 5 actually means that a monotonic behavior of the entropy is far more probable than a non-monotonic one. In the case of a weak broken symmetry, this is not necessarily so; it could very well be that the most probable scenario would be high entropy at both ends. Thus, this is definitely a weaker statement, but it can nevertheless be argued that it can be used to explain the time asymmetry that we observe, referring to a kind of anthropic principle: it is an obvious observational fact that we live in a universe with low entropy at at least one end. If the statement in Definition 4 is fulfilled, then clearly among such scenarios, the monotonic ones (LH and HL) are the by far most probable ones. Thus, since universes with high entropy at both ends would seem to be quite uninhabitable, one can argue that given the existence of an observer, then with almost certainty he must live in a universe with monotonic entropy.

Summing up, both limits above can be used to argue in favor of time asymmetry. Nevertheless, at least to the mind of the author, the strong broken symmetry is the preferable one. This alternative will be further studied in Section 9.

## 8. Numerical computations in the combinatorial multiverse

With the setup in Sections 6 and 7, we can now use Mathematica or MATLAB to generate instances of the combinatorial multiverse for small values of $m$ and $W$ and then compute the corresponding probability weights $P_{LL}$, $P_{LH}$, $P_{HL}$ and $P_{HH}$. It is important to note that the matrices here can be treated as sparse, rather than as full matrices, which make the computations considerably faster.

In particular, in the case $m = 2$ in Section 6 and with a randomly generated dynamics which is manifested by an adjacency matrix $A$, we can compute the power $A^4$ and read of the first row, which contains all the information we need about the paths from the state at $t = -2$ with $S = 0$. So what do we find?

In **Figure 3**, I have plotted the ratio $N_{LL}/(N_{LH} + N_{HL})$ for the cases $m = 2$ (light gray) and $m = 3$ (dark gray) for values of $W$ ranging from 3 to 30. What is actually displayed are the mean values of 1000 randomly generated matrices as above for each value of $W$. Although the picture clearly supports the claim that

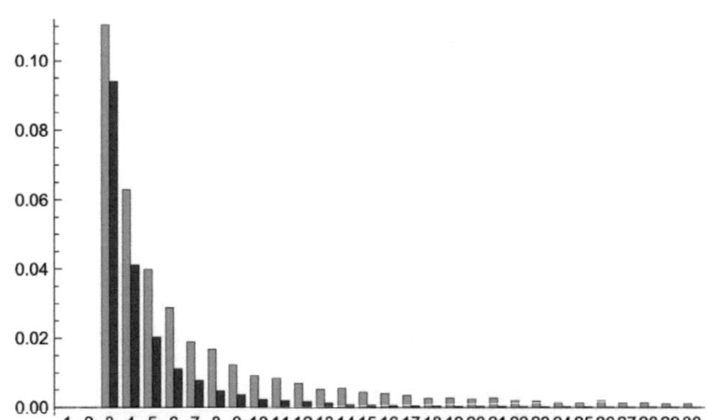


**Figure 3.**
*The ratio $N_{LL}/(N_{LH} + N_{HL})$ as a function of W for the cases $m = 2$ (light gray) and $m = 3$ (dark gray) [4].*

$N_{LL}/(N_{LH}+N_{HL}) \to 0$ when $W \to \infty$, there is not really enough support for a firm prediction about the more precise asymptotic behavior for large $W$. Having said this, the behavior seems to be rather close to a relationship of the form $\rho \sim 1/W$.

It should be possible, although perhaps not so easy, to prove exact limit theorems to confirm these kinds of predictions. The problem is that we use a large number of instances to model something much more complicated, namely, the full quantum mechanical development of the multiverse. For very special unlikely choices of these instances, the ratio $N_{LL}/(N_{LH}+N_{HL})$ may behave quite differently.

## 9. Can the dynamics be modified to generate a strong broken symmetry?

Obviously, the above model represents an extreme simplification. But from the point of view of the author, most of the simplifications can be said to be rather harmless for the purpose of explaining time's arrow.

However, there is one assumption which is somewhat problematic in the dynamics that we have discussed so far: the model can be said to exhibit a kind of Markov property in the sense that the probability for the entropy to go up or down at a certain step is completely independent of the prehistory of the state; it just depends on the state itself. This does not appear to be what is happening in our own universe: for instance, light emitted from (more or less) pointlike sources like stars continues to spread out concentrically for billions of years, and in this way it preserves a memory of the prehistory for a very long time.

A very interesting research project is therefore to try to find better models which do not exhibit this property. We can, for instance, attempt to construct models where the behavior of the entropy not only depends on the previous (or following) step but on a larger part of the prehistory (or post-history). As a particularly simple example one could let the probabilities for an increase (or decrease) of the entropy at a certain step, depend not only on the previous and following step but on the two previous (and following) steps. In fact, such dynamics would not only be more realistic but would in general also have a much better chance to exhibit a strong broken time symmetry.

I will now briefly discuss an example of such a modified model. In Section 6 it was noted that the number of paths between a state $i$ at time $-m$ and another state $j$ at time $m$ can be computed using the adjacency matrix $A$ as

$$\left(A^{2m}\right)_{ij} = \sum_{q_1}\sum_{q_2}\cdots\sum_{q_{2m-1}} a_{iq_1}a_{q_1q_2}\cdots a_{q_{2m-1}j}. \tag{14}$$

This sum can now be modified by introducing various weights depending on the path. An example of such a weight can be constructed as follows: given a path $U$ with vertices $v_{-m}, v_{-m+1}, v_{-m+2}, \dots, v_m$, we let $S_{-m}, S_{-m+1}, S_{-m+2}, \dots, S_m$ denote the corresponding entropies. We can now define

$$\xi = \sum_{k=-m+1}^{m} (S_k - S_{k-1})(S_{k+1} - S_k), \tag{15}$$

and note that periods of monotonic growth or decrease of the entropy will tend to make $\xi$ positive, whereas switches between growth and decrease tend to make it negative. In fact, if $S$ is monotonic on $[k-1, k+1]$, then $(S_k - S_{k-1})(S_{k+1} - S_k) = 1$ and if not, then $(S_k - S_{k-1})(S_{k+1} - S_k) = -1$.

Given a real number $\mu \geq 0$, we can then consider the probability measure which to each path $U$ assigns the (un-normalized) probability $\exp\{\mu\xi\}$ and replace the sum in (14) by

$$\sum_{q_1}\sum_{q_2}\cdots\sum_{q_{2m-1}} e^{\mu\xi} a_{iq_1} a_{q_1q_2} \cdots a_{q_{2m-1}j}. \tag{16}$$

With this definition, it is now again possible to compute the probability weights $P_{LL}, P_{LH}, P_{HL}$ and $P_{HH}$, and we can note that for $\mu = 0$, these will be exactly the same as in the case without weights in Section 8. Thus, this model is really a generalization of the previous theory.

**Conjecture 1**. If $\mu > 0$, then we have a strong broken time symmetry in the limit $m \to \infty$ (for a suitable fixed choice of $p$, $K$, and $W$ with $K \ll W$).

## 10. Conclusions

Clearly, there is a large gap between the extremely simplified dynamics in this paper and more realistic dynamics based on, say, ordinary Newtonian physics or quantum mechanics. This is, for better or for worse, both the strength and the weakness of the combinatorial method presented here: extreme simplification may be the price we have to pay in order to see the forest in spite of all the trees.

In any case, the few simple examples in this paper should only be considered as a first step toward more realistic models. And in fact, when the object of study is something as enormously large as the multiverse, one should not expect a single method of attack to give all the answers. Rather, it can be expected that future developments will have to combine computer computations, heuristics, and exact mathematical methods in completely new ways.

## Acknowledgements

Most of the computations presented here have be done using Mathematica and/or MATLAB. Many of them have been carried out on an ordinary mac, but the heavier ones have been processed using MATLAB at PDC, the center for parallel computing at the Royal Institute of Technology in Stockholm.

The author would like to thank PDC for the possibility to use their resources and also the friendly staff for helping to implement the computations in an economical way.

*DOI: http://dx.doi.org/10.5772/intechopen.90696*

## Author details

Martin Tamm
Department of Mathematics, University of Stockholm, Stockholm, Sweden

*Address all correspondence to: matamm@math.su.se

## References

[1] Tamm M. Probabilistic cosmology. Journal of Modern Physics. 2019;**10**: 1424-1438. DOI: 10.4236/jmp.2019.1012095

[2] Tamm M. Time's arrow from the multiverse point of view. Physics Essays. 2013;**26**(2):237-246

[3] Tamm M. Time's arrow in a finite universe. International Journal of Astronomy and Astrophysics. 2015;**5**: 70-78. DOI: 10.4236/ijaa.2015.52010

[4] Tamm M. A combinatorial approach to time asymmetry. Symmetry. 2016; **8**(3):11. DOI: 10.3390/sym8030011

[5] Eddington AS. The Nature of the Physical World. Cambridge: Cambridge University Press; 1928. Chapters 3 & 4

[6] Barbour J. The End of Time. Oxford: University Press; 1999

[7] Halliwell JJ, Perez-Mercander J, Zurek WH. Physical Origins of Time Asymmetry. Cambridge: Cambridge University Press; 1994

[8] Zeh HD. The Physical Basis of the Direction of Time. 4th ed. Berlin Heidelberg: Springer Verlag; 2001

[9] Price H. Time's Arrow and Archimedes' Point. Oxford: University Press; 1996

[10] Adams F, Laughlin G. A dying universe: The long-term fate and evolution of astrophysical objects. Reviews of Modern Physics. 1997;**69**(2): 337-372

[11] Egan C, Lineweaver C. A larger estimate of the entropy of the universe. The Astrophysical Journal. 2010;**710**(2): 1825-1834

[12] Boltzmann L. In: McGuinness B, editor. Theoretical Physics and Philosophical Problems. Dordrecht: Reidel Publishing Co.; 1974. Foulkes P (Trans.)

[13] Biggs N. Algebraic Graph Theory. Cambridge Mathematical Library. 2nd ed. Cambridge: Cambridge University Press; 1993

[14] Pemmaraju S, Skiena S. Computational Discrete Mathematics. Cambridge: Cambridge University Press; 2003. pp. 323-334